VOLUME ONE HUNDRED AND THIRTY FIVE

Advances in
COMPUTERS

Applications of Nature-Inspired Computing and Optimization Techniques

VOLUME ONE HUNDRED AND THIRTY FIVE

ADVANCES IN
COMPUTERS

Applications of Nature-Inspired Computing and Optimization Techniques

Edited by

ANUPAM BISWAS

*Department of Computer Science and Engineering,
National Institute of Technology Silchar, Silchar, Assam, India*

ALBERTO PAOLO TONDA

*UMR 518 MIA-PS, INRAE, Université Paris-Saclay,
Palaiseau; Institut des Systèmes Complexes de Paris
Île-de-France (ISC-PIF)—UAR 3611 CNRS, Paris, France*

RIPON PATGIRI

*Department of Computer Science and Engineering,
National Institute of Technology Silchar, Silchar, Assam, India*

KRISHN KUMAR MISHRA

*Assistant Professor, Department of Computer Science and
Engineering, Motilal Nehru National Institute of Technology
Allahabad, Prayagraj, Uttar Pradesh, India*

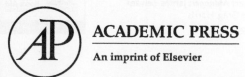

ACADEMIC PRESS

An imprint of Elsevier

ELSEVIER

Academic Press is an imprint of Elsevier
125 London Wall, London, EC2Y 5AS, United Kingdom
525 B Street, Suite 1650, San Diego, CA 92101, United States
50 Hampshire Street, 5th Floor, Cambridge, MA 02139, United States

First edition 2024

ISBN: 978-0-323-95768-7
ISSN: 0065-2458

For information on all Academic Press publications
visit our website at https://www.elsevier.com/books-and-journals

Publisher: Zoe Kruze
Editorial Project Manager: Palash Sharma
Production Project Manager: James Selvam
Cover Designer: Greg Harris

Typeset by STRAIVE, India

Working together
to grow libraries in
developing countries

www.elsevier.com • www.bookaid.org

Contents

Part II
Ecological and economic systems

Part III
Information and computational systems

Part IV
Communication and networking systems

Part V
Deep learning and neural networking systems

Contributors

R. Amith
Department of Information Science & Engineering, RV College of Engineering, Bengaluru, India

Milos Antonijevic
Singidunum University, Belgrade, Serbia

Harish Chandra Arora
CSIR-Central Building Research Institute, Roorkee, Uttarakhand; Academy of Scientific and Innovative Research (AcSIR), Ghaziabad, Uttar Pradesh, India

Nebojsa Bacanin
Singidunum University, Belgrade, Serbia

Arman Bath
Faculty of Engineering, Computer Engineering, Zeytinburnu, Zeytinburnu, Istanbul, Turkey; Electrical and Computer Engineering Faculty; Department of Political Science, Vancouver, British Columbia, Canada

Aniruddha Bhattacharya
Department of Electrical Engineering, NIT, Durgapur, India

Ajay Kumar Bhurjee
VIT Bhopal University, Sehore, Madhya Pradesh, India

Anupam Biswas
Department of Computer Science and Engineering, National Institute of Technology, Silchar, Silchar, Assam, India

Malaya Dutta Borah
Department of Computer Science and Engineering, National Institute of Technology Silchar, Silchar, Assam, India

Priyanath Das
Department of Electrical Engineering, NIT, Agartala, India

Prasenjit Dey
Cluster of Logistics and Rail Engineering, Mahidol University, Nakhon Pathom, Thailand

Milos Dobrojevic
Singidunum University, Belgrade, Serbia

A.J. Fernández-Ares
Department of Languages and Computer Systems, ETSIIT-CITIC, University of Granada, Granada, Spain

P. García-Sánchez
Department of Computer Engineering, Automatics, and Robotics, ETSIIT-CITIC, University of Granada, Granada, Spain

Diego Hernández
Department of Signal Theory, Telematics and Communications, ETSIIT-CITIC, University of Granada, Granada, Spain

Zakir Hussain
Department of Computer Science and Engineering, National Institute of Technology Silchar, Silchar, Assam, India

Deepali Jain
Department of Computer Science and Engineering, National Institute of Technology Silchar, Silchar, Assam, India

Suja Jayachandran
Department of Computer Engineering, Ramrao Adik Institute of Technology, D Y Patil Deemed to be University, Navi Mumbai; Vidyalankar Institute of Technology, Mumbai, Maharashtra, India

Bharti Joshi
Department of Computer Engineering, Ramrao Adik Institute of Technology, D Y Patil Deemed to be University, Navi Mumbai, Maharashtra, India

Haripriya V. Joshi
Department of Information Science & Engineering, RV College of Engineering, Bengaluru, India

Luka Jovanovic
Singidunum University, Belgrade, Serbia

Nishant Raj Kapoor
CSIR-Central Building Research Institute, Roorkee, Uttarakhand; Academy of Scientific and Innovative Research (AcSIR), Ghaziabad, Uttar Pradesh, India

Phumin Kirawanich
Cluster of Logistics and Rail Engineering; Department of Electrical Engineering, Mahidol University, Nakhon Pathom, Thailand

Rahul Kottath
School of Electrical and Electronics Engineering, VIT Bhopal University, Bhopal; Digital Tower, Bentley Systems India Private Limited, Pune, India

Aman Kumar
CSIR-Central Building Research Institute, Roorkee, Uttarakhand; Academy of Scientific and Innovative Research (AcSIR), Ghaziabad, Uttar Pradesh, India

Anuj Kumar
CSIR-Central Building Research Institute, Roorkee, Uttarakhand; Academy of Scientific and Innovative Research (AcSIR), Ghaziabad; Haldaur Group of College, Bijnor, Uttar Pradesh, India

Ashok Kumar
CSIR-Central Building Research Institute, Roorkee, Uttarakhand; Academy of Scientific and Innovative Research (AcSIR), Ghaziabad, Uttar Pradesh, India

Pankaj Kumar
DoMSC, National Institute of Technology Hamirpur, Hamirpur, Himachal Pradesh, India

Rahul Chandra Kushwaha
Department of Computer Science and Engineering, Rajiv Gandhi University (Central University), Itanagar, Arunachal Pradesh, India

Arunanshu Mahapatro
Veer Surendra Sai University of Technology, Burla, India

G.S. Mamatha
Department of Information Science & Engineering, RV College of Engineering, Bengaluru, India

Boonruang Marungsri
School of Electrical Engineering, Suranaree University of Technology, Nakhon Ratchasima, Thailand

P. Merino
Department of Signal Theory, Telematics and Communications, ETSIIT-CITIC, University of Granada, Granada, Spain

Swadhin Kumar Mishra
Veer Surendra Sai University of Technology, Burla, India

Jyoti Mohanty
Siksha O Anusandhan Deemed to be University, Bhubaneswar, India

A.M. Mora
Department of Signal Theory, Telematics and Communications, ETSIIT-CITIC, University of Granada, Granada, Spain

Miloš Nikolić
University of Belgrade, Faculty of Transport and Traffic Engineering, Belgrade, Serbia

Latha Parameswaran
Department of Computer Science and Engineering, Amrita Vishwa Vidyapeetham, Coimbatore, India

Ripon Patgiri
Department of Computer Science and Engineering, National Institute of Technology Silchar, Silchar, Assam, India

Prabina Pattanayak
National Institute of Technology Silchar, Silchar, India

Anubhav Kumar Prasad
Department of Computer Science and Engineering, United Institute of Technology, Naini, Prayagraj, Uttar Pradesh, India

Jasenka Rakas
University of California Berkeley, National Center of Excellence for Aviation Operations Research, Berkeley, United States

Gomathi Ramasamy
TCS Research and Innovation, IITM Research Park, Tata Consultancy Services Limited, Chennai, India

Mahdi Roshanzamir
Department of Computer Engineering, Faculty of Engineering, Fasa University, Fasa, Iran

Mohamad Roshanzamir
Department of Computer Engineering, Faculty of Engineering, Fasa University, Fasa, Iran

Anulekha Saha
Department of Electrical Engineering, Chulalongkorn University, Bangkok, Thailand

Meril Sakaria
Retail Strategic Initiatives, Tata Consultancy Services Limited, Chennai, India

Mohamed Salb
Singidunum University, Belgrade, Serbia

Sridharan Sankaran
TCS Research and Innovation, IITM Research Park, Tata Consultancy Services Limited, Chennai, India

Dharm Raj Singh
Department of Computer Applications, Jagatpur Post Graduate College, Varanasi, Uttar Pradesh, India

Priyanka Singh
Department of Computer Science and Engineering, Indian Institute of Information Technology, Raichur, India

Reetendra Singh
Indian Institute of Management Visakhapatnam, Visakhapatnam, India

Yeshwant Singh
School of Computer Science, UPES, Dehradun, Uttarakhand, India

Ivana Strumberger
Singidunum University, Belgrade, Serbia

Chaiyut Sumpavakup
Research Centre for Combustion Technology and Alternative Energy—CTAE and College of Industrial Technology, King Mongkut's University of Technology North Bangkok, Bangkok, Thailand

Shirmohammad Tavangari
Electrical and Computer Engineering Faculty; Department of Political Science, Vancouver, British Columbia, Canada

Dušan Teodorović
Serbian Academy of Sciences and Arts, Belgrade, Serbia

Bagyammal Thirumurthy
Department of Computer Science and Engineering, Amrita Vishwa Vidyapeetham, Coimbatore, India

Alberto Paolo Tonda
UMR 518 MIA-PS, INRAE, Université Paris-Saclay, Palaiseau; Institut des Systèmes Complexes de Paris Île-de-France (ISC-PIF)—UAR 3611 CNRS, Paris, France

Karthikeyan Vaiapury
TCS Research and Innovation, IITM Research Park, Tata Consultancy Services Limited, Chennai, India

Srihari Veeraraghavan
Retail Strategic Initiatives, Tata Consultancy Services Limited, Chennai, India

Vinay Yadav
Vinod Gupta School of Management, Indian Institute of Technology Kharagpur, Kharagpur, India; School of Civil and Environmental Engineering, Nanyang Technological University, Singapore, Singapore

Aref Yelghi
Faculty of Engineering, Computer Engineering, Zeytinburnu, Zeytinburnu, Istanbul, Turkey; Department of Political Science, Vancouver, British Columbia, Canada

Miodrag Zivkovic
Singidunum University, Belgrade, Serbia

CHAPTER ONE

A brief introduction to nature-inspired computing, optimization, and applications

Anupam Biswas[a], Alberto Paolo Tonda[b,c], and Ripon Patgiri[a]

[a]Department of Computer Science and Engineering, National Institute of Technology Silchar, Silchar, Assam, India
[b]UMR 518 MIA-PS, INRAE, Université Paris–Saclay, Palaiseau, France
[c]Institut des Systèmes Complexes de Paris Île-de-France (ISC-PIF)—UAR 3611 CNRS, Paris, France

Contents

Abstract

This chapter provides preliminary details about nature-inspired computing and optimization techniques. It starts by describing different components of optimization problems and their taxonomy. Nature-inspired computing techniques, created to solve such optimization problems, are then discussed. The chapter briefs the application prospects of nature-inspired optimization techniques in a few representative emerging domains that are covered in the book. Applications presented in this book cover five major domains, which include Controller and Power Systems, Ecological and Economic Systems, Information and Computational Systems, Communication and Networking Systems, and Deep Learning and Neural Networking Systems. The book primarily focuses on the practical challenges that are faced while applying nature-inspired algorithms to different problems in these domains, which include the feasibility of the problem, control parameters and constraints, representation of the solution space, and design of the objective function. Lastly, the chapter concluded by highlighting common challenges, criticisms, and future perspectives.

Advances in Computers, Volume 135
ISSN 0065-2458
https://doi.org/10.1016/bs.adcom.2023.11.010

1. Optimization problems

Technological advancements have led to ever-increasing complexities in the diverse problems associated with different application domains. Optimization problems [1–4] are one of the widely encountered problems across different domains. Optimization algorithms are computational techniques used to find the best solution to a problem, picked from a set of possible candidate solutions. Researchers and practitioners across the globe are constantly trying to develop models, processes, and application modules that are cost-effective, more reliable, and efficient. Despite differences in the nature of the problem depending on applications, the optimization problems have the following common components:

- Decision variables (X): The set of unknown parameters x_i of a problem, which are taken as inputs to the objective functions. Depending on the permissible values of x_i there will be multiple sets of values for the decision variable, which are referred to as candidate solutions. All sets of candidate solutions constitute the search space of the problem, also referred to as the solution space.
- Objective functions $f(X)$: It determines the quality of a candidate solution, i.e., a set of values of decision variables. The objective functions tell whether a set of values of decision variables is good or bad for the problem. Defining an appropriate objective function is crucial for an optimization problem.
- Constraints $C(X)$: These are the criteria used to define permissible values of decision variables. Normally, constraints are defined as inequalities on decision variables or as a function too.

An optimization problem can be formulated for any application with relevant decision variables, objective functions, and constraints. Objective function can be defined either as a maximization problem [5] (i.e., maximum value of the objective function will imply the best solution) or a minimization problem [6] (i.e., minimum value of the objective function will imply the best solution). A generic minimization problem is given as follows:

$$\text{Minimize} : f_i(x), \text{ for } i = 1, 2, 3..., I \tag{1}$$

$$\text{Subject to} : g_j(x) =\leq a_j, \text{ for } i = 1, 2, 3...,J \tag{2}$$

$$h_k(x) = b_k, \text{ for } i = 1, 2, 3..., K \tag{3}$$

where f_i shows the ith objective function, g_j indicates the jth inequality constraint, and h_k is the kth equality constraint, a_j is a constraint for jth in equality constraint, b_k is a constant for kth equality constraint. I is the number of objectives, J is the number of inequality constraints, and K is the number of equality constraints, which are dependent on the application.

Let us now understand the physical interpretation of different components of an optimization problem. Consider the following problem:

$$\text{Minimize} : f(x) = x^2 + 7 \, \sin(3x/2) \tag{4}$$

$$\text{Subject to} : g(x) = x + 2 \geq 0 \tag{5}$$

$$\text{Where} : \ -5 \leq x \leq 5 \tag{6}$$

The above problem is a minimization problem with only one objective function, one decision variable, and a constraint. The range of the decision variable is specified as $-5 \leq x \leq 5$. Fig. 1 shows the shape of the objective function, i.e., how the search space is defined by it for the range of decision variable x. However, the constraint has restricted the values beyond -2 as invalid, as indicated by the *dotted line*. If the same objective function is considered with two decision variables (without constraint), then the function will be as follows:

$$\text{Minimize} : f(x_1, x_2) = x_1^2 + 7 \, \sin(3x_1/2) + x_2^2 + 7 \, \sin(3x_2/2) \tag{7}$$

$$\text{Where} : -5 \leq x_1, x_2 \leq 5 \tag{8}$$

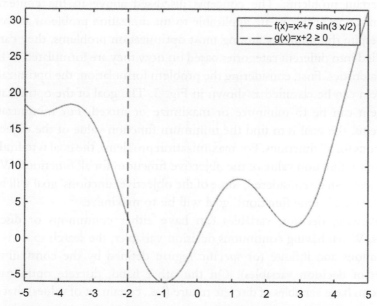

Fig. 1 Search space defined by an example objective function, constraint, and single decision variable.

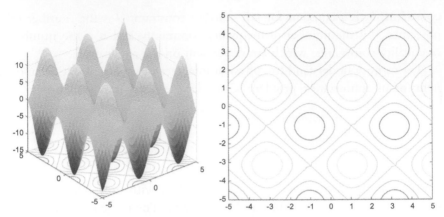

Fig. 2 Search space defined by example objective function with two decision variables.

In the above problem, two decision variables make the problem two dimensional, or, in other words, the number of decision variables determines the dimensionality of the optimization problem. The search space of the problem is presented in 3-D and 2-D as shown in Fig. 2. It is important to note that optimization problems can have multiple optima, among which the best one is referred to as global optima [7]. There can be multiple global optima for certain problems. The concepts discussed above in the context of minimization problems are applicable to maximization problems as well.

Despite the similarity among most optimization problems, they can be classified into different categories based on how they are formulated or their characteristics. First, considering the problem formulation, the optimization problem can be classified as shown in Fig. 3. The goal of the optimization problem can be to minimize or maximize or mixed. For minimization problems, the goal is to find the minimum function value of the objective functions for all functions. For maximization problems, the goal is to find the maximum function value of the objective functions for all functions. While both goals can be considered, some of the objective functions' goal will be to minimize and some functions' goal will be to maximize.

Likewise, decision variables can have either continuous or discrete values. When having continuous decision variables, the search space is also continuous and infinite (or specific region defined by the constraints or ranges of decision variables). On the other hand, discrete optimization problems have variables of discrete nature (i.e., certain set of values) that lead

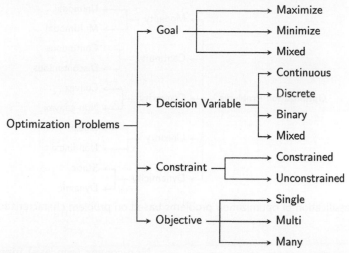

Fig. 3 Classification of optimization problems based on problem formulation.

to a finite search space. Mostly, the real-world optimization problems have mixed decision variables, which can also be divided into two classes: unconstrained and constrained.

In terms of the objective functions, optimization problems may have one or more than one objectives. A single-objective optimization problem has one objective function, so there is usually only one global optimum. Problems with multiple objectives, however, have more than one objective function, and multiple solutions can be found representing the best trade-offs between the objectives. The category of many-objective refers to problems with many objectives. Though both multiobjective and many-objective optimization have more than one objective function, they have a slight difference; in case of many-objective problems, the number of objectives is comparatively larger. The name many-objective was coined by researchers to highlight the importance of an ever-increasing number of objectives and the complexity of addressing them all simultaneously.

Optimization problems can also be categorized based on the characteristics of their objective function as shown in Fig. 4. The modality of optimization problem determines the number of optima present in the search space defined by the objective functions. Optimization problems can have one or more optima, which are referred to as unimodal or multimodal problems, respectively. In terms of continuity of the search space defined by the objective, optimization problems are classified into continuous and

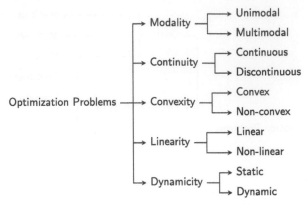

Fig. 4 Classification of optimization problems based on problem characteristics.

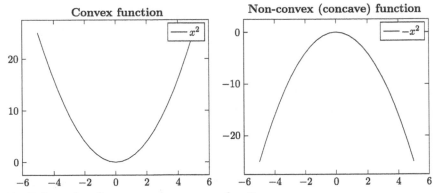

Fig. 5 Examples of convex and nonconvex functions.

discontinuous. If highest power of the decision variables with which the objective function is defined as well as constraints have highest power one then the function will be linear, otherwise it will be nonlinear. Linear optimization problems are comparatively simpler, while nonlinear optimization problems are complex and realistic problems where the objective function and/or constraints are nonlinear.

Again, depending on the objective function's convexity, the optimization problems can be categorized as convex or nonconvex optimization. A function is said to be convex at any interval if, for all pairs of points on the graph, the line segment that connects these two points passes above the curve. An example is shown in Fig. 5. All of the constraints as well as the objectives of a convex function for convex optimization problems.

If the problem is minimizing, function will be convex, where as it will be a concave function if maximizing. Mostly, linear optimization problems are convex, and nonlinear are nonconvex only.

Real-life optimization problems in general emerge from different applications, where execution of different phases requires different optimal settings of parameters. Thus, the optimal values for the same optimization problem change over time. The optimization problems that have this kind of characteristic are referred to as dynamic optimization problem. In contrast, optimal values of static optimization problems do not change over time or in different phases of the application.

2. Nature-inspired optimization techniques

The development of computers and advancements in computational methods have greatly expanded the scope and complexity of optimization problems that can be tackled. Numerous techniques have been developed for solving various types of optimization problems discussed above. Optimization techniques can be classified into two main categories: deterministic and stochastic. Deterministic optimization algorithms aim to find the best solution based on a set of rules or mathematical models, while stochastic algorithms use randomness to explore the solution space and find better solutions [8–11]. Deterministic algorithms, such as linear programming and dynamic programming, have been widely used in OR, while stochastic algorithms, such as genetic algorithms and particle swarm optimization, have emerged from the field of nature-inspired optimization. Our focus will be mainly on nature-inspired optimization techniques that are developed for solving optimization problems. A generic flow diagram of nature-inspired optimization techniques is presented in Fig. 6.

Nature-inspired optimization techniques mimic natural phenomena, such as natural selection, swarm behavior, and animal movements, to solve complex problems. These techniques are inspired by the observation of how biological systems adapt to their environment, and they seek to replicate the way living organisms behave or learn. Irrespective of the natural phenomena it is based on, the nature-inspired optimization techniques comprise four key components: Initialization, Terminate, Update, and Evaluate. The nature-inspired optimization algorithms are initialized with a population comprising a set of solutions, and in each iteration, the population is updated and evaluated until it meets the termination condition.

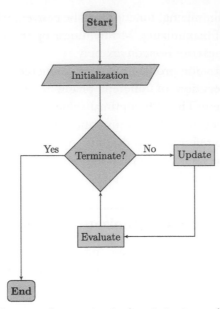

Fig. 6 Generic flow diagram of nature-inspired optimization techniques.

The development of nature-inspired optimization techniques was moti-vated by the need to find effective solutions to complex optimization prob-lems, typically those problems for which traditional techniques fail or cannot deliver a solution in a reasonable amount of time. Furthermore, traditional or optimization techniques rely on mathematical models that may not be able to capture the complex relationships between variables and constraints that occur in real-world problems. Nature-inspired optimization tech-niques, on the other hand, use heuristics that can explore the solution space more effectively, making them well suited to finding acceptable solutions for harder problems.

While nature-inspired optimization can feature a nonnegligible track record of success stories, these techniques still have known drawbacks. Being stochastic in nature, there is no guarantee of finding the best possible solution (global optimum), and two different runs of the same algorithm with the same parameters might lead to different solutions. When compared to techniques that make assumptions on the regularity of the function they aim to optimize, such as gradient-based optimization algorithms, nature-inspired optimization algorithms can be considerably more computationally expensive, as they typically need to evaluate each candidate solution produced during their search in the search space [12,13]

Despite these limitations, nature-inspired algorithms are currently con-sidered among the state of the art in several domains. Even more classical

OR approaches are seeing an increased hybridization with stochastic elements; for example, traditional gradient-based optimization techniques have largely been replaced by stochastic gradient descent and related algorithms, such as adaptive momentum, for the tuning of large neural network models. Nature-inspired optimization techniques have also seen the reverse, incorporating deterministic algorithms to perform local search, as in the case of memetic algorithms.

The *"Applications of Nature-inspired Computing and Optimization Techniques"* aims at portraying a selection of case studies where nature-inspired optimization has been successfully applied to solve problems from a variety of different real-world domains, ranging from summarizing legal documents to modeling air quality. The book primarily focuses on the practical challenges that are faced while applying nature-inspired algorithms to any specific problem, which include the feasibility of the problem, control parameters and constraints, representation of the solution space, and design of the objective function.

3. Application areas

Nature-inspired optimization techniques have emerged as powerful tools for solving complex problems in a wide range of practical applications. These methods draw inspiration from natural phenomena and biological systems to develop efficient algorithms that can find near-optimal solutions. Such techniques are especially useful in domains where exact optimization is not applicable due to the large search spaces or nonlinearity of the objective function. Applications presented in this book include domains such as Controller and Power Systems, Ecological and Economic Systems, Information and Computational Systems, Communication and Networking Systems, and Deep Learning and Neural Networking Systems. A considerable part of the applications in computer science are focused on deep learning, currently the dominant paradigm in machine learning, where nature-inspired algorithms are used to optimize the architecture of large neural networks. The use of nature-inspired optimization techniques has greatly contributed to these fields and has enabled researchers to tackle complex problems that were previously intractable.

3.1 Controller and power systems

Control problems are common in avionics and engineering and involve the design of control systems that can regulate the behavior of complex dynamic systems, such as aircraft and industrial processes and power systems, under uncertain and changing conditions while satisfying multiple constraints, such as safety, efficiency, and stability. Most of these problems fall into nonlinear categories. The book includes four chapters on such applications. Chapter "Overview of nonlinear interval optimization problems" by Bhurjee et al. starts with various kinds of nonlinear optimization problems, with special emphasis on nonlinear interval optimization problems. In chapter "Solving the aircraft landing problem using the bee colony optimization algorithm" by Nikolić et al., the authors propose a heuristic approach based on Bee Colony Optimization to solve the Aircraft Landing Problem in a static version, assuming that the information on scheduled aircraft is known in advance. The target objective is to minimize the total deviation of all aircraft from the target landing times while adhering to specific constraints. Chapter "Situation-based genetic network programming to solve agent control problems" by Roshanzamir and Roshanzamir deals with finding agent control strategies in complex environments, using the Tile-World benchmark as a case study. The authors employ an Evolutionary Algorithm, Situation-based Genetic Network Programming, to generate a strategy for each agent based on its situation rather than finding an optimal strategy for all agents. This approach improves the performance of traditional Genetic Network Programming and can be added to all versions of the optimization algorithm without additional overhead. Results show that the proposed method can achieve the goal more easily and quickly than finding a strategy that can guide all agents and generate more flexible solutions. Chapter "Small signal stability enhancement of large interconnected power system using grasshopper optimization algorithm tuned power system stabilizer" by Dey et al. tackles the optimization of a power system stabilizer used to damp potentially disruptive low-frequency oscillations in power systems and achieve small signal stability. The authors propose the use of a Grasshopper Optimization Algorithm to optimize an objective function consisting of eigenvalues and damping ratios, searching for optimal control parameters for the stabilizer.

3.2 Ecological and economic systems

Biological and ecological problems are among the most pressing challenges we face today, from climate change to the loss of biodiversity and habitat destruction. To address these complex issues, optimization techniques can play a critical role. Optimization approaches can help us model and simulate ecological systems and predict their behavior, identify optimal management strategies for conservation and resource allocation, and improve our understanding of how biological processes work. Furthermore, optimization techniques can help us design more sustainable and efficient systems, such as renewable energy grids and ecofriendly buildings. By using optimization techniques to address ecological and biological problems, we have the potential to greatly benefit the environment, society, and future generations. Chapter "Air quality modeling for smart cities of India by nature inspired AI: A sustainable approach" by Kapoor et al. tackles the issue of air quality modeling to estimate the correlation between pollution levels and their impacts on air quality. The authors propose the use of Particle Swarm Optimization merged as an optimizer for an Artificial Neural Networks tasked to predict air quality in seven Indian smart cities. The proposed approach has outstanding performance while evaluating high-dimensional data, which can help air quality managers forecast the effects of prospective new emissions and policymakers predict ambient air pollution concentrations under various scenarios.

Chapter "Genetic algorithm for the optimization of infectiological parameter values under different nutritional status" by Hussain and Borah proposes to use evolutionary optimization to find the parameters of mathematical equations describing the severity of infectious diseases. Optimizing these parameters can aid in identifying precautionary measures to mitigate losses in adverse situations. This chapter proposes mathematical equations for the infectiological parameters under normal-nutritional status and malnutrition and integrates nutritional status levels with a genetic algorithm to solve the formulated equations. Experimental results show that malnutrition negatively influences the optimization of infectiological parameters, and susceptibility, infection, and recovery are best optimized under normal-nutritional status, considering heuristic crossover for both sets of values. Chapter "A novel influencer mutation strategy for nature-inspired optimization algorithms to solve electricity price forecasting problem" by Singh and Kottath proposes a novel approach to the electricity price forecasting problem, an important issue for optimizing the management of energy grids, including different types of electricity sources, from renewable to carbon based to nuclear. The approach is based on a new mutation strategy that can be integrated with several nature-inspired optimization algorithms to improve their convergence rate and avoid local stagnation problems. The strategy involves a group of particles, called an influencer group, consisting of top-performing particles that guide the remaining particles toward the optimal point. Chapter "Recent trends in human- and bioinspired computing: Use-case study from a retail perspective" by Vaiapury et al., the authors apply nature-inspired optimization techniques to the task of extracting semantic descriptors from images and videos in the apparel and fashion industry. Specifically, the authors use Harris Hawk optimization and Progressive Spinalnet algorithms, obtaining satisfying results.

3.3 Information and computational systems

Nature-inspired optimization techniques are a popular approach for problems in the field of computer science. In particular, the recent developments in the field of information processing systems pose a large number of novel challenges, most of which cannot be tackled by resorting to classical optimization techniques or exhaustive search. Chapter "Domain knowledge-enriched summarization of legal judgment documents via grey wolf optimization" by Jain et al. proposes a nature-inspired approach to extractive summarization of legal documents by modeling the task as an optimization problem. The proposed objective function is infused with domain-specific knowledge and pretrained embeddings to better score candidate summaries. The experimental evaluation is conducted on an annotated Indian Legal Judgment document summarization dataset using ROUGE metrics, with promising results that can have practical utility in summarizing lengthy legal documents.

Chapter "Bio-intelligent computing and optimization techniques for developing computerized solutions" by Mamatha et al. presents a short survey on the use of nature inspired in different related domains: design of routing algorithms, sensors and visual systems, auditory systems, brain-controlled systems, and artificial intelligence. Chapter "Optimizing the feature selection methods using a novel approach inspired by the TLBO algorithm for student performance prediction" by Jayachandran and Joshi deals with a pivotal issue in machine learning, feature selection. In feature selection, an algorithm is tasked with selecting the minimal amount of informative input variables for a specific machine learning algorithm to improve the training speed and interpretability of the results. The novel approach to feature selection introduced in the chapter is tested on educational data, where it improves the performance of several different machine learning algorithms.

3.4 Communication and networking systems

Optimization techniques play an important role in tackling problems in communication and networking systems. With the increasing complexity and size of communication systems, optimization techniques can be used to improve their performance, reduce their cost, and enhance their reliability. In circuit design, optimization algorithms can help find the optimal configuration of components and their values, reducing power consumption and improving the overall efficiency. In addition, optimization techniques can be used in manufacturing processes to minimize defects, increase yield, and reduce the time and cost of production. Therefore, optimization techniques can significantly benefit the electronics industry by providing innovative solutions to challenging problems, improving the quality of electronic products related to networking systems, and optimizing their performance. In chapter "Applying evolutionary methods for the optimization of an intrusion detection system to detect anomalies in network traffic flows" by Mora et al., the authors propose the use of different evolutionary algorithms, including Particle Swarm Optimization, to optimize the weights and improve the performance of a Multivariate Statistical Network Monitoring, a state-of-the-art system for detecting different types of attacks with high performance. The proposed approach performs better than the classical optimization algorithm used for setting the system weights.

Chapter "Modified grey wolf optimization in user scheduling and antenna selection in MU–MIMO uplink system" by Mishra et al. tackles the challenges of user scheduling and antenna selection in multiuser multiple-input multiple-output wireless communication networks. Exhaustive search algorithms can deliver optimal solutions, but their computational cost is prohibitive. Techniques such as ant colony optimization, binary particle swarm optimization, and binary grey wolf optimization are employed to reduce the complexity of the scheduling algorithm, achieving high throughput with much less processing load. In chapter "Spectral efficiency optimization by the application of metaheuristic optimization techniques" by Mohanty and Pattanayak, the authors employ different optimization techniques, such as binary flower pollination algorithm, binary spider monkey optimization, and binary artificial bee colony optimization, to solve the combined user and receiver antenna scheduling problems in the downlink channel of multiuser multiple-input multiple-output systems. Such communication systems increase system throughput and spectral efficiency in networks, and their efficient scheduling of users is critical for optimal system capacity. In chapter "An effective genetic algorithm for solving traveling salesman problem with group theory" by Kushwaha et al., a novel variant of genetic algorithm has been proposed for solving traveling salesman problem that explores group theory for initial population generation. In the group tour construction method, each individual tour has a distinct start city, provided that the population size is equal to the total number of cities. In the initial population, each individual, i.e., tour, has a distinct starting city. The distinct starting cities of each tour provide genetic material for exploration for the whole search space. Therefore, a heterogeneous starting city of a tour in initial population is generated to have rich diversity. Also introduced crossover based on the greedy method of subtour connection, which drives the efficient local search, followed by 2-opt mutation for improvement of tour for optimal solution.

3.5 Deep learning and neural networking systems

With the recent emergence of the deep learning field in machine learning, nature-inspired techniques have found several niche optimization problems with vast search spaces where classical approaches fail. Most of these challenges lie in the domain of hyperparameter tuning, in other words, finding the best structure, activation function, number of neurons, and so on, for deep neural networks. Chapter "Adaptation of nature inspired optimization algorithms for deep learning" by Singh and Biswas discussed different architecture design-related adaptations required to use nature-inspired optimization techniques in deep learning models. The authors emphasize the need of optimal architectures for improving and enhancing the performance of deep learning models. Various nature-inspired optimization techniques are being explored in recent years for both optimal architecture design and weight adaptation for deep learning models. Various approaches that are followed for solution representation, objective function design, and constraint handling with respect to deep learning models are detailed as well as highlighted key challenges encountered during the incorporation of nature-inspired optimization techniques in deep learning models.

In chapter "Long short-term memory tuning by enhanced Harris hawks optimization algorithm for crude oil price forecasting" by Jovanovic et al., the authors propose the use of the Harris Hawk optimization algorithm to tune the hyperparameters of a long-short-term memory network, a deep learning architecture used for time series forecasting. The methodology is applied to real-world time series describing the price of crude oil, showing better results than the heuristics commonly used for this task. Chapter "Discovering the characteristic set of metaheuristic algorithm to adapt with ANFIS model" by Yelghi et al. focuses on the optimization of the parameters of Adaptive Neuro–Fuzzy Inference Systems, a hybrid machine learning architecture using both parts of Artificial Neural Network and Fuzzy Logic, used to create models for classification and regression tasks.

Six popular metaheuristic algorithms are assessed, and their results are compared. The study finds that metaheuristic algorithms based on evolutionary computation are more stable than swarm intelligence methods in tuning the parameters of Adaptive Neuro-Fuzzy Inference Systems. Chapter "Artificial neural network optimized with PSO to estimate the interfacial properties between FRP and concrete surface" by Kumar et al. tackles the problem of deterioration of concrete structures and the need for sustainable solutions such as strengthening with fibre-reinforced polymer composites. The bond between the concrete and the composite is crucial for effective strengthening, but predicting the bond strength is a complex task. A novel approach based on an Artificial Neural Network optimized with Particle Swarm Optimization is proposed, and the methodology is shown to deliver better results than analytical models and classic Artificial Neural Networks.

4. Concluding remarks

Nature-inspired computation techniques have been used in a wide range of applications to solve complex and large-scale optimization problems in various domains. This book focuses on a few representative application domains that have emerged in recent years, including controller and power systems, ecological and economic systems, information and computational systems, communication and networking systems, and lastly, deep learning and neural network models. In addition to other domains, a considerable part of the applications in computer science are nowadays focused on deep learning, which is currently the dominant paradigm in machine learning, whereas nature-inspired algorithms are often used to optimize the architecture of large neural networks. The use of nature-inspired optimization techniques has greatly contributed to these fields and has enabled researchers to tackle complex problems that were previously intractable. While nature-inspired optimization techniques are being applied to diverse domains, these techniques still have well-known drawbacks. It is also interesting to mention that the proliferation of novel nature-inspired metaheuristics has drawn severe criticism from researchers, highlighting how some of the newest proposed approaches have been poorly tested and are essentially slightly modified versions of established algorithms. Nevertheless, nature-inspired algorithms are enjoying growing popularity and are currently considered among the state of the art in numerous domains.

References

[1] C. Dimopoulos, A.M.S. Zalzala, Recent developments in evolutionary computation for manufacturing optimization: problems, solutions, and comparisons, IEEE Trans. Evol. Comput. 4 (2) (2000) 93–113.

[2] H. Tuy, Monotonic optimization: problems and solution approaches, SIAM J. Optim. 11 (2) (2000) 464–494.

[3] X. Hu, R. Eberhart, et al., Solving constrained nonlinear optimization problems with particle swarm optimization, in: Proceedings of the Sixth World Multiconference on Systemics, Cybernetics and Informatics, vol. 5, CiteSeer, 2002, pp. 203–206.

[4] A. Seyyedabbasi, F. Kiani, Sand Cat swarm optimization: a nature-inspired algorithm to solve global optimization problems, Eng. Comput. 39 (4) (2023) 2627–2651.

[5] R. Zhao, F. Zhu, M. Tang, L. He, Profit maximization in cache-aided intelligent computing networks, Phys. Commun. 58 (2023) 102065.

[6] S. Bartels, A. Kaltenbach, Explicit and efficient error estimation for convex minimization problems, Math. Comput. 92 (343) (2023) 2247–2279.

[7] T. Weise, Global optimization algorithms-theory and application, in: Self-Published Thomas Weise, vol. 361, 2009.

[8] M. Cavazzuti, M. Cavazzuti, Deterministic optimization, in: Optimization Methods: From Theory to Design Scientific and Technological Aspects in Mechanics, Springer, 2013, pp. 77–102.

[9] S. Uryasev, P.M. Pardalos, Stochastic Optimization: Algorithms and Applications, vol. 54, Springer Science & Business Media, 2013.

[10] J.C. Spall, Stochastic optimization, in: Handbook of Computational Statistics: Concepts and Methods, Springer, 2012, pp. 173–201.

[11] L. Liberti, S. Kucherenko, Comparison of deterministic and stochastic approaches to global optimization, Int. Trans. Oper. Res. 12 (3) (2005) 263–285.

[12] K. Sörensen, Metaheuristics-the metaphor exposed, Int. Trans. Oper. Res. 22 (1) (2013) 3–18, https://doi.org/10.1111/itor.12001.

[13] C. Aranha, C.L.C. Villalón, F. Campelo, M. Dorigo, R. Ruiz, M. Sevaux, K. Sörensen, T. Stützle, Metaphor-based metaheuristics, a call for action: the elephant in the room, Swarm Intell. 16 (1) (2021) 1–6, https://doi.org/10.1007/s11721-021-00202-9.

About the authors

Dr. Anupam Biswas is Assistant Professor in the Department of Computer Science & Engineering, the National Institute of Technology Silchar, Assam, Bharat. He received his PhD degree in Computer Science and Engineering from the Indian Institute of Technology (BHU), Varanasi, Bharat. His research interests include Machine learning, Social Networks, Computational music, Information retrieval, and Evolutionary computation. He is Principal Investigator/ Co-Investigator of five DST-SERB and Miety sponsored research projects in the domain of machine learning and evolutionary computation. He has served as Program Chair for BigDML 2019 and Publicity Chair of BigDML 2021, General Chair for 25th

International Symposium Frontiers of Research in Speech and Music (FRSM 2020) and coedited the proceedings book of FRSM 2020. He has edited seven books.

Dr. Alberto Paolo Tonda is a Permanent Researcher (CRCN) at the National Institute of Research for Agriculture and Environment (INRAE), and Université Paris-Saclay, Paris, France. His research interests include semi-supervised modeling of complex systems, evolutionary optimization and machine learning, with applications in food science and biology. He led the COST Action CA15118 FoodMC, a 4-year European networking project on in-silico modelling in food science. He has been program committee of International conferences including the *International Genetic and Evolutionary Computation Conference (GECCO, 2013–2020)*, the *European Conference on Evolutionary Computation* (EvoStar, 2010–2020), the *Bi-annual Conference on Artificial Evolution* (EA, 2013–2019), *and Nature Inspired Cooperative Strategies for Optimization* (NICSO, 2011). He is currently an editorial board member of the *Journal of Genetic Programming and Evolvable Machines*. He received his PhD degree in Computer Science Engineering from Politecnico di Torino, Italy.

Dr. Ripon Patgiri is Assistant Professor at the Department of Computer Science & Engineering, the National Institute of Technology Silchar, Bharat. His research interests include, big data, data-intensive computing, distributed systems, storage systems, file systems, Hadoop and MapReduce and bloom filter. He is a senior member of IEEE, a member of ACM and EAI, associate member of IETE and a lifetime member of ACCS, Bharat. He was General Chair of the 6th International Conference on Advanced Computing, Networking, and Informatics (ICACNI 2018, and the International Conference on Big Data, Machine Learning and Applications (BigDML 2019. He was organizing chair of the 25th International Symposium on Frontiers of Research in Speech and Music (FRSM 2020) and the International Conference on Modeling, Simulations and Applications (CoMSO 2020), and Organizing Chair and Program Chair of the 26th annual International Conference on Advanced Computing and Communications (ADCOM 2020). He was guest editor for the special issue "*Big Data: Exascale computation and beyond*" of the EAI Endorsed Transactions on Scalable Information Systems. He is also editor of "*Health Informatics: A Computational Perspective in Healthcare.*" He received his Doctor of Philosophy from the National Institute of Technology Silchar, Bharat.

Controller and power systems

PART 1

Controller and power systems

CHAPTER TWO

Overview of nonlinear interval optimization problems

Ajay Kumar Bhurjee[a], Pankaj Kumar[b], Reetendra Singh[c], and Vinay Yadav[d,e]
[a]VIT Bhopal University, Sehore, Madhya Pradesh, India
[b]DoMSC, National Institute of Technology Hamirpur, Hamirpur, Himachal Pradesh, India
[c]Indian Institute of Management Visakhapatnam, Visakhapatnam, India
[d]Vinod Gupta School of Management, Indian Institute of Technology Kharagpur, Kharagpur, India
[e]School of Civil and Environmental Engineering, Nanyang Technological University, Singapore, Singapore

Contents

Abstract

Variables and parameters in natural systems are often associated with uncertainty. There are broadly three approaches to tackle the uncertainty of such systems, i.e., stochastic, fuzzy, and interval/gray/inexact programming. Uncertainty is modeled as a probability distribution function in stochastic approach. In fuzzy programming, variables and parameters are governed by the membership functions. In many realistic data scarce

Advances in Computers, Volume 135
ISSN 0065-2458
https://doi.org/10.1016/bs.adcom.2023.11.011

21

scenarios, adequate data are not available to define a probabilistic or membership function. The model has to be developed and solved with only information of the extreme values of variables and parameters. Under such situations, interval analysis-based approaches pave the way for efficient model development and solutions. Nonlinearity in the modeling process is conspicuous in recent years due to multidisciplinary driven complexities. Therefore, decision makers have to address interval uncertainties of the models along with nonlinear objective function and constraints. Many techniques have been developed in recent past to solve such nonlinear interval optimization problem (NLIOP). In this chapter, we provide an overview of such techniques to solve NLIOP including nature inspired swarm algorithms.

1. Introduction

Optimization problems have a wide range of applications in different research fields [1–3]. The variables and parameters of some emerging applications of optimization models are often associated with uncertainty, which can be modeled as intervals. Optimization models with interval uncertainty provide flexibility to solve nonprobabilistic, and nonstatistical optimization problems [4] and are known as interval optimization. In addition to the uncertainty, nonlinearity in the modeling process has been conspicuous in recent years due to multidisciplinary driven complexities [5]. Interval optimization is one of the processes to tackle the uncertainty in the problem's parameters without any specified assumption, such as the membership function for the fuzzy technique and the probability density function for the probabilistic approach. Furthermore, the parameters' bounds can be easily estimated from past observations or expert knowledge and make the parameters as closed intervals. Also, it is seen in the real-world problem that one of the decision variables is proportional to itself or other variables, which raises a nonlinear interval optimization problem (NLIOP). In recent times, a number of literature exist on such domains, and repeatedly, new research is also going on in theoretical as well as application aspects. A survey on interval linear programming problems has been presented in [6] and provided a bibliography. However, a collection of all essential literature about NLIOP is not specifically available. Hence, this chapter intends to provide an overview of the various techniques provided in the literature to study NLIOP models through interval analysis using illustrative examples.

This survey is not intended to be exhaustive but merely represents the totality of results and explanations. In the beginning, a quadratic programming problem [7, 8] has been elaborated where coefficients of the objective functions and constraints are closed intervals. A nonlinear programming problem with interval coefficients has been subsequently illustrated, which has obtained the bounds of the objective function [9–13]. Due to the infeasibility of interval parameters based constraints on the solution obtained in the earlier mentioned model. Further, development has also been made in solution methodology [14–21]. Next, multiobjective NLIOP [22, 23] and evolutionary solution techniques [24] are discussed. Recently, NLIOP is extended by assuming the unknowns in the form of closed intervals, and it is applied in portfolio optimization; Kumar et al. [25–27] are also illustrated.

In the following sections, Section 2 starts with the basics of interval analysis. Section 3 discusses developed methods in chronological order. Finally, the conclusion and future work are in Section 4.

2. Interval analysis

This section provides some notations and definitions of interval analysis that are useful for studying the solutions to several types of interval optimization problems. In this chapter, a closed interval $[a^L, a^R]$ with $a^L \leq a^R$; $a^L, a^R \in \mathbb{R}$ is called a proper interval otherwise it is called improper interval. If $a^L = a^R = a$, then $[a, a]$ is called a degenerate interval or real number a. The set of proper interval on $\mathbb{R}$ is denoted by $\mathbb{I}(\mathbb{R})$, and $\mathbb{I}(\mathbb{R}) = \{[a^L, a^R] | a^L \leq a^R; a^L, a^R \in \mathbb{R}\}$. The set of improper interval on $\mathbb{R}$ is denoted by $\overline{\mathbb{I}(\mathbb{R})}$, and $\overline{\mathbb{I}(\mathbb{R})} = \{[a^L, a^R] | a^R \leq a^L; a^L, a^R \in \mathbb{R}\}$. An interval is said to be nonnegative if $a^L \geq 0$, $a^R \geq 0$ and set of all nonnegative intervals in $\mathbb{R}$ is represented by $\mathbb{I}(\mathbb{R}_+)$. Set of all negative intervals ($a^L < 0$, $a^R < 0$) in $\mathbb{R}$ is denoted by $\mathbb{I}(\mathbb{R}_-)$. Let $* \in \{+, -, \cdot, /\}$ be a binary operation on the set of real numbers. The binary operation $\circledast$ between two intervals $[a^L, a^R]$ and $[b^L, b^R]$ in $\mathbb{I}(\mathbb{R})$ can be expressed as

$$[a^L, a^R] \circledast [b^L, b^R] = \{a * b \,|\, a \in [a^L, a^R], b \in [b^L, b^R]\}.$$

In the case of division, it is assumed that $0 \notin [b^L, b^R]$.

An interval $A = [a^L, a^R]$ can also be expressed by its mean a^C and half width length (or radius) a^W. So $A = \langle a^C, a^W \rangle = \{a : (a^C - a^W) \leq a \leq (a^C + a^W), a \in R\}$, where $a^C = (a^L + a^R)/2$ and $a^W = (a^R - a^L)/2$. A precise parametric representation of an interval $[a^L, a^R]$ as $\{a_t \,|\, t \in [0, 1]\}$ with algebraic operations is defined by Bhurjee and Panda [14] and shown that the traditional algebraic operations of intervals defined either the forms of lower and upper bound or mean and spread of the intervals are equally performed with respect to parameters.

2.1 Partial ordering on $\mathbb{I}(\mathbb{R})$

In order to compare two closed intervals, it is necessary to define a partial ordering on $\mathbb{I}(\mathbb{R})$. Many partial order relations are available in the literature [14, 28, 29]. According to Moore [28], the order relation between two closed intervals A and B can be explained as an extension of "$<$" on the real line, which is "$A < B$ iff $a^R < b^L$," or as an extension of set inclusion, which is "$A \subseteq B$ iff $a^L \geq b^L$ and $a^R \leq b^R$." These order relations cannot explain ranking between two partially overlapping intervals. Ishibuchi and Tanaka [29] proposed some improved definitions on interval order relations for analyzing the mathematical programming with intervals coefficients. These order relations represent the decision makers preference between the intervals. They defined a partial order relation "$\preceq_{LR}$" in $\mathbb{I}(\mathbb{R})$ as,

$$A \preceq_{LR} B \quad \text{iff} \quad a^L \leq b^L \text{ and } a^R \leq b^R,$$
$$A \prec_{LR} B \quad \text{iff} \quad A \preceq_{LR} B \text{ and } A \neq B.$$

2.2 Interval-valued function

Interval-valued function is defined in different ways by many researchers in the literature. Moore [28] and Hansen [30] defined an interval function as the extension of real valued function with one or more interval arguments onto an interval. Ishibuchi and Tanaka [29] considered the interval-valued function, $\mathbf{F} : \mathbb{R}^n \to \mathbb{I}(\mathbb{R})$ as, $\mathbf{F}(x) = [F^L(x), F^R(x)]$, where $F^L, F^R : \mathbb{R}^n \to \mathbb{R}, F^L(x) \leq F^R(x) \ \forall x \in \mathbb{R}^n$.

3. Existing approaches for solving nonlinear interval optimization problem

This section provides a chronological overview of the methodologies that have been developed to deal with the nonlinear programming problem with interval uncertainty.

3.1 Numeric method to interval quadratic programming

A quadratic programming problem has been discussed by Liu and Wang [7] with coefficients of the objective function and constraints are closed intervals, while quadratic part of the objective function is free from interval uncertainty. Quadratic problem is the one of the best and powerful techniques increasing the productivity and rectifying the efficiency of any organization. The aim of this model is to find conventional one-level quadratic programs where the objective function have to a convex and the constraints have to a linear. They formulated a pair of two-level mathematical programs to calculate the bounds for the objective values of the problem and using the duality theorem according to Dorn [31], and transformation technique two-level mathematical programming technique can be transformed into a pair to one-level quadratic programs in which numerical techniques can be applied. They also explained the requirements of probability distribution to construct a posterior frequency determination. And alternative that apply to interval estimation to represent the uncertain parameters, instead of single values.

Consider the interval quadratic programming of the following form:

$$\min \quad \mathbf{Z} = \mathbf{C}_v^{nT} x \oplus \frac{1}{2} x^T Q x,$$

$$\text{subject to} \quad \left. \begin{array}{l} \sum_{j=1}^{n} [a_{ij}^L, a_{ij}^R] x_j \preceq \mathbf{B}_i, \ i = 1, 2, \ldots, m, \\ x_j \geq 0, \ j = 1, 2, \ldots, n, \end{array} \right\} \tag{1}$$

where $\mathbf{C}_v^n = (\mathbf{C}_1, \mathbf{C}_2, \ldots, \mathbf{C}_n)^T, \mathbf{C}_j \in \mathbb{I}(\mathbb{R})$; Q is a symmetric matrix of order $n \times n$, semidefinite, and $\mathbf{B}_i \in \mathbb{I}(\mathbb{R})$.

The Lagrangian duality say that if one problem is unbounded, then other is feasible. Next statement say that if both problems are feasible, then they have optimal solution with same objective value.

The deterministic lower level problem of interval quadratic programming is stated as follows.

$$z^L = \min \mathbf{Z} = \sum_{j=1}^{n} c_j^L x_j + \frac{1}{2} \sum_{i=1}^{n} \sum_{j=1}^{n} q_{ij} x_i x_j,$$

$$\text{subject to} \quad \left. \begin{array}{l} \sum_{j=1}^{n} a_{ij}^L x_j \leq b_i^R, \ i = 1, 2, \ldots, m, \\ x_j \geq 0, \ j = 1, 2, \ldots, n. \end{array} \right\} \tag{2}$$

This is a general quadratic programming problem. The Karush–Kuhn–Tucker (KKT) conditions for quadratic problem can be written in form of a linear complementary problem.

The deterministic upper level problem of interval quadratic programming is stated as follows.

$$z^R = \max_{x, \lambda, \delta} Z = -\frac{1}{2} \sum_{i=1}^{n} \sum_{j=1}^{n} q_{ij} x_i x_j - \sum_{i=1}^{m} b_i^L \lambda_i,$$

$$\text{subject to} \quad \left. \begin{array}{l} \sum_{i=1}^{n} q_{ij} x_i + \sum_{i=1}^{m} r_{ij} - \delta_j = -c_j, \; j = 1, 2, \ldots, n, \\[2mm] c_j^L \leq c_j \leq c_j^R, \; j = 1, 2, \ldots, n, \\[2mm] a_{ij}^L \lambda_i \leq r_{ij} \leq a_{ij}^R \lambda_i, \; i = 1, 2, \ldots, m, \; j = 1, 2, \ldots, n, \\[2mm] \lambda_i, \delta_j \geq 0, \; i = 1, 2, \ldots, m, \; j = 1, 2, \ldots, n, \end{array} \right\} \quad (3)$$

where $r_{ij} = a_{ij} \lambda_i$, $a_{ij} \in [a_{ij}^L, a_{ij}^R]$. This is a general quadratic programming problem. Both the above problems can be solved using general quadratic programming technique. Optimal values of the lower and higher level problems are declared as the optimal bounds of the original problem.

3.2 Numeric method for general interval quadratic programming

Li and Tian [8] generalized Liu and Wang's [7] method for solving a general interval quadratic programming problem, where all coefficients in the objective function and constraints are interval numbers. The advantage of this model is that it requires less computational compared to Liu and Wang's [7] methods. Further, they formulate a numerical example that give a whole idea about general quadratic problem. They used an interval coefficient to describe the imprecise and uncertain elements of real life problem. Because it is not easy to specify the membership function or the probability distribution in an inexact environment.

Consider the interval quadratic programming of the following form:

$$\min \quad \mathbf{Z} = \mathbf{C}_v^{nT} x \oplus \frac{1}{2} x^T \mathbf{Q}_m x,$$

$$\text{subject to} \quad \left. \begin{array}{l} \sum_{j=1}^{n} [a_{ij}^L, a_{ij}^R] x_j \preceq \mathbf{B}_i, \; i = 1, 2, \ldots, m, \\[2mm] x_j \geq 0, \; j = 1, 2, \ldots, n, \end{array} \right\} \quad (4)$$

where $\mathbf{C}_v^n = (\mathbf{C}_1, \mathbf{C}_2, \ldots, \mathbf{C}_n)^T$, $\mathbf{C}_j \in \mathbb{I}(\mathbb{R})$; $\mathbf{Q}_m$ is an interval matrix of order $n \times n$, which is symmetric positive semidefinite. $\mathbf{Q}_m = ([q_{ij}^L, q_{ij}^R])_{n \times n}$; and $\mathbf{B}_i \in \mathbb{I}(\mathbb{R})$. Denote

$S = \{(c_j, b_i, a_{ij}, q_{ij}) | c_j \in \mathbf{C}_j, \; b_i \in \mathbf{B}_i, \; a_{ij} \in [a_{ij}^L, a_{ij}^R], \; q_{ij} \in [q_{ij}^L, q_{ij}^R], i = 1, 2,$
$\ldots, m; j = 1, 2, \ldots, n\}$. For inequality constraint of the "$\geq$" form, and it can be transformed into the form of "$\leq$" just multiplying the both side by -1.

If the objective function is "max," then it can be changed to "min" and vice-versa.

The objective is to find the upper and lower bound of the range of the objective values by employing the two-level mathematical programming technique.

The deterministic lower level problem of interval quadratic programming is stated as follows.

$$z^L = \min \mathbf{Z} = \sum_{j=1}^{n} c_j^L x_j + \frac{1}{2} \sum_{i=1}^{n} \sum_{j=1}^{n} q_{ij}^L x_i x_j,$$

subject to
$$\left. \begin{array}{l} \sum_{j=1}^{n} a_{ij}^L x_j \leq b_i^R, \ i = 1, 2, ..., m, \\ x_j \geq 0, \ j = 1, 2, ..., n. \end{array} \right\} \tag{5}$$

This is a general quadratic programming problem.

The deterministic upper level problem of interval quadratic programming is stated as follows.

$$z^R = \max_{x, \lambda, \delta} \mathbf{Z} = -\frac{1}{2} \sum_{i=1}^{n} \sum_{j=1}^{n} q_{ij}^L x_i x_j - \sum_{i=1}^{m} b_i^L \lambda_i,$$

subject to
$$\left. \begin{array}{l} \sum_{i=1}^{n} q_{ij}^R x_i + \sum_{i=1}^{m} r_{ij} - \delta_j = -c_j^R, \ j = 1, 2, ..., n, \\ a_{ij}^L \lambda_i \leq r_{ij} \leq a_{ij}^R \lambda_i, \ i = 1, 2, ..., m, \ j = 1, 2, ..., n, \\ \lambda_i, \delta_j \geq 0, \ i = 1, 2, ..., m, \ j = 1, 2, ..., n, \end{array} \right\} \tag{6}$$

where $r_{ij} = a_{ij} \lambda_i$, $a_{ij} \in [a_{ij}^L, a_{ij}^R]$. This is a general quadratic programming problem. Both the above problems can be solved using general quadratic programming technique. Optimal values of the lower and higher level problems are declared as the optimal bounds of the original problem.

Proposed Approach: The following steps describe the proposed approach:

Step 1. Consider a general interval quadratic programming problem and write its matrix form stated as (4).

Step 2. Construct the deterministic lower and upper level problems of interval quadratic programming problems are stated as (5) and (6), respectively.

Step 3. Solve these deterministic lower and upper level problems using any existing optimization software; and find the lower and upper level objective values as Z^L and Z^R, respectively.

Step 4. Finally, one can obtain the optimal bounds of the objective function of the original considered problem as $[Z^L, Z^R]$.

This methodology can be well understood through the following example.

Example 1 (Numerical example due to Li and Tian [8]).

$$\min \ \mathbf{Z} = [-10, \ -6]x_1 \oplus [2, 3]x_2 \oplus [4, 10]x_1^2 \oplus [-1, 1]x_1x_2 \oplus [10, 20]x_2^2$$

$$\text{subject to} \quad [1, 2]x_1 \oplus [3, 3]x_2 \preceq [1, 10],$$

$$[-2, 8]x_1 \oplus [4, 6]x_2 \preceq [4, 6], x_1, x_2 \geq 0.$$

Here the objective function can be expressed as

$$\mathbf{Z} = [-10, -6]x_1 \oplus [2,3]x_2 \oplus [4,10]x_1^2 \oplus [-1,1]x_1x_2 \oplus [10,20]x_2^2$$

$$= ([-10, -6], [2,3]) \begin{pmatrix} x_1 \\ x_2 \end{pmatrix} \oplus \frac{1}{2}(x_1, x_2) \begin{pmatrix} [8,20] & [-1,1] \\ [-1,1] & [20,40] \end{pmatrix} \begin{pmatrix} x_1 \\ x_2 \end{pmatrix}.$$

The upper bound of the objective value z^R can be formulated as

$$z^R = \max_{x, \lambda, \delta} \mathbf{Z} = -10x_1^2 - x_1x_2 - 20x_2^2 - \lambda_1 - 4\lambda_2$$

$$\left. \begin{array}{c} 20x_1 + x_2 + r_{11} + r_{21} - \delta_1 = 6, \\[4pt] x_1 + 40x_2 + r_{12} + r_{22} - \delta_2 = -3, \\[4pt] \text{subject to} \quad \lambda_1 \leq r_{11} \leq 2\lambda_1, \ -2\lambda_2 \leq r_{21} \leq 8\lambda_2, \\[4pt] r_{12} = \lambda_1, 4\lambda_2 \leq r_{22} \leq 6\lambda_2, \\[4pt] \lambda_1, \lambda_2, \delta_1, \delta_2 \geq 0. \end{array} \right\} \tag{7}$$

This model is a general quadratic programming problem. The optimum solution is

$$z^R = -0.9, \quad x_1 = 0.3, \quad x_2 = 0.$$

The lower bound of the objective value z^L can be formulated as

$$z^L = \max_x \mathbf{Z} = -10x_1 + 2x_2 + 4x_1^2 - x_1x_2 + 10x_2^2$$

$$\left. \begin{array}{c} x_1 + 3x_2 \leq 10, \\[4pt] \text{subject to} \quad -2x_1 - 4x_2 \leq 6, \\[4pt] x_1, x_2 \geq 0. \end{array} \right\} \tag{8}$$

This model is also a general quadratic programming problem. The optimum solution is

$$z^L = -6.25, \quad x_1 = 1.25, \quad x_2 = 0.$$

Combining these two results, they have concluded that the optimal bounds of objective values of this interval quadratic programming problem are -6.25 and -0.9.

One may observe that the lower bound of the given problem, which is $z^L = -6.25$ occurs at the point $(x_1, x_2) = (1.25, 0)$ and the upper bound of the given problem, which is $z^R = -0.9$ occurs at the point $(x_1, x_2) = (0.3, 0)$. Points $(1.25, 0)$ and $(0.3, 0)$ are not the optimal solutions of this problem. These are the points which provide the optimal bounds. One of these points lie in the maximum feasible region and other lies in the minimum feasible region.

3.3 Nonlinear interval method for uncertain optimization

Jiang et al. [9] solved NLIOP by transforming the uncertain single-objective optimization into a deterministic two-objective optimization programming problem, where the midpoint and radius of the uncertain objective function simultaneously. In this paper they have shown that the two-objective functions are converted into a single-objective problem through the linear combination method, and the deterministic inequality constraints will treated with the penalty function technique. The constraints and objective function are linked to the stochastic processes, fuzzy sets and their membership function [32]. These kind of approaches play an important roles. Some we are not able to apply these type of approaches in uncertain environment. For the most of real-world problem, the objective function and constraints are nonlinear, they can be solve by some of strong algorithms such as finite element method (FEM) instead as explicit form Quan et al. [33]. They also used the IP-GA for optimize the penalty function and obtain the intervals of objective function and constraints.

A general NLIOP with interval coefficients in both of the objective function and constraints can be given as follows:

$$\begin{aligned}
\min \; & f(X, U) \\
\text{subject to} \; & g_i(X, U) \geq (=, \leq)[v_i^L, v_i^R], \; i = 1, 2, \ldots, l, \\
& X \in \Omega^n, U = [U^L, U^R],
\end{aligned} \tag{9}$$

where X is an n-dimensional decision vector and Ω^n is its range.

The NLIOP problem (9) is transformed into the following deterministic two-objective programming problem:

$$\min \ \{f_m(X, U), f_w(X, U)\}$$
$$\text{subject to} \quad P_{M_i \geq N_i} \geq \lambda_i, \ i = 1, 2, \ldots, k,$$
$$f_m(X, U) = \frac{1}{2} \left(\min_{U \in \Gamma} f(X, U) + \max_{U \in \Gamma} f(X, U) \right),$$
$$f_w(X, U) = \frac{1}{2} \left(\max_{U \in \Gamma} f(X, U) - \min_{U \in \Gamma} f(X, U) \right),$$
$$X \in \Omega^n, U \in \Gamma = \{U : U \in [U^L, U^R], \lambda_i \in [0, 1],$$

where M_i and N_i can be both intervals or one of them is a real number, k is the number of the uncertain inequality constraints after transformation of the uncertain equality constraints. (For detail see the paper of Jiang et al. [9]).

3.4 Optimal value bound method for nonlinear programming

Hladik [10] obtained optimal bounds of the objective function of NLIOP by solving two deterministic optimization problem namely, lower and upper bounds of optimization problem using some assumptions. He explained that cases of better approximation, where duality gap become zero due to independence relation between the interval quantities. He discussed the approach of convex quadratic programs and posynomial programs in very easy manner. Alefeld and Herzberger [34] have introduced some notion from interval analysis. further we will use this technique.

An interval nonlinear program means the family of nonlinear programs

$$f(A, c) = \inf f_c(x) \text{ subject to } F_A(x) \leq 0, \tag{10}$$

where $f_c : R^n \rightarrow R$ is a real function with input data c varying inside an interval vector $\mathbf{c}$, and $F_A : R^n \rightarrow R^m$ is a vector function with input data A varying inside an interval matrix $\mathbf{A}$. We suppose that $f_c(x)$ and $F_A(x)$ do not have an interval parameter in common.

We consider any dual problem to (10) having the form of

$$g(A, c) = \sup g_{A,c}(y) \text{ subject to } G_{A,c}(y) \leq 0. \tag{11}$$

Here in $g_{A,c} : R^k \rightarrow R$, and $G_{A,c} : R^k \rightarrow R^l$ are functions depending on $A \in \mathbf{A}$ and $c \in \mathbf{c}$. The set

$$M = \{x \in R^n \, | \, F_A(x) \leq 0, A \in \mathbf{A}\},$$
$$N = \{y \in R^m \, | \, G_{A,c}(y) \leq 0, A \in \mathbf{A}, c \in \mathbf{c}\}$$

Next, we introduce the functions

$$f(x) = \inf f_c(x) \text{ subject to } c \in \mathbf{c},$$
$$g(y) = \sup g_{A,c}(y) \text{ subject to } A \in \mathbf{A}, \, c \in \mathbf{c}.$$

Now, we compute the bounds of optimal value of the objective function as follows.

Case 1. Consider a convex quadratic programming problem

$$\min \; x^T C x + d^T x \text{ subject to } Ax \leq b, \, x \geq 0, \tag{12}$$

where C, A, b, and d vary in given interval matrices $\mathbf{C}, \mathbf{A}$ and interval vectors $\mathbf{b}$ and $\mathbf{d}$. Suppose that C is positive semidefinite for all $C \in \mathbf{C}$. Then the range $[f^L, f^R]$ in which the optimal value of (12) varies is calculated as follows.

$$\left.\begin{array}{l} f^L = \inf x^T C^L x + \left(d^L\right)^T x \text{ subject to } A^L x \leq b^R, \, x \geq 0, \\ f^R = \inf x^T C^R x + \left(d^R\right)^T x \text{ subject to } A^R x \leq b^L, \, x \geq 0. \end{array}\right\} \tag{13}$$

Case 2. Consider a convex general quadratic programming problem

$$\min \; x^T C x + d^T x \text{ subject to } Ax \leq b, \, Ex = h, \, x \geq 0, \tag{14}$$

where $C \in \mathbf{C}, A \in \mathbf{A}, b \in \mathbf{b}, d \in \mathbf{d}, E \in \mathbf{E}$, and $h \in \mathbf{h}$. Suppose that C is positive semidefinite for all $C \in \mathbf{C}$. Then the range $[f^L, f^R]$ in which the optimal value of (14) varies is calculated as follows.

$$f^L = \inf x^T C^L x + \left(d^L\right)^T x$$
$$\text{subject to } A^L x \leq b^R, \, E^R x \leq h^L, \, E^L x \leq h^R, \, x \geq 0,$$
$$f^R \geq \sup -x^T C^R x - \left(b^L\right)^T u - h_c^T v + h_\triangle^T |v|$$
$$\text{subject to } 2C^R x + \left(A^R\right)^T u + E_c^T v + E_\triangle^T |v| + d^R \geq 0, \, u \geq 0,$$

and equality holds if the system

$$2C^R x + \left(A^R\right)^T u + \left(E^L\right)^T v^1 - \left(E^R\right)^T v^2 + d^R \geq 0, \, u \geq 0,$$
$$v^1 \geq 0, \, v^2 \geq 0 \tag{15}$$

is solvable. System (15) is particularly true if C^R is positive definite.

Example 2. Consider the interval optimization problem,

$$(IOP) \; \min\{[-10, -6]x_1 + [2, 3]x_2 + [4, 10]x_1^2 + [-1, 1]x_1 x_2 + [10, 20]x_2^2\}$$
$$\text{s.t.} \quad [1, 2]x_1 + 3x_2 \preceq [1, 10], [-2, 8]x_1 + [4, 6]x_2 \preceq [4, 6], x_1, x_2 \geq 0.$$

Thus the corresponding interval matrices and vectors are

$$\mathbf{C} = \begin{pmatrix} [4,10] & [-0.5,0.5] \\ [-0.5,0.5] & [10,20] \end{pmatrix}, \quad \mathbf{d} = \begin{pmatrix} [-10,-6] \\ [2,3] \end{pmatrix},$$

$$\mathbf{A} = \begin{pmatrix} [1,2] & [3,3] \\ [-2,8] & [4,6] \end{pmatrix}, \quad \mathbf{d} = \begin{pmatrix} [1,10] \\ [4,6] \end{pmatrix}.$$

Using (13), we determine the lower bound of the optimal value function by computing the convex quadratic program

$$f^L = \inf(-4x^2 - x_1x_2 + 10x_2^2 - 10x_1 + 2x_2)$$

$$\text{subject to} \quad x_1 + 3x_2 \le 10, \; -2x_1 + 4x_2 \le 6, x_1 \ge 0, x_2 \ge 0.$$

and

$$f^R = \inf(-10x^2 + x_1x_2 + 20x_2^2 - 6x_1 + 3x_2)$$

$$\text{subject to} \quad 2x_1 + 3x_2 \le 1, 8x_1 + 6x_2 \le 4, x_1 \ge 0, x_2 \ge 0.$$

Using Lingo software, the solutions of these programs are $x_1 = 1.25, x_2 = 0$, $f^L = -6.25$ and $x_1 = 0.30, x_2 = 0, f^R = -0.90$, respectively. Hence the range of optimal value of the problem lies in the interval $[-6.25, -0.90]$.

3.5 Efficient solution method for interval optimization problem

Bhurjee and Panda [14] obtained efficient solution of the interval optimization problem using duality theory. They explained interval-valued function, its convex property, and solution of convex interval-valued programming problem developing a methodology for general optimization problem [35]. The linear interval value objective function are independent from interval uncertainty in collection of constraints [36]. This methodology applied in the interval-valued convex quadratic programming problem to find the optimal value of interval and it corresponding decision variables. Convexity play an important role to prove the existence of the solution of general optimization problem. There is nice application of this paper is that it can be applicable any type of optimization problems like convex, nonconvex, linear, nonlinear.

Consider a general interval optimization problem as follows.

$$(IOP) \qquad \min \; \mathbf{F}_{\mathbf{C}_v^k}(x)$$

$$\text{subject to} \quad \mathbf{G}_{\mathbf{D}_v^k}^j(x) \preceq [b_j^L, \, b_j^R], \, j \in \Lambda_p,$$

where $[b_j] \in \mathbb{I}(\mathbb{R})$, the interval-valued functions $\mathbf{F}_{\mathbf{C}_v^k}, \mathbf{G}_{\mathbf{D}_v^k}^j : \mathbb{R}^n \to \mathbb{I}(\mathbb{R})$ are the sets,

$$\mathbf{F}_{\mathbf{C}_v^k}(x) = \left\{ f_{\mathbf{c}_t}(x) \mid f_{\mathbf{c}_t} : \mathbb{R}^n \to \mathbb{R}, \ \mathbf{c}_t \in \mathbf{C}_v^k \right\},$$

$$\mathbf{G}_{\mathbf{D}_v^k}^j(x) = \left\{ g_{\mathbf{d}_{t_j}}^j(x) \mid g_{\mathbf{d}_{t_j}}^j : \mathbb{R}^n \to \mathbb{R}, \ \mathbf{d}_{t_j} \in \mathbf{D}_v^k \right\}.$$

Following the partial orderings as discussed in Section 2, the feasible region of (*IOP*) can be expressed as the set,

$$S = \left\{ x \in \mathbb{R}^n : \mathbf{G}_{\mathbf{D}_v^k}^j(x) \preceq [b_j^L, b_j^R], \ j \in \Lambda_p \right\}$$

$$= \left\{ x \in \mathbb{R}^n : g_{\mathbf{d}_{t_j}}^j(x) \leq b_{t_j}, \ j \in \Lambda_p \right\}$$

$$\equiv \left\{ x \in \mathbb{R}^n : \max_{\mathbf{d}_{t_j} \in [\mathbf{d}_j]} g_{\mathbf{d}_{t_j}}^j(x) \leq \max_{t_j} b_{t_j}, \ \min_{\mathbf{d}_{t_j} \in [\mathbf{d}_j]} g_{\mathbf{d}_{t_j}}^j(x) \leq \min_{t_j} b_{t_j}, \ j \in \Lambda_p \right\}$$

Definition 1. A point $x^* \in S$ is called an efficient solution of *IOP*, if there exist no $x \in S$ such that for every $t \in [0, 1]$,

$$f_{\mathbf{c}_t}(x) \leq f_{\mathbf{c}_t}(x^*) \text{ and for at least one } t' = t, \ f_{c_{t'}}(x) < f_{c_{t'}}(x^*).$$

For any choice of preference functions $p_j : [0, 1]^{k_j} \to \mathbb{R}_+$ such as $\int_k p(t) dt = 1$, consider the following multiobjective optimization problem which is free from the interval uncertainty.

$$(IOP_I) \qquad \min_{x \in S} \int_k p(t) f_{\mathbf{c}_t}(x) dt,$$

where $\int_k \equiv \underbrace{\int_0^1 \int_0^1 \dots \int_0^1}_{k_j \text{ times}}, \ dt = dt_1 dt_2 \dots dt_k.$

The following theorem establishes the relationship between the solution of the transformed problem IOP_I and the original problem IOP. This methodology is explained in the following numerical example in Li and Tian [8].

Example 3. Consider the interval optimization problem,

$(IOP) \ \min\{[-10, -6]x_1 + [2, 3]x_2 + [4, 10]x_1^2 + [-1, 1]x_1 x_2 + [10, 20]x_2^2\}$

$\quad$ s.t. $[1, 2]x_1 + 3x_2 \preceq [1, 10], [-2, 8]x_1 + [4, 6]x_2 \preceq [4, 6], x_1, x_2 \geq 0.$

According to Li and Tian [8], the lower level problem has an optimal value -6.25 corresponding to the solution $(1.25, 0)$ and the upper level problem has an optimal value -0.9 corresponding to the solution $(0.3, 0)$. By the methodology described in this paper, we get an efficient solution.

Here $t = (t_1, t_2, t_3, t_4, t_5)^T$, $t_j \in [0, 1]$. $f_{c(t)}(x) = (-10 + 4t_1)x_1 + (2 + t_2)x_2 + (4 + 6t_3)x_1^2 + (-1 + 2t_4)x_1x_2 + (10 + 10t_5)x_2^2$ and $F = \{(x_1, x_2) : 2x_1 + 3x_2 \leq 1, 8x_1 + 6x_2 \leq 4, x_1, x_2 \geq 0\}$. For some $w : [0, 1]^5 \rightarrow R$, the corresponding (IOP_l) becomes:

$$\min_F \int_0^1 \cdots \int_0^1 w(t)\{(-10 + 4t_1)x_1 + (2 + t_2)x_2 + (4 + 6t_3)x_1^2$$

$$+ (-1 + 2t_4)x_1x_2 + (10 + 10t_5)x_2^2\}\mathbf{dt},$$

where $\mathbf{dt} = dt_1dt_2dt_3dt_4dt_5$. For a particular weight function $w(t) = t_1 + t_3$, the solution of the above problem is $x^* = (x_1^*, x_2^*) = (0.5, 0)$. The corresponding compromising optimal value of (IOP) is $[-4, -0.5]$.

3.6 Alternative method for interval objective constrained optimization problems

Karmakar and Bhunia [11] obtained an alternative optimization technique for interval objective constrained optimization problems via multiobjective programming. They have solved the optimal value of considered problems in the form of an interval with minimum uncertainty. The objective of this technique is to find such solutions with higher accuracy and lower computational cost with an adequate number of examples using Refs. [37–41]. They used this technique because it is very useful to tackle the uncertainty in different branches of Operation Research and Management Science.

Consider a special form of a constrained optimization problem with interval-valued objective function as follows:

$$\max Z = F(x, U)$$
$$\text{subject to } g_\lambda(x) \leq 0, \quad \lambda = 1, 2, ..., k,$$
$$h_\mu(x) = 0, \quad \mu = 1, 2, ..., m,$$
$$x \in D \subseteq \mathbb{R}^n,$$

where $F : \mathbb{R}^n \rightarrow \mathbb{I}(\mathbb{R})$ be an interval-valued function, $x = (x_1, x_2, ..., x_n)$ be an n-dimensional decision vector, $U = (U_1, U_2, ..., U_q)$ be a q-dimensional interval vector whose components are all intervals; D is the n-dimensional interval (or box) and is given by $D = \{x \in \mathbb{R}^n : l \leq x \leq u\}$. Here $l, u \in \mathbb{R}^n$ be two vectors given by $l = (l_1, l_2, ..., l_n)$ and $u = (u_1, u_2, ..., u_n)$ such that $l_j \leq x_j \leq u_j, (j = 1, 2, ..., n)$.

Definition 2. A decision vector $x^* \in D$ is a minimum point if $F_m(x^*) \leq F_m(x)$ (maximum if $F_m(x^*) \geq F_m(x)$ for maximization problem) and $F_w(x^*) \leq F_w(x)$ for any $x \in D$. In this case, the minimum value is denoted by F^* and the minimizer point by x^*, i.e., $F^* = \min_{x \in D} F(x, U) = F(x^*, U)$.

Solution procedure: The general structure of the proposed optimization technique is comprised by the following steps:

- Representation of an interval function in its mean and width form:

$$F(x, U) = U_1 f_1(x) \pm U_2 f_2(x) \pm \cdots \pm U_n f_n(x)$$
$$= \langle U_{1m}, U_{1w} \rangle f_1(x) \pm \langle U_{2m}, U_{2w} \rangle f_2(x) \pm \cdots \pm \langle U_{nm}, U_{nw} \rangle f_n(x)$$

- Construction of biobjective optimization problem: The problems for which the interval objective function is explicitly expressible in terms of mean and width, the biobjective optimization problem can be constructed directly as follows.

$$\max_{x \in S} \{F_m, F_w\},$$

where $F_m = U_{1m} f_1(x) \pm U_{2m} f_2(x) \pm \cdots \pm U_{nm} f_n(x)$ and $F_w = U_{1w} |f_1(x)| \pm U_{2w} |f_2(x)| \pm \cdots \pm U_{nw} |f_n(x)|$.

- Solution of biobjective problem: The Pareto optimal solution for the constructed biobjective optimization problem is obtained by the Global Criterion Method (GCM) or any other suitable methods, depending on the problem consideration and requirement of the decision maker.

Example 4.

$$\min \quad F(U,x) = U_1 \log(x_2 + 1) + U_2 \log(x_1 - x_2 + 1) + U_3 x_4$$
$$+ U_4 x_5 + U_5 x_6 + U_6 x_1 - 7x_3 + U_7$$
$$\text{subject to} \quad 0.8 \log(x_2 + 1) - 0.96 \log(x_1 - x_2 + 1) + 0.8x_3 \leq 0,$$
$$- \log(x_2 + 1) - 1.2 \log(x_1 - x_2 + 1) + x_3 + 2x_2 - 2 \leq 0,$$
$$x_2 - x_1 \leq 0, -x_2 + x_1 - 2x_5 \leq 0, x_2 - 2x_4 \leq 0, x_4$$
$$+ x_5 - 1 \leq 0,$$

where $U_1 = [-18.5, -17.5]$, $U_2 = [-19.5, -19]$, $U_3 = [4.5, 5.7]$, $U_4 = [5.7, 6.3]$, $U_5 = [7.5, 8.3]$, $U_6 = [9.9, 10.5]$, $U_7 = [9.9, 10.9]$ and $x_1, x_2 \in [0, 2]$, $x_3, x_4, x_5, x_6 \in [0, 1]$. Hence, the corresponding biobjective optimization problem is as follows:

$$\min \ F_c = -18 \log(x_2 + 1) - 19.25 \log(x_1 - x_2 + 1) + 5.1x_4 + 6x_5$$
$$+ 7.9x_6 + 10.2x_1 - 7x_3 + 10.4,$$
$$\min \ F_w = 0.5|\log(x_2 + 1)| - 0.25|\log(x_1 - x_2 + 1)| + 0.6|x_4| + 0.3|x_5|$$
$$+ 0.4|x_6| + 0.3|x_1| + 0.5$$

subject to the constraints as given in the original problem.

Here, the ideal objective vector is $(1.39204, 0.0)$ and the Pareto optimal solution is

$$x^* = (1.13795, 0.435103, 1.0, 0.217551, 0.351424, 0.0)^T$$

with $F_{\min} = [0.085043, 2.686498]$.

According to Mahato and Bhunia [42] order relation, the above solutions are incomparable. The expected value of new solution is worse than the previous one, but uncertainty our this solution is very less than the Inuiguchi and Sakawa [43].

3.7 Method of generalized interval vector spaces and optimization

Costa et al. [15] formulated new generalized interval optimization problems and relate them to classic multiobjective optimization problems (see [44]). This technique was obtained help of the concepts of algebraic and order relations, and of a bijection mapping between the set of generalized interval and the set R^{2n} and a vector structures. In this process they used unifying approach to generate interval optimization problem from algorithm. An interval optimization problem is considered as follows. They provided Neumann's theorem to solve interval-valued problems because it help us to find optimality and bijective function helped us to get results between M^n and subset $\mathbb{I}(\mathbb{R})^n$, according to Wu [45, 46].

Consider the interval optimization problem

$$(IOP) \quad \min_{x \in X} \ F(x),$$

where $X \subset \mathbb{R}^n$, and $F : X \to \mathbb{I}(\mathbb{R})$ are interval-valued functions such that $F(x) = [f^L(x), f^R(x)]$, with $f^L(x) \leq f^R(x)$ for all $x \in X$.

Then, they show that IOP is equivalent to a vector interval optimization problem $V IOP$, which is defined as follows.

$$(VIOP) \quad \min_{x \in X} \ \phi(F(x)),$$

where $\phi(F(x)) : X \to \mathbb{R}^2$ is a function. If we consider an order relation $\leqq_{\mathbb{R}^2}$ in $\mathbb{R}^2$ then we have the following concept of solution.

Definition 3. $x^* \in X$ is a solution of $V\ IOP$ if and only if there is no $x \in X$, $x^* \neq x$ such that

$$\phi(F(x)) \leqq \phi(F(x^*))$$

Theorem 1. $x^* \in X$ is an efficient solution of $V\ IOP$ if and only if $x^* \in X$ is a solution of IOP.

Example 5. We consider the following interval optimization problem

$$(IOP) \quad \min_{x \in [0,\,2]} \ F(x) = [x^2 - x, 3x^2 - x]. \tag{16}$$

Consider the bijective function ϕ_1 defined in Ref. [15], then the problem (16) is equivalent to the following biobjective programming problem as

$$(VIOP) \quad \min_{x \in [0,\,2]} \ \{3x^2 - x, x^2 - x\}$$

The efficient solution of $V\ IOP$ is $x^* = 1/4 \in [0, 2]$. From Theorem 1, $x^* = 1/4$ is one of the efficient solution of IOP.

3.8 Numerical method for interval quadratic problems and Swarm algorithms

Elsisy et al. [24] formulated the solution of interval quadratic programming problem help of numerical methods and swarm algorithms. The aim of author is to find optimal solution of feasible region. Advantages of these techniques are this is very fast in decision making for optimal solution of interval, effectiveness is more reliable compare to others, very useful to solve multiobjective linear programming with interval coefficients, this is more general technique. Approaches and methods have taken from help of these following papers [24, 47–57], etc.

An interval quadratic programming problem (IQPP) is considered as follows:

$$\min \ \sum_{i=1}^{k} \gamma_i^f f_i(x)$$

$$\text{subject to} \ \sum_{i=1}^{l} \gamma_{ij}^f g_{ij}(x) \leq \gamma_j^R, \quad j = 1, 2, \ldots, m,$$

where $\gamma_i^f = [\gamma_i^{fL}, \gamma_i^{fU}] \forall i$, $f_i(x)$ is quadratic function, $g_{ij}(x), i = 1, \ldots, l$, $j = 1, \ldots, m$ are the linear functions, $\gamma_{ij}^f = [\gamma_{ij}^{fL}, \gamma_{ij}^{fU}]$, and $\gamma_j^R = [\gamma_j^{RL}, \gamma_j^{RU}]$. It is assumed that feasible region is nonempty.

Proposed Approach: The following steps describe the proposed approach:

Step 1. Replacing all intervals in IQPP by additional variables which is called modified quadratic programming problem (MQPP) and obtaining the dual form of MQPP.

Step 2. Constructing KKT for MQPP, and solving KKT equations by the numerical algorithm.

Step 3. Using the solutions of the numerical algorithm as a start point of chaotic particle swarm optimization algorithm (CPSO) and chaotic firefly algorithm (CFA).

Step 4. Solving MQPP and its dual form by CPSO and CFA.

Step 5. The values of the objective function which are obtained from solving the problem by SA and its dual form are compared. If their values are the same, the global optimal solution of our problem is found. If there is a difference between the outputs from the problem and its dual, we solve the problem and its dual form again until the difference between them is ϵ, where ϵ can be computed a

$$\epsilon = \frac{\delta}{\text{the optimal value of the problem}}, \tag{17}$$

where δ is the difference between the optimal value of the problem and the optimal value of its dual problem. This solution is a local optimal solution. This comparison is used as a new stopping criterion. The suggested method is suitable for convex and nonconvex problems.

Example 6. Consider the following interval quadratic programming problem:

$$\begin{aligned}
\min \quad & [2,3]x_1^2 + 2x_2^2 - 2x_1x_2 + [-5, -3]x_1 + [1,2]x_2, \\
\text{subject to} \quad & [1,2]x_1 + x_2 \leq [2,4], \\
& [2,3]x_1 + [-1, -0.5]x_2 \leq [3,4], \\
& x_1, x_2 \geq 0.
\end{aligned} \tag{18}$$

By replacing all intervals by additional parameters, the problem becomes

$$\begin{aligned}
\min \quad & a_1x_1^2 + 2x_2^2 - 2x_1x_2 + a_2x_1 + a_3x_2, \\
\text{subject to} \quad & b_1x_1 + x_2 \leq b_2, \\
& b_3x_1 + b_4x_2 \leq b_5, \\
& x_1, x_2 \geq 0,
\end{aligned} \tag{19}$$

where $a_1 = [2, 3]$, $a_2 = [-5, -3]$, $a_3 = [1, 2]$, $b_1 = [1, 2]$, $b_2 = [2, 4]$, $b_3 = [2, 3]$, $b_4 = [-1, -0.5]$, and $b_5 = [3, 4]$. The dual problem of the problem (19) is

$$\max \quad a_1 x_1^2 + 2x_2^2 - 2x_1 x_2 + a_2 x_1 + a_3 x_2 + u_1(b_1 x_1 + x_2 - b_2)$$
$$+ u_2(b_3 x_1 + b_4 x_2 - b_5) - u_3 x_1 - u_4 x_2,$$

$$\text{subject to} \quad u_1 \geq 0, u_2 \geq 0, u_3 \geq 0, u_4 \geq 0.$$

$$(20)$$

The KKT conditions of the problem (19) are

$$2a_1 x_1 - 2x_2 + a_2 + u_1 b_1 + u_2 b_3 - u_3 = 0,$$
$$4x_2 - 2x_1 + a_3 + u_1 + u_2 b_4 - u_4 = 0,$$
$$u_1(b_1 x_1 + x_2 - b_2) = 0,$$
$$u_2(b_3 x_1 + b_4 x_2 - b_5) = 0,$$
$$u_3 x_1 = 0, u_4 x_2 = 0, \qquad (21)$$
$$b_1 x_1 + x_2 \leq b_2,$$
$$b_3 x_1 + b_4 x_2 \leq b_5,$$
$$x_1 \geq 0, x_2 \geq 0,$$
$$u_1 \geq 0, u_2 \geq 0, u_3 \geq 0, u_4 \geq 0.$$

Problem (16) is divided into two problems. The first problem is

$$\min \quad 2x_1^2 + 2x_2^2 - 2x_1 x_2 - 5x_1 + x_2,$$
$$\text{subject to} \quad x_1 + x_2 \leq 4,$$
$$2x_1 - x_2 \leq 4, \qquad (22)$$
$$x_1, x_2 \geq 0.$$

Its solution is $(x_1, x_2) = (1.5, 0.5)$ and $f(x) = -3.5$. The second problem is

$$\min \quad 3x_1^2 + 2x_2^2 - 2x_1 x_2 - 3x_1 + 2x_2,$$
$$\text{subject to} \quad 2x_1 + x_2 \leq 2,$$
$$3x_1 - 0.5x_2 \leq 3, \qquad (23)$$
$$x_1, x_2 \geq 0.$$

Its solution is $(x_1, x_2) = (0.5, 0)$ and $f(x) = -0.75$. The solution of KKT conditions in (21) can be expressed as

$$\left\{ \begin{array}{l} \left(\dfrac{2a_2 + a_3}{2 - 4a_1}, \dfrac{a_2 + a_1 a_3}{2 - 4a_1} \right), \left(\dfrac{a_2}{2a_1}, 0 \right), \left(\dfrac{-b_2 b_4 + b_5}{b_3 - b_1 b_4}, \dfrac{b_2 b_3 - b_1 b_5}{b_3 - b_1 b_4} \right), \left(\dfrac{b_5}{b_3}, 0 \right), \left(\dfrac{b_2}{b_1}, 0 \right), \\[4mm] \left(\dfrac{2b_2(2b_1 + 1) - a_2 + a_3 b_1}{2a_1 + 4b_1(b_1 + 1)}, \dfrac{a_2 b_1 - a_3 b_1^2 + 2b_2(a_1 + b_1)}{2a_1 + 4b_1(b_1 + 1)} \right) \end{array} \right\}$$

$$(24)$$

In addition, the numerical solution provides the boundaries of the basic variables as $x_1 = [0.4, 4.08333]$ and $x_2 = [-1.6, 3.6]$.

3.9 Solving nonlinear interval optimization problems using stochastic programming

In this approach, a relation has been established between interval and random variable according to the 3 sigma-rule [58]. Using this relation an interval function is associated with function of random variables and interval inequality is converted into chance constraint. The interval optimization problem is then transformed into a nonlinear stochastic programming problem [59–62]. Further, the solution methodology is developed using chance constrained programming technique.

A nonlinear interval programming problem is considered, wherein the parameters of the objective function and constraints are closed interval. The problem is mathematically defined as follows:

$$\textbf{NLIP}: \qquad \min \ \bar{\mathbf{F}}_{\bar{\mathbf{C}}_\nu^k}(\mathbf{x}), \tag{25}$$

$$\text{subject to} \quad \bar{\mathbf{G}}_{i\bar{\mathbf{A}}_{\nu(i)}^k}(\mathbf{x}) \succcurlyeq \bar{\mathbf{B}}_i, \, i = 1, 2, \ldots, m, \tag{26}$$

where $\bar{\mathbf{C}}_\nu^k = (\bar{\mathbf{C}}_1, \bar{\mathbf{C}}_2, \ldots, \bar{\mathbf{C}}_k)^T$, $\bar{\mathbf{A}}_{\nu(i)}^k = (\bar{\mathbf{A}}_{1i}, \bar{\mathbf{A}}_{2i}, \ldots, \bar{\mathbf{A}}_{ki})^T$, $\bar{\mathbf{B}}_{\nu(i)}^k = (\bar{\mathbf{B}}_{1i}, \bar{\mathbf{B}}_{2i}, \ldots, \bar{\mathbf{B}}_{ni})^T$ and for $i = 1, 2, \ldots, m$,

$$\bar{\mathbf{G}}_{i\bar{\mathbf{A}}_{\nu(i)}^k}(\mathbf{x}) = \left\{ g_i(\mathbf{x}; \mathbf{a}_i) \mid g_i : \mathbb{R}^n \to \mathbb{R}, \mathbf{a}_i \in \bar{\mathbf{A}}_{\nu(i)}^k \right\}.$$

Transformation of NLIP to deterministic form: Consider first objective function $\bar{\mathbf{F}}_{\bar{\mathbf{C}}_\nu^k}(\mathbf{x})$ is a interval-valued function. Here $\bar{\mathbf{C}}_\nu^k = (\bar{\mathbf{C}}_1, \bar{\mathbf{C}}_2, \ldots, \bar{\mathbf{C}}_k)^T$, and $\bar{\mathbf{C}}_j, j = 1, 2, \ldots, k$ are interval, so each $\bar{\mathbf{C}}_j$ can be associated with random variable $C_j \sim \mathcal{N}\left(m(\bar{\mathbf{C}}_j), \left(\frac{w(\bar{\mathbf{C}}_j)}{3}\right)^2\right)$ for all $j = 1, 2, \ldots, k$. Denote $\mathbf{C} = (C_1, C_2, \ldots, C_k)^T$, $m_{\mathbf{C}} = (m(\bar{\mathbf{C}}_1), m(\bar{\mathbf{C}}_2), \ldots, m(\bar{\mathbf{C}}_k))^T$ and $\sigma_{\mathbf{C}} = (w(\bar{\mathbf{C}}_1)/3, w(\bar{\mathbf{C}}_2)/3, \ldots, w(\bar{\mathbf{C}}_k)/3)^T$. Correspondingly, $\bar{\mathbf{F}}(\mathbf{x}; \bar{\mathbf{C}}_\nu^k)$ can be associated to a function of random variables $f(\mathbf{x}; \mathbf{C})$ with mean $\mathbb{E}(f(\mathbf{x}; \mathbf{C}))$ and variance $\text{Var}(f(\mathbf{x}; \mathbf{C}))$.

Now consider constraint $\bar{\mathbf{G}}_{i\bar{\mathbf{A}}_{\nu(i)}^k}(\mathbf{x}) \succcurlyeq \bar{\mathbf{B}}_i \Rightarrow \bar{\mathbf{G}}_{i\bar{\mathbf{A}}_{\nu(i)}^k}(\mathbf{x}) \ominus_M \bar{\mathbf{B}}_i \succcurlyeq [0, 0]$. Suppose that $\bar{\mathbf{G}}_{i\bar{\mathbf{A}}_{\nu(i)}^k}(\mathbf{x}) \ominus_M \bar{\mathbf{B}}_i \triangleq \bar{\mathbf{H}}_{i\bar{\mathbf{A}}_{\nu(i)}^{k+1}}(\mathbf{x}) \, i = 1, 2, \ldots, m$, where $\bar{\mathbf{A}}_{\nu(i)}^{k+1} = (\bar{\mathbf{A}}_{1i}, \bar{\mathbf{A}}_{2i}, \ldots, \bar{\mathbf{A}}_{ki}, \bar{\mathbf{B}}_i)^T$. Now every interval $\bar{\mathbf{A}}_{ji}$ is associated by a random variable $A_{ji} \sim \mathcal{N}\left(m(\bar{\mathbf{A}}_{ji}), (w(m(\bar{\mathbf{A}}_{ji}))/3)^2\right), j = 1, 2, ..., k,$ and $\bar{\mathbf{B}}_i$ is associated by $B_i \sim \mathcal{N}(m(\bar{\mathbf{B}}_i), (w(\bar{\mathbf{B}}_i)/3)^2), i = 1, 2, ..., m.$ Denote, $\mathbf{A}_i = (A_{1i}, A_{2i}, ..., A_{ki}, B_i)^T,$ $m_{\mathbf{A}_i} = (m(\bar{\mathbf{A}}_{1i}), m(\bar{\mathbf{A}}_{2i}), ..., m(\bar{\mathbf{A}}_{ki}), m(\bar{\mathbf{B}}_i))^T,$ and $\sigma_{\mathbf{A}_i} = (w(\bar{\mathbf{A}}_{1i}), w(\bar{\mathbf{A}}_{2i}), ..., w(\bar{\mathbf{A}}_{ki}), w(\bar{\mathbf{B}}_i))^T.$ Correspondingly, the interval-valued function $\bar{\mathbf{H}}_{i\bar{\mathbf{A}}_{v(i)}^{k+1}}(\mathbf{x})$ can be associated to a function of random variables $h_i(\mathbf{x}; \mathbf{A}_i)$ with mean $\mathbb{E}(h_i(\mathbf{x}; \mathbf{A}_i))$ and variance $\text{Var}(h_i(\mathbf{x}; \mathbf{A}_i))$ for all $i = 1, 2, ..., m.$

The feasible set of **NLIP** is

$$\mathcal{F} = \{\mathbf{x} = (x_1, x_2, ..., x_n) \in \mathbb{R}^n \,|\, \bar{\mathbf{G}}_{i\bar{\mathbf{A}}_{v(i)}^k}(\mathbf{x}) \succcurlyeq \bar{\mathbf{B}}_i, i = 1, 2, ..., m\}. \quad (27)$$

If we assign a bound $\lambda \in [0, 1]$ for the degree of closeness of the interval inequalities in (26)

$\left\{(x_1, x_2, ..., x_n) \,|\, \bar{\mathbf{H}}_{i\bar{\mathbf{A}}_{v(i)}^{k+1}}(\mathbf{x}) \succcurlyeq [0, 0]\right.$ satisfies with at least degree of closeness $\lambda, i = 1, 2, ..., m\}$

$$= \{(x_1, x_2, ..., x_n) \,|\, \text{P}(h_i(\mathbf{x}; \mathbf{A}_i) \geq 0) \geq \lambda, i = 1, 2, ..., m\}. \quad (28)$$

Hence the feasible region of **NLIP** in (27), with acceptability λ can be expressed as

$$\mathcal{F}_\lambda = \{(x_1, x_2, ..., x_n) \,|\, \text{P}(h_i(\mathbf{x}; \mathbf{A}_i) \geq 0) \geq \lambda, i = 1, 2, ..., m\}. \quad (29)$$

Using the above transformation **NLIP** can be transformed to

$$(\textbf{NLIP}_\lambda) \quad \min \ f(\mathbf{x}; \mathbf{C}) \quad (30)$$
$$\text{subject to } \text{P}(h_i(\mathbf{x}; \mathbf{A}_i) \geq 0) \geq \lambda, i = 1, 2, ..., m, \quad (31)$$

with λ as preassign value and $\lambda \in [0, 1]$. $(\textbf{NLIP}_\lambda)$ is a stochastic programming problem. Using the chance constraint programming technique, $(\textbf{NLIP}_\lambda)$ can be converted into a nonlinear programming problem as described below.

The objective function $f(\mathbf{x}; \mathbf{C})$ can be expended around the vector of mean values of the random variables, $m_{\mathbf{C}}$, as

$$f(\mathbf{x}; \mathbf{C}) \simeq f(\mathbf{x}; m_{\mathbf{C}}) - \sum_{j=1}^{4}\left(\left.\frac{\partial f(\mathbf{x}; \mathbf{C})}{\partial C_j}\right|_{m_{\mathbf{C}}}\right)m(\bar{\mathbf{C}}_j)$$
$$+ \sum_{j=1}^{4}\left(\left.\frac{\partial f(\mathbf{x}; \mathbf{C})}{\partial C_j}\right|_{m_{\mathbf{C}}}\right)C_j \quad (32)$$

From this equation, the mean value, $\mathbb{E}(f(\mathbf{x}; \mathbf{C}))$, and variance, $\text{Var}(f(\mathbf{x}; \mathbf{C}))$, of $f(\mathbf{x}; \mathbf{C})$ can be obtained as

$$\mathbb{E}(f(\mathbf{x}; \mathbf{C})) = f(\mathbf{x}; m_{\mathbf{C}}) \triangleq m_f \tag{33}$$

$$\text{Var}(f(\mathbf{x}; \mathbf{C})) = \sum_{j=1}^{k} \left(\frac{\partial f(\mathbf{x}; \mathbf{C})}{\partial C_j} \bigg|_{m_{\mathbf{C}}} \right)^2 \left(\frac{w(\bar{C}_j)}{3} \right)^2 \triangleq \sigma_f^2 \tag{34}$$

Since all C_j are independent. For the purpose of optimization, a new objective function can be constructed as follows

$$\alpha_1 m_f + \alpha_2 \sigma_f \tag{35}$$

where $\alpha_1 \geq 0$ and $\alpha_2 \geq 0$, and their numerical values indicate the relative importance of m_f and σ_f for minimization.

Next, the constraint function $h_i(\mathbf{x}; \mathbf{A}_i)$ can be expended around the vector of the mean values of the random variables, $m_{\mathbf{A}_i}$, as

$$h_i(\mathbf{x}; \mathbf{A}_i) \simeq h_i(\mathbf{x}; m_{\mathbf{A}_i}) + \left\{ \sum_{j=1}^{k} \left(\frac{\partial h_i(\mathbf{x}; \mathbf{A}_i)}{\partial A_{ji}} \bigg|_{m_{\mathbf{A}_i}} \right) (\bar{A}_{ji} - m(A_{ji})) \right. \tag{36}$$

$$\left. + \left(\frac{\partial h_i(\mathbf{x}; \mathbf{A}_i)}{\partial B_i} \bigg|_{m_{\mathbf{A}_i}} \right) (B_i - m(\bar{B}_i)) \right\} \tag{37}$$

From the above equation, the mean value, $\mathbb{E}(h_i(\mathbf{x}; \mathbf{A}_i))$, and variance Var $(h_i(\mathbf{x}; \mathbf{A}_i))$ of $h_i(\mathbf{x}; \mathbf{A}_i)$ can be obtained as follows

$$\mathbb{E}(h_i(\mathbf{x}; \mathbf{A}_i)) = h_i(\mathbf{x}; m_{\mathbf{A}_i}) \triangleq m_{h_i}, \tag{38}$$

$$\text{Var}(h_i(\mathbf{x}; \mathbf{A}_i)) = \sum_{j=1}^{k} \left(\frac{\partial h_i(\mathbf{x}; \mathbf{A}_i)}{\partial A_{ji}} \bigg|_{m_{\mathbf{A}_i}} \right)^2 \left(\frac{w(\bar{A}_{ji})}{3} \right)^2 + \left(\frac{\partial h_i(\mathbf{x}; \mathbf{A}_i)}{\partial B_i} \bigg|_{m_{\mathbf{A}_i}} \right)^2 \left(\frac{w(\bar{B}_i)}{3} \right)^2$$

$$\triangleq \sigma_{h_i}^2 \tag{39}$$

Using the values of $\mathbb{E}(h_i(\mathbf{x}; \mathbf{A}_i))$ and $\text{Var}(h_i(\mathbf{x}; \mathbf{A}_i))$ from (38) and (39), the inequality (31) can be expressed as

$$\text{P} \left(\frac{h_i(\mathbf{x}; \mathbf{A}_i) - m_{h_i}}{\sigma_{h_i}} \geq \frac{-m_{h_i}}{\sigma_{h_i}} \right) \geq \lambda,$$

$$\Phi \left(\frac{-m_{h_i}}{\sigma_{h_i}} \right) \geq \lambda, \tag{40}$$

$$\frac{-m_{h_i}}{\sigma_{h_i}} \leq \Phi^{-1}(\lambda),$$

$$m_{h_i} + \Phi^{-1}(\lambda)\sigma_{h_i} \geq 0, \quad i = 1, 2, \dots, m.$$

With this background the original nonlinear interval programming problem **NLIP** can be stated as a deterministic nonlinear programming problem as follows, which is free from interval uncertainty.

$$(\textbf{NLIP}'_\lambda) \qquad \min \quad \alpha_1 m_f + \alpha_2 \sigma_f,$$
$$\text{subject to} \quad m_{h_i} + \Phi^{-1}(\lambda)\sigma_{h_i} \geq 0, \quad i = 1, 2, \dots, m.$$

For preassigned values of $\lambda \in [0, 1]$, α_1 and α_2 by decision maker, suppose the solution of the model **NLIP**$_\lambda$ is $\mathbf{x}^*$.

Example 7. Consider the interval optimization problem

$$\textbf{NLIP}: \quad \min \ \bar{\mathbf{F}}_{\bar{\mathbf{C}}^4_\nu}(x_1, x_2) = [1.5, 3.5]e^{[1,3]x_1} + [1.5, 2.5]e^{[0.6, 1.2]x_2} \quad (41)$$

$$\text{subject to} \ \bar{\mathbf{G}}_{\bar{\mathbf{A}}^2_\nu}(x_1, x_2) = [0.8, 2]x_1 + [0.5, 1.5]x_2 \geqslant [6, 10] \quad (42)$$

$$x_1, x_2 \geq 0, \quad (43)$$

where $\bar{\mathbf{C}}^4_\nu = ([1.5, 3.5], [1, 3], [1.5, 2.5], [0.6, 1.2])^T$, $\bar{\mathbf{A}}^2_\nu = ([0.5, 1.5],$ $[0.5, 1.5])^T$ and $\bar{\mathbf{B}} = [6, 10]$, i.e., $\mathbf{C} = (C_1, C_2, C_3, C_4)^T$, $m_{\mathbf{C}} = (2.5, 2,$ $2, 0.9)^T$, $\sigma_{\mathbf{C}} = \left(\frac{1}{3}, \frac{1}{3}, \frac{1}{3}, 0.1\right)^T$, $\mathbf{A} = (A_1, A_2, B)^T$, $m_{\mathbf{A}} = (1.4, 1, 8)^T$ and $\sigma_{\mathbf{A}} = \left(0.2, \frac{0.5}{3}, \frac{2}{3}\right)^T$

The feasible set of **NLIP** problem is

$$\mathcal{F} = \{\mathbf{x} = (x_1, x_2) | [0.8, 2]x_1 + [0.5, 1.5]x_2 \geqslant [6, 10], x_1, x_2 \geq 0\} \quad (44)$$

Since inequality (42) is an interval inequalities, therefore, these can be transformed to algebraic inequality as follows:

The variable intervals $[0.8, 2]x_1 + [0.5, 1.5]x_2 \ominus_M [6, 10] = \bar{\mathbf{H}}_{\bar{\mathbf{A}}^3_\nu}$ is associated with random variable $h(x_1, x_2; \mathbf{A})$ and

$$\left.\frac{\partial h(x_1, x_2; \mathbf{A})}{\partial A_1}\right|_{m_\mathbf{A}} = x_1, \qquad \left.\frac{\partial h(x_1, x_2; \mathbf{A})}{\partial A_2}\right|_{m_{\mathbf{A}_1}} = x_2.$$

$$\left.\frac{\partial h(x_1, x_2; \mathbf{A})}{\partial B_1}\right|_{m_{\mathbf{A}_1}} = 1.$$

Using Taylor's approximation, we obtain

$$h(x_1, x_2; \mathbf{A}_1) = m(\bar{\mathbf{A}}_1)x_1 + m(\bar{\mathbf{A}}_2)x_2 - m(\bar{\mathbf{B}}) - x_1 m(\bar{\mathbf{A}}_1) - x_2 m(\bar{\mathbf{A}}_2) + m(\bar{\mathbf{B}})$$
$$+ x_1 A_1 + x_2 A_2 - B$$
$$= x_1 A_1 + x_2 A_2 - B.$$

$$(45)$$

Accordingly,

$$\mathbb{E}(h(x_1, x_2; \mathbf{A})) = m(\bar{\mathbf{A}}_1)x_1 + m(\bar{\mathbf{A}}_2)x_1 - m(\bar{\mathbf{B}})$$
$$1.4x_1 + x_2 - 8$$
$$\mathrm{Var}(h(x_1, x_2; \mathbf{A})) = x_1^2 \left(\frac{w(\bar{\mathbf{A}}_1)}{3}\right)^2 + x_2^2 \left(\frac{w(\bar{\mathbf{A}}_2)}{3}\right)^2 + \left(\frac{w(\bar{\mathbf{B}})}{3}\right)^2$$
$$= 0.04x_1^2 + \frac{2.5}{9}x_2^2 + \frac{4}{9}$$

and $h(x_1, x_2; \mathbf{A}_1) \backsim \mathcal{N}\left(1.4x_1 + x_2 - 8, 0.04x_1^2 + \frac{2.5}{9}x_2^2 + \frac{4}{9}\right)$.

Correspondingly, feasible set (44) is transformed to the feasible set with degree of acceptability $\lambda = 0.9015$ and is given by

$$\mathcal{F}_\lambda = \{\mathbf{x} = (x_1, x_2) | P(h(x_1, x_2; \mathbf{A}) \geq 0) \geq 0.9015, x_1, x_2 \geq 0\} \quad (46)$$

Since the objective function (41) is also a variable interval, so this can also be associated to a random variable $f(x_1, x_2; \mathbf{C})$ and

$$\left.\frac{\partial f(x_1, x_2; \mathbf{C})}{\partial C_1}\right|_{m_C} = e^{m(\bar{\mathbf{C}}_1)x_1}, \quad \left.\frac{\partial f(x_1, x_2; \mathbf{C})}{\partial C_2}\right|_{m_C} = m(\bar{\mathbf{C}}_1)e^{m(\bar{\mathbf{C}}_2)x_2}x_1,$$

$$\left.\frac{\partial f(x_1, x_2; \mathbf{C})}{\partial C_3}\right|_{m_C} = e^{m(\bar{\mathbf{C}}_4)x_2}, \quad \left.\frac{\partial f(x_1, x_2; \mathbf{C})}{\partial C_4}\right|_{m_C} = m(\bar{\mathbf{C}}_3)e^{m(\bar{\mathbf{C}}_4)x_2}x_2.$$

Accordingly, Taylor's approximation of $f(x_1, x_2; \mathbf{C})$ is as follows:

$$f(x_1, x_2; \mathbf{C}) = m(\bar{\mathbf{C}}_1)e^{m(\bar{\mathbf{C}}_2)x_1} + m(\bar{\mathbf{C}}_3)e^{m(\bar{\mathbf{C}}_4)x_2} - e^{m(\bar{\mathbf{C}}_2)x_1}m(\bar{\mathbf{C}}_1)$$

$$- m(\bar{\mathbf{C}}_1)e^{m(\bar{\mathbf{C}}_2)x_1}x_1 m(\bar{\mathbf{C}}_2) - e^{m(\bar{\mathbf{C}}_4)x_2}m(\bar{\mathbf{C}}_3)$$

$$- m(\bar{\mathbf{C}}_3)e^{m(\bar{\mathbf{C}}_4)x_2}x_2 m(\bar{\mathbf{C}}_4) + e^{m(\bar{\mathbf{C}}_2)x_1}C_1 + m(\bar{\mathbf{C}}_1)e^{m(\bar{\mathbf{C}}_2)x_1}x_1 C_2$$

$$+ e^{m(\bar{\mathbf{C}}_4)x_2}C_3 + m(\bar{\mathbf{C}}_3)e^{m(\bar{\mathbf{C}}_4)x_2}x_2 C_4$$

$$= -m(\bar{\mathbf{C}}_1)e^{m(\bar{\mathbf{C}}_2)x_1}x_1 m(\bar{\mathbf{C}}_2) - m(\bar{\mathbf{C}}_3)e^{m(\bar{\mathbf{C}}_4)x_2}x_2 m(\bar{\mathbf{C}}_4) + C_1 e^{m(\bar{\mathbf{C}}_2)x_1}$$

$$+ m(\bar{\mathbf{C}}_1)e^{m(\bar{\mathbf{C}}_2)x_1}x_1 C_2 + e^{m(\bar{\mathbf{C}}_4)x_2}C_3 + m(\bar{\mathbf{C}}_3)e^{m(\bar{\mathbf{C}}_4)x_2}x_2 C_4$$

Correspondingly,

$$\mathbb{E}(f(x_1, x_2; \mathbf{C})) = m(\bar{\mathbf{C}}_1)e^{m(\bar{\mathbf{C}}_2)x_1} + m(\bar{\mathbf{C}}_3)e^{m(\bar{\mathbf{C}}_4)x_2}$$

$$= 2.5e^{2x_1} + 2e^{0.9x_2}$$

$$\mathrm{Var}(f(x_1, x_2; \mathbf{C})) = \left(e^{m(\bar{\mathbf{C}}_2)x_1}\right)^2 \left(\frac{w(\bar{\mathbf{C}}_1)}{3}\right)^2 + \left(m(\bar{\mathbf{C}}_1)e^{m(\bar{\mathbf{C}}_2)x_1}x_1\right)^2 \left(\frac{w(\bar{\mathbf{C}}_2)}{3}\right)^2$$

$$+ \left(e^{m(\bar{\mathbf{C}}_4)x_2}\right)^2 \left(\frac{w(\bar{\mathbf{C}}_3)}{3}\right)^2 + \left(m(\bar{\mathbf{C}}_3)e^{m(\bar{\mathbf{C}}_4)x_2}x_2\right)^2 \left(\frac{w(\bar{\mathbf{C}}_4)}{3}\right)^2$$

$$= (e^{2x_1})^2 \left(\frac{1}{3}\right)^2 + (2.5e^{2x_1}x_1)^2 \left(\frac{1}{3}\right)^2 + (e^{0.9x_2})^2 \left(\frac{1}{3}\right)^2 + (2e^{0.9x_2}x_2)^2 \left(\frac{0.3}{3}\right)^2$$

$$= \frac{1}{9}\left\{e^{4x_1} + 6.25e^{4x_1}x_1^2 + e^{1.8x_2} + 0.18e^{1.8x_2}x_2^2\right\}$$

and $f(x_1, x_2; \mathbf{C}) \backsim \mathcal{N}\left(2.5e^{2x_1} + 2e^{0.9x_2}, \frac{1}{9}\left\{e^{4x_1} + 6.25e^{4x_1}x_1^2 + e^{1.8x_2} + 0.18e^{1.8x_2}x_2^2\right\}\right)$

Using the above transformation, **NLIP** can be transformed to

$$\mathbf{NLIP}'_\lambda : \min f(x_1, x_2; \mathbf{C}) \tag{47}$$

$$\text{subject to} \quad \mathrm{P}(h(x_1, x_2; \mathbf{A}_1) \geq 0) \geq 0.9015, \tag{48}$$

$$x_1, x_2 \geq 0. \tag{49}$$

Using the chance constraint programming technique, **NLIP**$'_\lambda$ can be converted into a nonlinear programming problem as given below.

$$\min \quad \alpha_1(2.5e^{2x_1} + 2e^{0.9x_2}) + \alpha_2\sqrt{\frac{1}{9}\left\{e^{4x_1} + 6.25e^{4x_1}x_1^2 + e^{1.8x_2} + 0.18e^{1.8x_2}x_2^2\right\}}$$

$$\text{subject to} \quad 1.4x_1 + x_2 - 8 + \Phi^{-1}(0.9015)\sqrt{0.04x_1^2 + \frac{2.5}{9}x_2^2 + \frac{4}{9}} \geq 0,$$

$$x_1 \geq 0, x_2 \geq 0.$$

For the value of $\alpha_1 = 1.0$, $\alpha_2 = 1.0$ and $\Phi^{-1}(0.9015) = 1.29$, the optimal solution is obtained using Karush–Kuhn–Tucker conditions as $x_1 = 0.116$ and $x_2 = 4.596$ with objective value 0.

3.10 Recent study of NLIOP in brief

Bhurjee and Panda [16] derived the necessary and sufficient optimality conditions for the NLIOP. The efficient solutions of an interval optimization problem with the help of saddle point are studied by Ghosh et al. [17] in (2018). Wu [63] explained the concept of null set in the space of all bounded closed intervals, transforming the interval-valued optimization problems into the conventional vector optimization problem. Bhurjee and Panda [22] illustrate a multiobjective NLIOP problem and establish a relation between the solution of two different forms of multiobjective programming problems with interval parameters. Roy and Panda [18] discussed the conditions for optimization problems with interval parameters using gH-differentiability. Kumar and Bhurjee [26] have made an extension of NLIOP by assuming the decision variables as closed intervals in an interval optimization problem, called the Enhanced interval optimization problem, and developed the solution methodology based on the parametric representation of a closed interval. They have verified the efficacy and applicability of the problem by applying it to the portfolio optimization problem. Kumar and Bhurjee [21] have developed a method to solve the multiobjective enhanced nonlinear programming problem simultaneously. Rahman [20] established necessary and sufficient optimality conditions for a parametric form of unconstrained interval optimization problem.

4. Conclusions

The interval optimization problem is one of the techniques to study the uncertainty in real-world issues without any specified assumption. This chapter provides an overview of the different approaches to dealing with uncertainty in nonlinear optimization problems using interval analysis, including nature inspired techniques. For example, quadratic interval programming problems, general NLIOPs in different considerations of the researchers, an evolutionary process for quadratic interval programming problems, and a scholastic programming approach for nonlinear programming problems. The satisficing and optimizing techniques have been reviewed and illustrated employing small numerical examples.

References

[1] W.L. Winston, J.B. Goldberg, Operations Research: Applications and Algorithms, vol. 3, Thomson Brooks/Cole Belmont, 2004.
[2] V. Yadav, A.K. Bhurjee, S. Karmakar, A.K. Dikshit, A facility location model for municipal solid waste management system under uncertain environment, Sc. Total Environ. 603 (2017) 760–771.
[3] V. Yadav, S. Karmakar, A.K. Dikshit, A.K. Bhurjee, Interval-valued facility location model: an appraisal of municipal solid waste management system, J. Clean. Prod. 171 (2018) 250–263.
[4] I. Aguirre-Cipe, R. López, E. Mallea-Zepeda, L. Vásquez, A study of interval optimization problems, Optim. Lett. 15 (3) (2021) 859–877.

[5] L. Wang, G. Yang, Z. Li, F. Xu, An efficient nonlinear interval uncertain optimization method using Legendre polynomial chaos expansion, Appl. Soft Comput. 108 (2021) 107454.

[6] M. Hladík, Interval linear programming: a survey, in: Linear Programming-New Frontiers in Theory and Applications, Nova Science Publishers, New York, 2012, pp. 85–120.

[7] S.-T. Liu, R.-T. Wang, A numerical solution method to interval quadratic programming, Appl. Math. Comput. 189 (2) (2007) 1274–1281.

[8] W. Li, X. Tian, Numerical solution method for general interval quadratic programming, Appl. Math. Comput. 0096-3003, 202 (2) (2008) 589–595.

[9] C. Jiang, X. Han, G.R. Liu, G.P. Liu, A nonlinear interval number programming method for uncertain optimization problems, Eur. J. Oper. Res. 188 (1) (2008) 1–13.

[10] M. Hladík, Optimal value bounds in nonlinear programming with interval data, TOP 19 (1) (2011) 93–106.

[11] S. Karmakar, A.K. Bhunia, An alternative optimization technique for interval objective constrained optimization problems via multiobjective programming, J. Egypt. Math. Soc. 22 (2) (2014) 292–303.

[12] A.K. Bhurjee, P. Kumar, Optimality conditions and duality theory for multi-objective interval optimization problems, in: Recent Advances in Nonlinear Analysis and Optimization with Applications, Cambridge Scholars Publishing, 2020, p. 81.

[13] P. Kumar, Multi-objective interval linear programming problem with the bounded solution, in: AIP Conference Proceedings, vol. 2277, AIP Publishing, 2020.

[14] A.K. Bhurjee, G. Panda, Efficient solution of interval optimization problem, Math. Meth. Oper. Res. 76 (3) (2012) 273–288.

[15] T.M. Costa, Y. Chalco-Cano, W.A. Lodwick, G.N. Silva, Generalized interval vector spaces and interval optimization, Inf. Sci. 311 (2015) 74–85.

[16] A.K. Bhurjee, G. Panda, Sufficient optimality conditions and duality theory for interval optimization problem, Ann. Oper. Res. 243 (1) (2016) 335–348.

[17] D. Ghosh, D. Ghosh, S.K. Bhuiya, L.K. Patra, A saddle point characterization of efficient solutions for interval optimization problems, J. Appl. Math. Comput. 58 (1) (2018) 193–217.

[18] P. Roy, G. Panda, Existence of solution of constrained interval optimization problems with regularity concept, RAIRO-Oper. Res. 55 (2021) S1997–S2011.

[19] A.K. Bhurjee, P. Kumar, S. Panigrahi, G. Panda, Existence of the solutions of an interval linear complementarity problem and its application, in: AIP Conference Proceedings, vol. 2277, AIP Publishing LLC, 2020, p. 200001.

[20] M.S. Rahman, Optimality theory of an unconstrained interval optimization problem in parametric form: its application in inventory control, Results Control Optim. 7 (2022) 100–111.

[21] P. Kumar, B.S. Bishakha Rani, A.K. Bhurjee, Multi-objective portfolio selection problem using admissible order vector space, in: AIP Conference Proceedings, 2516, AIP Publishing, 2022. vol.

[22] A.K. Bhurjee, G. Panda, Parametric multi-objective fractional programming problem with interval uncertainty, Int. J. Oper. Res. 35 (1) (2019) 132–145.

[23] A.K. Bhurjee, P. Kumar, Calculus for interval valued function on real space, in: AIP Conference Proceedings, 2516, AIP Publishing, 2022. vol.

[24] M.A. Elsisy, D.A. Hammad, M.A. El-Shorbagy, Solving interval quadratic programming problems by using the numerical method and swarm algorithms, Complexity 2020 (2020) 6105952.

[25] P. Kumar, G. Panda, U.C. Gupta, Generalized quadratic programming problem with interval uncertainty, in: 2013 IEEE International Conference on Fuzzy Systems (FUZZ-IEEE), IEEE, 2013, pp. 1–7.

[26] P. Kumar, A.K. Bhurjee, An efficient solution of nonlinear enhanced interval optimization problems and its application to portfolio optimization, Soft Comput. 25 (7) (2021) 5423–5436.

[27] P. Kumar, A.K. Bhurjee, Multi-objective enhanced interval optimization problem, Ann. Oper. Res. 311 (2022) 1035–1050.

[28] R.E. Moore, Interval Analysis, Prentice-Hall, 1966.

[29] H. Ishibuchi, H. Tanaka, Multiobjective programming in optimization of the interval objective function, Eur. J. Oper. Res. 48 (2) (1990) 219–225.

[30] E. Hansen, G.W. Walster, Global Optimization Using Interval Analysis, Marcel Dekker, Inc., 2004.

[31] W.S. Dorn, Duality in quadratic programming, Q. Appl. Math. 18 (2) (1960) 155–162.

[32] A. Sengupta, T.K. Pal, D. Chakraborty, Interpretation of inequality constraints involving interval coefficients and a solution to interval linear programming, Fuzzy Set. Syst. 119 (1) (2001) 129–138.

[33] Z. Quan, F. Zhiping, P. Dehui, A ranking approach with possibilities for multiple attribute decision making problems with intervals, Control Decis. 6 (56) (1999) 56.

[34] G. Alefeld, J. Herzberger, Introduction to Interval Computation, Academic Press, 2012.

[35] J. Rohn, Positive definiteness and stability of interval matrices, SIAM J. Matrix Anal. Appl. 15 (1) (1994) 175–184.

[36] H. Ishibuchi, H. Tanaka, Multiobjective programming in optimization of the interval objective function, Eur. J. Oper. Res. 48 (2) (1990) 219–225.

[37] A. Sengupta, T.K. Pal, Fuzzy Preference Ordering of Interval Numbers in Decision Problems, vol. 238, Springer, 2009.

[38] S. Karmakar, A.K. Bhunia, On constrained optimization by interval arithmetic and interval order relations, Opsearch 49 (1) (2012) 22–38.

[39] S. Karmakar, S.K. Mahato, A.K. Bhunia, Interval oriented multi-section techniques for global optimization, J. Comput. Appl. Math. 224 (2) (2009) 476–491.

[40] S. Karmakar, A.K. Bhunia, An efficient interval computing technique for bound-constrained uncertain optimization problems, Optimization 63 (11) (2014) 1615–1636.

[41] S. Karmakar, A.K. Bhunia, Uncertain constrained optimization by interval-oriented algorithm, J. Oper. Res. Soc. 65 (1) (2014) 73–87.

[42] S.K. Mahato, A.K. Bhunia, Interval-arithmetic-oriented interval computing technique for global optimization, Appl. Math. Res. Express 2006 (2006) 69642.

[43] M. Inuiguchi, M. Sakawa, Minimax regret solution to linear programming problems with an interval objective function, Eur. J. Oper. Res. 86 (3) (1995) 526–536.

[44] S. Chanas, D. Kuchta, Multiobjective programming in optimization of interval objective functions—a generalized approach, Eur. J. Oper. Res. 94 (3) (1996) 594–598.

[45] H.-C. Wu, Evaluate fuzzy optimization problems based on biobjective programming problems, Comput. Math. Appl. 47 (6–7) (2004) 893–902.

[46] H.-C. Wu, The Karush-Kuhn-Tucker optimality conditions in multiobjective programming problems with interval-valued objective functions, Eur. J. Oper. Res. 196 (1) (2009) 49–60.

[47] M.A. El-Shorbagy, A.E. Hassanien, Particle swarm optimization from theory to applications, Int. J. Rough Sets Data Anal. 5 (2) (2018) 1–24.

[48] S. Verma, V. Mukherjee, Firefly algorithm for congestion management in deregulated environment, Eng. Sci. Technol. Int. J. 19 (3) (2016) 1254–1265.

[49] M. Marinaki, Y. Marinakis, A glowworm swarm optimization algorithm for the vehicle routing problem with stochastic demands, Expert Syst. Appl. 46 (2016) 145–163.

[50] A.L. Bolaji, M.A. Al-Betar, M.A. Awadallah, A.T. Khader, L.M. Abualigah, A comprehensive review: Krill Herd algorithm (KH) and its applications, Appl. Soft Comput. 49 (2016) 437–446.

[51] Y. Zhou, X. Chen, G. Zhou, An improved monkey algorithm for a 0-1 knapsack problem, Appl. Soft Comput. 38 (2016) 817–830.

[52] S. Saremi, S. Mirjalili, A. Lewis, Grasshopper optimisation algorithm: theory and application, Adv. Eng. Soft. 105 (2017) 30–47.

[53] A.W. Mohamed, A.A. Hadi, A.K. Mohamed, Gaining-sharing knowledge based algorithm for solving optimization problems: a novel nature-inspired algorithm, Int. J. Mach. Learn. Cybern. 11 (2019) 1–29.

[54] Y. Cao, H. Zhang, W. Li, M. Zhou, Y. Zhang, W.A. Chaovalitwongse, Comprehensive learning particle swarm optimization algorithm with local search for multimodal functions, IEEE Trans. Evol. Comput. 23 (4) (2018) 718–731.

[55] W. Dong, M. Zhou, A supervised learning and control method to improve particle swarm optimization algorithms, IEEE Trans. Syst. Man Cybern. Syst. 47 (7) (2016) 1135–1148.

[56] Q. Kang, C. Xiong, M. Zhou, L. Meng, Opposition-based hybrid strategy for particle swarm optimization in noisy environments, IEEE Access 6 (2018) 21888–21900.

[57] Z. Lv, L. Wang, Z. Han, J. Zhao, W. Wang, Surrogate-assisted particle swarm optimization algorithm with Pareto active learning for expensive multi-objective optimization, IEEE/CAA J. Autom. Sin. 6 (3) (2019) 838–849.

[58] P. Kumar, G. Panda, Solving nonlinear interval optimization problem using stochastic programming technique, Opsearch 54 (4) (2017) 752–765.

[59] I. Stancu-Minasian, Stochastic programming with multiple fractile criteria, Rev. Roum. Math. Pures Appl. 37 (1992) 939–941.

[60] I.M. Stancu-Minasian, S. Tigan, On some fractional programming models occurring in minimum-risk problems, in: Generalized Convexity and Fractional Programming With Economic Applications, Springer, 1990, pp. 295–324.

[61] I.M. Stancu-Minasian, M.J. Wets, A research bibliography in stochastic programming, 1955-1975, Oper. Res. 24 (6) (1976) 1078–1119.

[62] I.M. Stancu-Minasian, Overview of different approaches for solving stochastic programming problems with multiple objective functions, in: Stochastic Versus Fuzzy Approaches to Multiobjective Mathematical Programming Under Uncertainty, Springer, 1990, pp. 71–101.

[63] H.-C. Wu, Solving the interval-valued optimization problems based on the concept of null set, J. Ind. Manage. Optim. 14 (3) (2018) 1157.

About the authors

Dr. Ajay Kumar Bhurjee is an assistant professor a VIT Bhopal University. Ajay has worked as an assistant professor at NIST Berhampur, Odisha, India. He earned his bachelor's and master's degrees from Bundelkhand University, Jhansi, India, and a doctoral degree from IIT Kharagpur, India. His expertise includes optimization under uncertainty, interval analysis-based operations research techniques, and game theory.

Dr. Pankaj Kumar, currently an Assistant Professor at NIT Hamirpur, previously held roles as a Research Assistant Professor at SRM Institute of Science and Technology and as a Visiting Faculty at IFMR Sricity. He earned his PhD from IIT Kharagpur, M.Tech from NIT Durgapur, and MSc/BSc from Ranchi University. Dr. Kumar specializes in optimization methods in finance, interval analysis-based operations research techniques, and machine learning.

Reetendra Singh is a PhD candidate in the Decision Sciences Area at the Indian Institute of Management, Visakhapatnam, Andhra Pradesh (AP), India. He holds a master's in mathematics from LNM IIT Jaipur and a bachelor's in mathematics from the University of Rajasthan, Jaipur. He is working on complex networks, game theory, and optimization.

Dr. Vinay Yadav is an Assistant Professor at IIT Kharagpur. Vinay worked as an assistant professor at IIM Jammu, Visakhapatnam, and Marie Curie Postdoctoral Fellow at the Technical University of Denmark prior to joining IIT Kharagpur. He earned his master's and doctoral degrees from IIT Bombay, Mumbai, and a bachelor's degree from Banaras Hindu University, Varanasi. His expertise includes optimization under uncertainty, interval analysis-based operations research techniques, and environmental management.

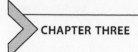

CHAPTER THREE

Solving the aircraft landing problem using the bee colony optimization (BCO) algorithm

Miloš Nikolić[a], Jasenka Rakas[b], and Dušan Teodorović[c]
[a]University of Belgrade, Faculty of Transport and Traffic Engineering, Belgrade, Serbia
[b]University of California Berkeley, National Center of Excellence for Aviation Operations Research, Berkeley, United States
[c]Serbian Academy of Sciences and Arts, Belgrade, Serbia

Contents

Abstract

This study examines a static version of the Aircraft Landing Problem (ALP) when information on scheduled aircraft is known in advance. We assume that for any scenario each aircraft must land on one of the given runways, and every aircraft must land within the minimal prescribed separation time between all pairs of aircraft. The objective is to minimize the total deviation of all aircraft from the target landing times. To solve the ALP, we propose a heuristic approach based on the Bee Colony Optimization (BCO) metaheuristic. The BCO algorithm is a nature-inspired optimization technique. This algorithm is based on the similarity between how bees search for food, and how optimization algorithms search for an optimum of known combinatorial optimization problems. We also propose a linear programming model for solution improvement. The proposed linear model solves a sub-problem, and we applied it to improve the initial and the final solution. We tested the BCO algorithm and the linear programming model on benchmark examples from the literature. We solved 48 examples with different numbers of

Advances in Computers, Volume 135
ISSN 0065-2458
https://doi.org/10.1016/bs.adcom.2023.11.002

aircraft (from 10 to 500) and runways (from 1 to 5). Our results show that the proposed algorithm can easily find optimal solutions for small instances (with 10 to 50 aircraft, and for up to four runways). In the case of larger-sized instances (with 100 to 500 aircraft, and up to five runways) the BCO approach was capable of finding high-quality solutions, making this approach competitive with other approaches.

1. Introduction

Following the Wright brothers' first powered flight in 1903 and the beginning of airline operations in the early 1920s, scheduled commercial aviation has experienced enormous development, accompanied by the rapid growth in organizational problems. In 2019, the International Air Transport Association (IATA) estimated that airlines transported 4.5 billion passengers globally and carried 57.6 million metric tons of cargo across a network comprising nearly 22,000 air routes [1]. Before the COVID-19 pandemic, about 1400 airlines operated more than 25,000 aircraft worldwide [2]. Although in 2020 the COVID-19 pandemic severely impacted the air transportation sector globally, it is estimated that by the end of 2023, most regions will be at, or exceeding the pre-pandemic level of air travel demand [1].

As a result of such enormous development of commercial aviation, runways have become a bottleneck at many airports around the world. Because of a shortage of land or environmental constraints, many busy airports operate with only one or two runways. When demand for landing during a specific time period exceeds the capacity of a runway system, some aircraft landings must be delayed, creating additional costs to airlines and passengers. In addition, passengers could miss their connecting flights, aircraft delays could propagate over the air transport network, and crew members might need to be rescheduled. Although these problems can partially be solved by building additional runways, expansion of runways is often limited because of land constraints, large capital costs and environmental damages. Hence, the Air Traffic Control (ATC) system and airport authorities usually try to use the existing airport infrastructure optimally; they develop and implement a range of tactical traffic control measures, methods, and strategies, while using sophisticated ATC automation tools and systems for communication, navigation and surveillance. During peak arrival periods, when aircraft arrival demand exceeds runway arrival capacity, ATC often applies tactical traffic control measures, such as holding patterns, vectoring, and speed control, ensuring safety by providing adequate landing time intervals.

However, such measures might not ensure the optimal reduction in arrival delays. Thus, landing time intervals can vary greatly, driven by decisions made by air traffic controllers, which in turn are based on visibility conditions, aircraft size, wind conditions and other factors [3,4]. The quandary about reducing arrival delays, known as the Aircraft Landing Problem (ALP), has been the object of some very important research studies. In the ALP, the common challenge is to minimize arrival delays by optimally reducing the total deviation of aircraft actual landing times from target landing times. Solving this problem would also improve the predictability of landing times, thereby significantly increasing efficiency and safety of operations.

To address the issue of deviations between aircraft actual landing times and target landing times, we base our study on the static version of the Aircraft Landing Problem (ALP), where all information related to the aircraft to be scheduled is known in advance. In the ALP, each aircraft requesting landing must be assigned to one of several runways, and every landing must be performed within its prescribed time window. At the same time, the prescribed time separation (i.e., headway) between all pairs of aircraft must be satisfied. Then, the objective is to minimize the total deviation of all aircraft actual landing times from their target landing times. Previous research on the ALP showed that the exact algorithms are capable of discovering an optimal solution of the ALP within acceptable central processing unit (CPU) time for instances that consider up to 50 aircraft. For the larger instances, the authors mainly use metaheuristic approaches. To solve the ALP, we propose a heuristic approach that is based on the Bee Colony Optimization (BCO) metaheuristic. The BCO metaheuristic is an algorithm inspired by nature. The artificial bees try to find the best solution to the combinatorial optimization problem in a way similar to how bees in nature look for food sources. Our motivation for applying this metaheuristic to the ALP was inspired by our positive experiences in applying this approach to many other combinatorial optimization problems. Although at first it was very challenging, we successfully combined the BCO metaheuristics with the exact approach to obtain solution improvements. To the best of our knowledge, this is the first attempt to combine the BCO algorithm with one exact approach for solving the sub-problem. The proposed algorithm in this study is tested on benchmark examples from the literature. The test examples are divided into two groups. The first group has a number of aircraft between 10 and 50, and the number of runways between 1 and 3. The second group contains instances with the number of aircraft between 100 and 500, and the number of runways between 1 and 5. Our results show that the proposed algorithm is competitive with other approaches suggested in the literature.

The rest of the article is organized as follows. Section 2 states the problem. Our implementation of the BCO metaheuristic is presented in Section 3. Section 4 presents the quantitative experiments, and in Section 5 we present our conclusions.

2. Problem statement

Understanding the roles of the airport zone, terminal airspace and Air Traffic Control (ATC) can help the reader appreciate the complexity of the ALP. It is commonly known that ATC monitors and controls flights from takeoff to landing. Takeoffs and landings occur within the airport zone, which usually has a radius of 2.0 to 2.5 nautical miles (nm) and has an altitude of 2000 ft. (600 m) Mean Sea Level (MSL). Depending on the airport, the zone can take various shapes – circle, semi-circle, ellipse, square, rectangle, or trapezoid, for example – as local conditions dictate. Aircraft fly along their prescribed arrival and departure trajectories within the terminal airspace, which spreads around and above the airport zone; the terminal airspace is established around a single busy airport or several airports of the same airport system, within a radius of 40 to 50 nm and altitude up to Flight Level (FL) 100 (each FL is determined by 10^3 ft $(1 \text{ ft.} \sim 0,305 \text{ m})$) [5]. An aircraft that wants to land needs information on landing time from the ATC, as well as information about an assigned runway. The ATC must coordinate flights to maximize safety, reliability and efficiency – the basis of the Airport Landing Problem.

The ALP was first considered by Beasley et al. [6]. The ALP is a non-deterministic polynomial time (NP)-hard combinatorial optimization problem [7,8], and many approaches to solving this problem have already been proposed in the literature. Beasley et al. [6] proposed a mixed integer programming mathematical formulation, as well as a heuristic approach, while Beasley et al. [9] considered a dynamic version of this problem. A scatter search metaheuristic approach for this problem was proposed in Pinol and Beasley [10]. Salehipour et al. [11] proposed a mixed-integer goal programming mathematical formulation, as well as a heuristic algorithm based on Variable Neighbor Search (VNS) and Simulated Annealing (SA) metaheuristics. Sabar and Kendall [12] considered implementation of an iterated local search algorithm. This algorithm uses multiple perturbation operators as well as time perturbation strength. Faye [13] developed an approach based on an approximation of the time separation matrix and on time discretization.

Herein, we consider the static version of the ALP. All relevant information about the aircraft to be scheduled is known in advance. Each of the considered aircraft must land on one of the given runways. Every aircraft landing must be done inside its given time window. At the same time, given time separations between all pairs of aircraft must be satisfied. The objective function to be minimized represents the total deviation of all aircraft actual landing times from the target landing times. The ALP that we consider can be defined in the following way: for each arriving aircraft determine an available runway and landing time in such a way as to minimize the total deviation of all aircraft actual landing times from the target landing times.

The ALP consists of two subsequent tasks: 1) assignment of each aircraft to one of the runways; and 2) scheduling aircraft landings. Depending on the number of runways that can be used for landing, we distinguish single- and multiple-runway systems for aircraft landings. For the purpose of this study we assume that the number of runways for landing is input data in this problem formulation, and there is no difference between runways. Scheduling of aircraft landings must be determined according to time windows given for each aircraft, as well as minimal safe time separation between any pairs of aircraft. In the input data we have information for each aircraft, consisting of the earliest time for landing and its lateness. Considering a target landing time for the aircraft, we can calculate the cost of deviation from this time. Depending on the actual landing time of aircraft, which can occur before or after the target landing time, we obtain the corresponding cost. The objective function is to minimize the total landing time deviation cost.

Predefined minimal landing time separations must be satisfied for each pair of aircraft that use the same runway. Depending on the aircrafts' size, minimal safety time separation (i.e., headway) takes different values. These values should be given in advance. If aircraft do not use the same runway for landing, in previous studies [6,10,11] it was assumed that the minimal time separation between aircraft must be equal to one-time unit. The following is a mixed-integer programming mathematical formulation of the ALP [11]:

Minimize

$$z = \sum_{i=1}^{n} \left(a_i c_i^{+} + b_i c_i^{-} \right) \tag{1}$$

s.t.

$$E_i \leq x_i \leq L_i \qquad i = 1, ..., n \tag{2}$$

$$x_i - T_i = a_i - b_i \qquad i = 1, ..., n \tag{3}$$

$$(x_j - x_i) \geq s_{ij}\delta_{ij} + t_{ij}(1 - \delta_{ij}) - My_{ji} \qquad i, j = 1, ..., n; \quad i \neq j \tag{4}$$

$$y_{ij} + y_{ji} = 1 \qquad i, j = 1, ..., n; \quad i \neq j \tag{5}$$

$$\delta_{ij} \geq \gamma_{ir} + \gamma_{jr} - 1 \qquad i, j = 1, ..., n; \quad i \neq j; \quad r = 1, ..., m \tag{6}$$

$$\sum_{r=1}^{m} \gamma_{ir} = 1 \qquad i = 1, ..., n \tag{7}$$

$$y_{ir}, \gamma_{ir}, \delta_{ij} \in \{0, 1\} \qquad i, j = 1, ..., n; \quad i \neq j; \quad r = 1, ..., m \tag{8}$$

$$x_i, a_i, b_i \geq 0 \qquad i = 1, ..., n \tag{9}$$

where:

s_{ij} – the time separation between two aircraft i and j when landing on a same runway

t_{ij} – the time separation between two aircraft i and j when landing on different runways

T_i – the target landing time of aircraft i

E_i – earliest landing time of aircraft i

L_i – latest landing time of aircraft i

c_i^+ – cost of late landing of aircraft i

c_i^- – cost of early landing of aircraft i

and decision variables are:

x_i – scheduled landing time of aircraft i ($i = 1, ..., n$)

y_{ij} – takes 1 if aircraft i lands before aircraft j and otherwise 0

γ_{ir} – takes 1 if aircraft i is allocated to runway r ($r = 1, ..., m$) and otherwise 0

δ_{ij} – takes 1 if aircraft i and j land on a same runway and otherwise 0

a_i – delay of landing aircraft i (landing after the target time)

b_i – earliness of landing aircraft i (landing before the target time)

We minimize the objective function (1), which represents the total cost of landing deviation from target times. Constraint (2) guarantees that every aircraft starts landing within a defined time window. Constraint (3) defines calculation of landing time deviations. Time separation between two aircraft must be satisfied according to Constraint (4). If aircraft i lands before aircraft j, then decision variable y_{ji} must take the value 0, as defined by Constraint (5). Constraint (6) defines that if aircraft i and j lend on the same runway, decision variable δ_{ij} must take value 1. According to Constraint (7) every aircraft must use exactly one runway for landing. Constraint (8) defines decision variables as binary. Constraint (9) defines decision variables as nonnegative.

3. The BCO metaheuristic for the aircraft landing problem

The Bee Colony Optimization (BCO) metaheuristic was proposed by Lučić and Teodorović [14]. This biologically inspired method explores collective intelligence applied by the honey bees during a nectar collecting process. The BCO technique uses a similarity between the way in which bees search for food and the way in which optimization algorithms search for an optimum of known combinatorial optimization problems. The BCO has been used for solving various combinatorial optimization problems [15–21], [22–31]. The broad literature review of the BCO can be found in recent review articles by Teodorović et al. [32,33].

The major idea behind the BCO metaheuristic is to construct a multi-agent system (colony of artificial bees) able to solve difficult combinatorial optimization problems effectively. In nature every bee makes a decision to arrive at the nectar source by following a nestmate who has previously discovered a patch of flowers. Each hive has a so-called dance floor area in which the bees that have discovered nectar sources dance. By dancing, bees try to encourage their nestmates to follow them. If a bee decides to depart from the hive to get nectar, it goes behind one of the bee dancers to one of the nectar areas. Upon arrival, the foraging bee takes a load of nectar and comes back to the hive, relinquishing the nectar to a food-storing bee. After it relinquishes the food, the bee can abandon the food source and become once more an uncommitted follower, or dance and thus recruit nestmates before it returns to the food source. The bee selects one of these alternatives with a certain probability. Within the dance area, the bee dancers promote different food areas.

Within the BCO algorithm, a population of artificial bees searches for the optimal solution. Artificial bees represent agents, which collaboratively solve complex combinatorial optimization problems. This algorithm has two versions: constructive and improvement. In this article we use the improvement version of the BCO algorithm. At the end of the search process, every artificial bee generates one solution to the problem. The BCO algorithm consists of two alternating phases, forward pass and backward pass, which represent the main steps for generating the initial solution. At the beginning of the search process, the generated initial solution is assigned to every artificial bee. During the forward pass, bees try to improve their solutions. Every artificial bee makes the #C solution modification. The backward pass starts after the end of the forward pass. During the forward pass, artificial bees exchange information about their solutions. The new forward pass starts after the end of the backward pass, etc.

The BCO algorithm parameters whose values must be set prior the algorithm execution are as follows:

#B - number of bees involved in the search,

#P - number of forward and backward passes in a single iteration,

#C - number of changes in one forward pass.

The analyst also has to define stopping criteria. Some of the most frequently used stopping criteria are the number of iterations and the CPU time. Also, let us introduce the following notations used for the BCO algorithm:

S_{best} – the best solution

$S_{initial}$ – the initial solution

S_b – solution of the bee b ($b = 1, ..., $ #B)

T_b – bee's b objective function value

O_b – normalized bee's b objective function value

p_b – the probability that bee b will stay loyal to its solution

The pseudo code of the improvement version of the BCO algorithm is given as Algorithm 1. As one can notice at the beginning of the algorithm, we have to determine the initial solution. For that purpose, we can use some simple greedy heuristic algorithm (step 1). That solution should be saved as the best solution (step 2). At the beginning of iterations one solution (usually the best solution that was discovered) should be assigned to the bees (steps 4–6). During the iterations the bees repeat the forward and backward passes (steps 7–10). The process continues while the stopping criteria are not satisfied (step 11).

ALGORITHM 1 Bee Colony Optimization

1. Generate $S_{initial}$
2. $S \leftarrow S_{initial}$
3. **do**
4.   **for** $b = 1$ **to** #B **do**
5. $S_b \leftarrow S$
6.   **end next**
7.   **for** $i = 1$ **to** #P **do**
8.   Forward pass
9.   Backward pass
10.   **end next**
11. **while** (stopping criteria are not satisfied)

> **ALGORITHM 2 Forward pass**
> 1. **for** $i=1$ **to** #C **do**
> 2.   **for** $b=1$ **to** #B **do**
> 3.   Modify S_b
> 4.   **end next**
> 5.   **if** the best S_b is better than S_{best} **then do**
> 6. $S_{best} \leftarrow S_b$
> 7.   **end if**
> 8. **end next**

The pseudo code of the forward pass is given as Algorithm 2. During the forward pass all bees modify their solutions (steps 2–4), then in the steps 6 and 7 we compare new solutions with the best-known solution. If the best new solution is better than previously best-known, it will be saved (steps 5 and 6).

In the backward pass of the BCO algorithm, the bees announce the quality of the generated solutions. When all solutions are evaluated, every bee decides with a certain probability whether it will stay loyal to its solution. The bees with better solutions have more chances to keep and advertise their solutions. Unlike bees in nature, the artificial bees that are loyal to their solutions are at the same time recruiters (their solutions would be considered by other bees). Once its solution is abandoned, a bee becomes uncommitted and must select one of the advertised solutions. This decision is taken with a probability that those selecting the better among the advertised solutions have a greater opportunity to be chosen for further exploration. The pseudo-code of the backward pass is given as Algorithm 3.

At the beginning of the backward pass we have to normalize the objective function value for each bee (steps 1–3). If we consider the problem where the objective function should be maximized, the normalized value of bee's b solution equals:

$$O_b = \frac{T_b - \min_{i=1,\,\ldots,\,\#B}\{T_i\}}{\max_{i=1,\,\ldots,\,\#B}\{T_i\} - \min_{i=1,\,\ldots,\,\#B}\{T_i\}} \tag{10}$$

In the opposite case, when the objective function should be minimized, the normalized objective function value equals:

$$O_b = \frac{\max_{i=1,\,\ldots,\,\#B}\{T_i\} - T_b}{\max_{i=1,\,\ldots,\,\#B}\{T_i\} - \min_{i=1,\,\ldots,\,\#B}\{T_i\}} \tag{11}$$

ALGORITHM 3 Backward pass
1. **for** $b=1$ to #B
2.   Calculate O_b
3. **end for**
4. **for** $b=1$ to #B
5.   Calculate p_b
6.   **if** *random number between 0 and* $1 \leq p_b$ **then**
7.   bee b is loyal
8.   **else**
9.   bee b is an uncommitted follower
10. **end for**
11. **for** $b=1$ to #B
12.   **if** bee b is the uncommitted follower then
13.   Determine loyal bee r that bee b will follow.
14. $S_b \leftarrow S_r$
15.   **end if**
16. **end for**

In the fifth step of algorithm 3 we calculate the probability that b^{th} bee (at the beginning of the new forward pass) will stay loyal to its previously generated solution. This probability p_b equals:

$$p_b = e^{-(O_{max}-O_b)} \tag{12}$$

where:

O_{max}– maximum over all normalized values of the solutions to be compared.

It is obvious that a bee with the better solution (i.e., the higher O_b value) has the higher probability to stay loyal to its solution. Using the probability p_b and the generated random numbers r_b, a decision is made for every artificial bee about whether to convert into an uncommitted follower (if $r_b > p_b$), or to continue exploring a previously generated path (if $r_b \leq p_b$). For every uncommitted bee with a certain probability we simulate which recruiter it would follow. The probability q_b that b's solution would be chosen by any uncommitted bee is equal to:

$$q_b = \frac{O_b}{\sum_{l \in L} O_l} \tag{13}$$

where L is a set of the loyal bees.

Taking a random number between 0 and 1 for each uncommitted bee, the bees make a decision as to which loyal bee(s), from the set L, they are going to follow (step 13 in the algorithm 3). Let us suppose that the bee r is chosen. This means that the solution of the bee r will be assigned to the bee b (step 14 in the algorithm 3).

3.1 Solution representation

In each solution, we must have the following information for each aircraft: the start time of landing, and the assigned runway it will use for landing. Also, we assign the order of the aircraft according to the start landing time. The first aircraft in the array will be the one with the smallest landing time value, and the last one will be the aircraft with the highest landing time value. In that way, we represent the generated solutions of the ALP by using the following three arrays: the array of the aircraft (AA), the array of the runways (AR), and the array of the landing times (AT). An example of the solution representation is given in Table 1. As one can see, Aircraft 1 starts landing first $(AA(1)=1)$. It uses runway 1 $(AR(1)=1)$. The aircraft starts landing at the moment 10 time units $(AT(1)=10)$. Aircraft 3 starts landing as the second one $(AA(2)=$ aircraft 3). It lands at 20 time units $(AT(2)=20)$ on runway 2 $(AR(2)=2)$. The third one is Aircraft 2 $(AA(3)=$ aircraft 2). It lands at the 30 time units $(AT(3)=30)$ on runway 1 $(AR(3)=1)$.

3.2 Generating the initial solution

We propose the following heuristic algorithm that generates the initial solution (Algorithm 4):

Table 1 An example of the solution with three aircraft.

	Landing scheduling index		
Array	1	2	3
AA	$AA(1)=$ aircraft 1	$AA(2)=$ aircraft 3	$AA(3)=$ aircraft 2
AR	$AR(1)=1$	$AR(2)=2$	$AR(3)=1$
AT	$AT(1)=10$	$AT(2)=20$	$AT(3)=30$

ALGORITHM 4 Initial solution generation
1.   Generate the list *A* of aircraft from the smallest arrival time value to the highest arrival time value.
2.   **while** (list *A* is not empty) **do**
3.   Take the first element from a list *A*.
4.   Determine the best landing time and runway for aircraft taking into consideration previously scheduled aircraft.
5.   Remove the aircraft from a list *A*.
6.   **end while**

Table 2 The example of the temporal solution.

			Landing scheduling index					
Array	**1**	**2**	**...**	**i - 1**	**i**	**$i+1$**	**...**	**m - 1**	**m**
AA	$AA(1)$	$AA(2)$		$AA(i\text{-}1)$	$AA(i)$	$AA(i+1)$		$AA(m\text{-}1)$	$AA(m)$
AR	$AR(1)$	$AR(2)$	...	$AR(i\text{-}1)$	$AR(i)$	$AR(i+1)$	...	$AR(m\text{-}1)$	$AR(m)$
AT	$AT(1)$	$AT(2)$	...	$AT(i\text{-}1)$	$AT(i)$	$AT(i+1)$	...	$AT(m\text{-}1)$	$AT(m)$

Here we explain in more detail how we determined the best runway and landing time for a particular aircraft (step 4). Let us assume that we have already determined runways and landing times for *m* aircraft as shown in Table 2.

We must check if it is possible to insert the aircraft in each of the places of arrays $AA(i)$, $i = 1, \ldots, m + 1$ and for each of the available runways ($r = 1, \ldots, R$). For each place in array, and for each available runway, we determine the earliest t^e_{kir} and the latest t^l_{kir} landing time of the aircraft *k*. The earliest landing time is:

$$t^e_{kir} = \max_{j=1,\ldots,i-1} \left\{ AT(j) + s_{AA(j),k} \cdot \delta_{AA(j),k} + t_{AA(j),k} \cdot \left(1 - \delta_{AA(j),k} \right) \right\} \tag{14}$$

where:

$s_{k,AA(j)}$ – the time separation between two aircraft *k* and $AA(j)$ when they are landing on the same runway (if $AR(j)$ is equal *r*)

$\delta_{AA(j),k}$ – takes value 1 if aircraft $AA(j)$ and *k* will use the same runway (if $AR(j)$ is equal *r*), otherwise it takes 0

$t_{AA(j),k}$ – the time separation between two aircraft k and $AA(j)$ when they are landing on different runways (if $AR(j)$ is not equal r)

In the similar way we define the latest landing time:

$$t^l_{kir} = \min_{j=i,\ldots,n}\left\{AT(j) - s_{k,AA(j)} \cdot \delta_{k,AA(j)} - t_{k,AA(j)} \cdot \left(1 - \delta_{k,AA(j)}\right)\right\} \tag{15}$$

We suppose that the insertion is possible in the following three cases:

Case 1. Target landing time (T_k) is between t_e and t_l ($t^e_{kir} \leq T_k \leq t^l_{kir}$) as depicted in Fig. 1. In this case we can see that aircraft k can be inserted in the position i and the landing time is: $t_{kir} = T_k$, and the cost of insertion is equal $c_{kir} = 0$.

Case 2. Target landing time value (T_k) is smaller than t^e_{kir} and the latest landing time of aircraft k (L_k) is between t^e_{kir} and t^l_{kir} ($T_k < t^e_{kir}$ and $t^e_{kir} \leq L_k \leq t^l_{kir}$) as depicted in Fig. 2. In this case $t_{kir} = t^e_{kir}$ and $c_{kir} = (t_{kir} - T_k) \cdot c^+_i$

Case 3. Target landing time value (T_k) is higher than t^l_{kir}, and the earliest landing time of aircraft k (E_k) is between t^e_{kir} and t^l_{kir} ($T_k > t^l_{kir}$ and $t^e_{kir} \leq E_k \leq t^l_{kir}$) as depicted in Fig. 3. In this case $t_{kir} = t^l_{kir}$ and $c_{kir} = (t_{kir} - T_k) \cdot c^-_i$.

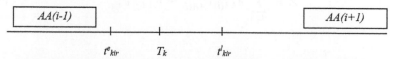

Fig. 1 The target landing time (T_k) for aircraft k in position i on runway r is between the earliest landing time t^e_{kir} and the latest landing time t^l_{kir}.

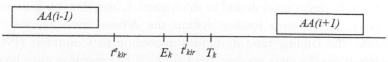

Fig. 2 The target landing time (T_k) is smaller than the earliest landing time t^e_{kir}, and the latest landing time of aircraft k (L_k) is between the earliest landing time t^e_{kir} and the latest landing time t^l_{kir}.

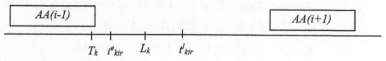

Fig. 3 The target landing time (T_k) is higher than the latest landing time t^l_{kir}, and the earliest landing time of aircraft k (E_k) is between the earliest landing time t^e_{kir} and the latest landing time t^l_{kir}.

For the aircraft k, the best place in the array and the best runway are related to the minimal costs:

$$c_{kir}^{*} = \min_{i=1,\,\ldots,\,m+1,\,r=1,\,\ldots,\,R} \{c_{kir}\} \tag{16}$$

When c_{kir}^{*} is determined, the aircraft can be added to the solution. The arrays AA, AR and AT get one more element ($m+1$). The new aircraft will be set at the i position ($AA(i)=k$, $AR(i)=r$, and $AT(i)=t_{kir}$). Elements that were previously at the positions from i to m are moved to the positions from $i+1$ to $m+1$. Their characteristics are the same as the previous ones.

3.3 Solution improvements

We improve the initial solution and the final solution by solving one linear programming model. By solving this model, we want to find the best possible array of time (AT). As inputs we use array of aircraft (AA) and array of runways (AR). Mathematical formulation for this linear programming model can be given as follows:

Minimize

$$Z = \sum_{i=1}^{n} a_{AA(i)} c_{AA(i)}^{+} + b_{AA(i)} c_{AA(i)}^{-} \tag{17}$$

s.t.

$$E_{AA(i)} \leq x_{AA(i)} \leq L_{AA(i)} \qquad i = 1, \ldots, n \tag{18}$$

$$x_{AA(i)} - a_{AA(i)} + b_{AA(i)} = T_{AA(i)} \tag{19}$$

$$x_{AA(j)} - x_{AA(i)} \geq t_{ij} \qquad i = 1, \ldots, n-1;$$
$$j = i+1, \ldots, n; AR(i) = AR(j) \tag{20}$$

$$x_{AA(j)} - x_{AA(i)} \geq s_{ij} \qquad i = 1, \ldots, n-1;$$
$$j = i+1, \ldots, n; AR(i) \neq AR(j) \tag{21}$$

$$x_{AA(i)} \geq 0 \qquad i = 1, \ldots, n \tag{22}$$

The objective function (17) that represents the total cost of landing deviations from the target times should be minimized. Constraint (18) guarantees that every aircraft starts lending within the defined time interval. We determine the landing time deviations according to Constraint (19). If two aircraft use the same runway, the landing time separation must be satisfied (Constraint (20)). Otherwise, when they use different runways, the time separation is described by Constraint (21). Constraint (22) defines the decision variable x as nonnegative. Unlike previous mathematical formulations (1)–(9), here we do not have any integer programming variables, which make the model easier to solve.

3.4 Bees' modification of the solution

To modify the bees' solutions, we use a two-step approach. In the first step, the bee discards a part of its solution. In the second step, the bee generates a feasible solution. All bees discard part of their solutions in the same way. To repair broken solutions, bees use two ways to make them feasible. We considered the situation with two types of bees, where each type chooses one of these ways to obtain a feasible solution. The concept of heterogeneous bees was first proposed by Nikolić and Teodorović [23,24]. We determine the number of aircraft (between 1 and 10) that will be removed from the bee's solution in a random manner. Let us denote this number by k. We remove randomly chosen k successive aircraft. All k removed aircraft are treated as unassigned. In the second step, we must determine the landing runway and landing time for each of the k aircraft. The bees of the first type generate feasible solutions by using the heuristic for the initial solution generation (explained in Section 3.2). The only difference is that we do not have to determine a runway and landing time for all aircraft, but only for k aircraft that are removed in the previous step. The bees of the second type use an approach that is similar to the one described for initial solution generation. The main difference is in the way the landing time is calculated for the first case, as described in Section 3.2. Now, if the target time is between the earliest (t_e) and the latest (t_l) time, the landing time T is determined randomly as:

$$T = t_e + rand() \cdot (t_l - t_e) \tag{23}$$

where:

$rand()$ is a function that returns a random number between 0 and 1.

4. Numerical experiments

We use a set of benchmark instances[a] to test the proposed algorithm. The total set contains 13 instances, with the number of aircraft from 10 to 500. For each instance a different number of runways was tested, varying

---

[a] The benchmark instances are given at: http://people.brunel.ac.uk/~mastjjb/jeb/orlib/airlandinfo. html [34]

from 1 to 5. The whole set of instances can be divided into two subsets: the subset of small size instances (where the number of aircraft is between 10 and 50) and the large size set (where the number of aircraft is between 100 and 500). We used, in all experiments, a computer with the following characteristics: Intel(R) Core(TM) i7-4700MQ CPU, 2.4 GHz, Installed memory (RAM): 8 GB. To solve the small size instances, we allocated CPU time of 1 min, and to solve large size instances we allocated 5 min of CPU time. We performed a small test to determine the best BCO parameters for the considered problem. According to our previous experience with the BCO, we tested the following values for the number of bees and the number of passes:

- #B = 10 (5 bees of each type), #B = 20 (10 bees of each type) and #B = 30 (15 bees of each type)
- #P = 5; #P = 10 and #P = 15

We set up in each test the number of changes per forward pass (#C) to be equal to one. For this test we use instance with 500 aircraft and 1 runway, and 5 min CPU time. We perform 3 runs for each combination of the parameters. The average values of the BCO parameters are given in Table 3. As one can notice, the best value is obtained for #B = 10 and #P = 5 (the corresponding values are denoted in Table 3 in bold and italic). These values of the BCO parameters are further used for the final experiment.

For the small-size instances we compared the BCO results with the results obtained by solving mixed-integer programs ((1)–(9) mathematical formulation) by CPLEX 12.6. With the bold numbers are denoted by the values of objective functions when the solutions are optimal. The obtained results are given in Table 4. As it can be noticed, the BCO algorithm was capable of discovering optimal solutions for all considered instances.

For the large-size instances, where the number of aircraft varies between 100 and 500, the BCO results are compared with the heuristic approaches given in the literature. We made 3 runs for each instance. Our results for each run are given in Table 5. The best obtained results from 3 runs are given

Table 3 Average results for different values of the BCO parameters.

Number of bees (#B)	Number of passes (#P)		
	5	10	15
10	*39,985.53*	41,317.00	41,996.16
20	40,315.94	41,638.57	42,495.81
30	40,577.83	42,273.99	42,869.82

Table 4 Comparison results obtained by the BCO algorithm with the optimal solutions obtained by CPLEX.

Instance	Number of aircraft	Number of runways	CPLEX Objective function value	CPU time [s]	BCO objective function value
airland1	10	1	700	0.03	700
		2	90	0.05	90
		3	0	0.01	0
airland 2	15	1	1480	0.08	1480
		2	210	0.09	210
		3	30	0.06	30
airland 3	20	1	820	0.08	820
		2	70	0.09	70
		3	10	0.09	10
airland 4	20	1	2520	1.00	2520
		2	660	2.31	660
		3	160	0.34	160
		4	30	0.09	30
airland 5	20	1	3100	3.20	3100
		2	650	4.17	650
		3	170	0.36	170
		4	30	0.13	30
airland 6	30	1	24,442	0.00	24,442
		2	554	0.22	554
		3	0	0.02	0
airland 7	44	1	1550	0.08	1550
		2	0	0.02	0
airland 8	50	1	1950	0.36	1950
		2	135	1.00	135
		3	10	0.58	10

Table 5 BCO results on large size instances.

Instance	Number of aircraft	Number of runways	BCO						gap (%)
			run 1	run 2	run 3	min	max		
airland 9	100	1	5666.64	5666.64	5659.68	5659.68	5666.64		0.12
		2	444.1	444.1	444.1	444.1	444.1		0.00
		3	75.75	75.75	75.75	75.75	75.75		0.00
		4	0	0	0	0	0		0.00
airland 10	150	1	12,957.64	13,070.07	12,709.27	12,709.27	13,070.07		2.84
		2	1144.04	1144.04	1144.04	1144.04	1144.04		0.00
		3	206.27	206.27	206.27	206.27	206.27		0.00
		4	35.28	35.28	35.28	35.28	35.28		0.00
		5	1.06	1.06	1.06	1.06	1.06		0.00
airland 11	200	1	12,593.27	12,545.88	12,628.63	12,545.88	12,628.63		0.66
		2	1330.91	1330.91	1330.91	1330.91	1330.91		0.00
		3	253.07	253.07	253.07	253.07	253.07		0.00
		4	54.53	54.53	54.53	54.53	54.53		0.00
		5	0	0	0	0	0		0.00

airland 12	250	1	17,081.15	16,894.15	16,988.35	16,894.15	17,081.15	1.11
		2	1696.59	1696.59	1696.59	1696.59	1696.59	0.00
		3	222.53	222.61	222.53	222.53	222.61	0.04
		4	2.44	2.44	2.44	2.44	2.44	0.00
		5	0	0	0	0	0	0.00
airland 13	500	1	40,059.97	39,387.93	40,513.93	39,387.93	40,513.93	2.86
		2	3951.43	3949.11	3944.11	3944.11	3951.43	0.19
		3	675.06	675.06	675.06	675.06	675.06	0.00
		4	89.95	89.95	89.95	89.95	89.95	0.00
		5	0	0	0	0	0	0.00

in column 7 (column name "min"), and the worst results are found in column 8 (column name "max"). The gap between these two values is given in the column 9 (column name "gap%"). In 17 considered cases, the gap between these two values is equal to 0, and in 7 cases the gap is between 0.12 and 2.86%.

The comparisons between the best BCO solutions and the results given in the literature are shown in Table 6. From the literature we take into consideration results obtained by a Scatter Search (SS) [10], a hybrid Simulated Annealing and Variable Neighborhood Descent (SA + VND) [11], a hybrid Simulated Annealing and Variable Neighborhood Search (SA + VNS) [27], and Iterated Local Search (ILS) [12]. The best-known values are denoted by the bold numbers. The cases when the BCO produced better results than the other techniques are denoted by bold and italic. In 9 cases BCO had slightly worse solutions. In three cases the BCO produced the same quality solutions as the other approaches. Finally, in 12 cases the BCO generated solutions of the better quality. The most significant improvements of the solutions are obtained for the largest instances (with 250 and 500 aircraft).

5. Conclusions

The ALP we have considered is an NP-hard combinatorial optimization problem. When solving this problem, the goal is to minimize the total deviation cost from target landing times. Various heuristics and metaheuristic approaches have been proposed in the literature for the ALP. To solve it, we tailored the BCO metaheuristic. The proposed method was tested on benchmark instances from the literature. Our results show that the proposed algorithm can easily find optimal solutions for small instances (with 10 to 50 aircraft, and up to four runways). In the case of larger-sized instances (with 100 to 500 aircraft, and up to five runways), the BCO approach was capable of outperforming some of the well-known algorithms from the literature.

This article presents the first attempt to combine BCO with an exact approach for a sub-problem. The results in this article show that this solution approach is capable of obtaining a high-quality solution. Nonetheless, this approach has two drawbacks: 1) the quality of the solution could depend on the assigned CPU time (most probably the algorithm will be capable of finding a better solution if it has more CPU time), and 2) the CPU time for solving a sub-problem's linear program could be very different depending on the software used (some software need much more CPU time than others).

Table 6 Comparison results for BCO and four approaches given in the literature.

Instance	Number of aircraft	Number of runways	SS	SA+VND	SA+VNS	ILS	BCO
airland 9	100	1	**5611.70**	6091.88	6091.88	**5611.70**	5659.68
		2	452.92	450.26	452.92	445.04	***444.1***
		3	75.75	75.75	75.75	**74.00**	75.75
		4	**0**	**0**	**0**	**0**	**0**
airland 10	150	1	12,329.31	12,329.31	12,329.31	**12,321.91**	12,709.27
		2	1288.73	1219.26	1288.73	1271.07	***1144.04***
		3	220.79	206.45	220.79	**200.01**	206.27
		4	34.22	35.28	35.28	**32.11**	35.28
		5	**0**	1.06	1.06	**0**	1.06
airland 11	200	1	12,418.32	12,418.32	12,418.32	**12,417.70**	12,545.88
		2	1540.84	1416.83	1540.84	1410.02	***1330.91***
		3	280.82	272.92	280.82	271.10	253.07
		4	54.53	54.53	54.53	**51.00**	54.53
		5	**0**	**0**	**0**	**0**	**0**

Continued

Table 6 Comparison results for BCO and four approaches given in the literature.—cont'd

Instance	Number of aircraft	Number of runways	SS	SA+VND	SA+VNS	ILS	BCO
airland 12	250	1	**16,209.78**	**16,209.78**	**16,209.78**	16,209.78	16,894.15
		2	1961.39	1961.39	1961.39	1961.39	*1696.59*
		3	290.04	279.7	290.04	272.03	*222.53*
		4	3.49	3.49	3.49	3.40	*2.44*
		5	**0**	**0**	**0**	0	0
airland 13	500	1	44,832.38	41,448.16	44,832.38	41,380.29	*39,387.93*
		2	5501.96	5475.81	5501.96	5458.49	*3944.11*
		3	1108.51	744.97	1108.51	1108.51	*675.06*
		4	188.46	100.6	188.46	92.89	*89.95*
		5	7.35	3.81	7.35	3.00	*0*

In future work we plan to improve the proposed algorithm and to obtain better results for larger instances. We will also try to apply this concept to other combinatorial optimization problems such as the Traveling Salesman Problem (TSP), the Vehicle Routing Problem with Time Windows (VRPTW), and the Gate Assignment Problem. For such combinatorial optimization problems, it will be interesting to use an exact approach for improving previously generated solutions. For example, in the TSP problem, instead of using 2-OPT or 3-OPT algorithms, we will try to improve an obtained solution by solving an appropriate linear (or mix-integer) program.

Acknowledgments

The authors would like to thank the U.S. Federal Aviation Administration (Award No. DTFAWA-11-D-00017), The Ministry of Education, Science and Technological Development of the Republic of Serbia, and the Serbian Academy of Sciences and Arts, for partial funding of this research.

References

[1] International air transport association (IATA), Annual Review 2022, 2022. http://www.iata.org/publications/Pages/annual-review.aspx. (accessed 29.09.22).
[2] International air transport association (IATA), Annual Review 2014, 2014. http://www.iata.org/publications/Pages/annual-review.aspx. (accessed 30.09.22).
[3] J. Rakas, F. Yin, Statistical modeling and analysis of landing time intervals: case study of Los Angeles international airport, California, Transp. Res. Rec. 2005 (1915) 69–78.
[4] H.F. Vandevenne, M.A. Lippert, Using maximum likelihood estimation to determine statistical model parameters for landing time separations, ATC Project Memorandum No. 92PM-AATT-0006, 2000.
[5] R. de Neufville, A. Odoni, Airport Systems: Planning, Design and Management, first ed., McGraw-Hill, New York, 2003 (Chapter 13).
[6] J.E. Beasley, M. Krishnamoorthy, Y.M. Krishnamoorthy, D. Krishnamoorthy, Scheduling aircraft landings - the static case, Transp. Sci. 34 (2) (2000) 180–197.
[7] J.A. Bennell, M. Mesgrarpour, C.N. Potts, Airport runway scheduling, 4OR 9 (2011) 115–138.
[8] J.A. Bennell, M. Mesgrarpour, C.N. Potts, Airport runway scheduling, Ann. Oper. Res. 204 (2013) 249–270.
[9] J.E. Beasley, M. Krishnamoorthy, Y.M. Krishnamoorthy, D. Krishnamoorthy, Displacement problem and dynamically scheduling aircraft landings, J. Oper. Res. Soc. 55 (2004) 54–64.
[10] H. Pinol, J.E. Beasley, Scatter search and bionomic algorithms for the aircraft landing problem, Eur. J. Oper. Res. 171 (2006) 439–462.
[11] A. Salehipour, M. Modarres, L.M. Naeni, An efficient hybrid meta-heuristic for aircraft landing problem, Comput. Oper. Res. 40 (2013) 207–213.
[12] N. Sabar, G. Kendall, An iterated local search with multiple perturbation operators and time varying perturbation strength for the aircraft landing problem, Omega 56 (2015) 88–98.
[13] A. Faye, Solving the aircraft landing problem with time discretization approach, Eur. J. Oper. Res. 242 (2015) 1028–1038.

[14] P. Lučić, D. Teodorović, Bee system: modeling combinatorial optimization transportation engineering problems by swarm intelligence, in: Preprints of the TRISTAN IV Triennial Symposium on Transportation Analysis, Sao Miguel, Azores Islands, Portugal, 13-19 June, 2001, pp. 441–445.

[15] T. Davidović, D. Ramljak, M. Šelmić, D. Teodorović, Bee colony optimization for the *p*-center problem, Comput. Oper. Res. 38 (2011) 1367–1376.

[16] T. Davidović, M. Šelmić, D. Teodorović, D. Ramljak, Bee Colony optimization for scheduling independent tasks to identical processors, J. Heuristics 18 (4) (2012) 549–569.

[17] B. Dimitrijević, D. Teodorović, V. Simić, M. Šelmić, A bee Colony optimization approach to solving the anti-covering location problem, J. Comput. Civ. Eng. 26 (6) (2011) 759–768.

[18] A. Jovanović, D. Teodorović, Fixed-time traffic control at superstreet intersections by bee Colony optimization, Transp. Res. Rec. 2676 (4) (2021) 228–241.

[19] A. Jovanović, D. Teodorović, M. Nikolić, Area-wide urban traffic control: a bee Colony optimization approach, Transp. Res. Part C 77 (2017) 329–350.

[20] A. Jovanović, A. Stevanović, N. Dobrota, D. Teodorović, Ecology based network traffic control: a bee colony optimization approach, Eng. Appl. Artif. Intel. 115 (2022) 105262.

[21] I. Jovanović, M. Šelmić, M. Nikolić, Metaheuristic approach to optimize placement of detectors in transport networks – case study of Serbia, Can. J. Civ. Eng. 46 (3) (2018) 176–187.

[22] M. Nikolić, D. Teodorović, Empirical study of the bee Colony optimization (BCO) algorithm, Expert Syst. Appl. 40 (11) (2013) 4609–4620.

[23] M. Nikolić, D. Teodorović, Transit network design by bee Colony optimization, Expert Syst. Appl. 40 (15) (2013) 5945–5955.

[24] M. Nikolić, D. Teodorović, A simultaneous transit network design and frequency setting: computing with bees, Expert Syst. Appl. 41 (16) (2014) 7200–7209.

[25] M. Nikolić, D. Teodorović, Vehicle rerouting in the case of unexpectedly high demand in distribution systems, Transp. Res. Part C 55 (2015) 535–545.

[26] M. Nikolić, D. Teodorović, K. Vukadinović, Disruption Management in Public Transit: the bee Colony optimization (BCO) approach, Transp. Plan. Technol. 38 (2) (2015) 162–180.

[27] M. Nikolić, D. Teodorović, Mitigation of disruptions in public transit by bee Colony optimization, Transp. Plan. Technol. 42 (6) (2019) 573–586.

[28] M. Nikolić, M. Šelmić, D. Macura, J. Ćalić, Bee Colony optimization metaheuristic for fuzzy membership functions tuning, Expert Syst. Appl. 158 (2020) 113601.

[29] M. Šelmić, D. Teodorović, K. Vukadinović, Locating inspection facilities in traffic networks: an artificial intelligence approach, Transp. Plan. Technol. 33 (6) (2010) 481–493.

[30] D. Teodorovic, M. Dell' Orco, Mitigation traffic congestion: solving the ride-matching problem by bee colony optimization, Transp. Plan. Technol. 31 (2) (2008) 135–152.

[31] D. Teodorović, M. Šelmić, L.J. Mijatović-Teodorović, Combining case-based reasoning with bee Colony optimization for dose planning in well differentiated thyroid cancer treatment, Expert Syst. Appl. 40 (6) (2012) 2147–2155.

[32] D. Teodorović, T. Davidović, M. Šelmić, M. Nikolić, Bee Colony optimization and its applications, handbook of AI-based Mataheuristics, in: P. Siarry, A. Kulkarni (Eds.), Handbook of AI-Based Metaheuristics, CRC Press, Boca Raton, FL, 2021, pp. 301–321.

[33] D. Teodorović, M. Nikolić, M. Šelmić, I. Jovanović, Bee Colony optimization with applications in transportation engineering, in: A. Biswas, C.B. Kalayci, S. Mirjalili (Eds.), Advances in Swarm Intelligence, Studies in Computational Intelligence, vol. 1054, Springer, Cham, 2023, pp. 135–152.

[34] J.E. Beasley, Benchmark Instances, 2016. http://people.brunel.ac.uk/~mastjjb/jeb/orlib/airlandinfo.html. (accessed 30.09.2022).

About the authors

Miloš Nikolić was born in 1984 in Belgrade. He finished his doctoral studies at the University of Belgrade—Faculty of Transport and Traffic Engineering, with the thesis "Disruption management in transportation by the Bee Colony Optimization metaheuristic." He has worked at the University of Belgrade since 2011. Currently, he is an Associate Professor at the Faculty of Transport and Traffic Engineering, University of Belgrade. Professor Nikolić was Visiting Scholar at the University of California, Berkeley (2013, 2015/2016, and 2017). His research areas are applications of metaheuristic algorithms and operational research techniques for solving combinatorial optimization problems in transportation, transport network design, disruption management in transportation, and vehicle routing problems. He is a coauthor of the book "Quantitative Methods in Transportation", two other book chapters, and many research articles published in scientific journals and conference proceedings.

Jasenka Rakas is faculty in the Civil and Environmental Engineering Department, affiliated faculty with the Jacobs Institute for Design Innovation, and founder of the Airport Design Studio and Aviation Futures Lab at the University of California, Berkeley (UCB). She is the deputy director of the UCB FAA Consortium in Aviation Operations Research (NEXTOR III) and its lead aviation researcher. Her research interests are in the areas of advanced concepts in aviation, aviation infrastructure and facility performance, and their interaction with the environment. Rakas' teaching interests are in the areas of air transport facility design, sustainability, planning and operations. She has authored a large number of publications that include scientific articles, conference proceedings, technical reports and two book chapters. She is also a recipient of numerous awards, including seven national awards in the FAA/ACRP Airport Design Competition for Universities.

Dušan Teodorović (Belgrade, 1951) is a professor at the University of Belgrade (retired) and Professor Emeritus at Virginia Polytechnic Institute and State University. He is an elected member of the Serbian Academy of Sciences and Arts and the European Academy of Sciences and Arts. He was visiting professor at Technical University of Denmark, University of Delaware, USA. and National Chiao Tung University, Taiwan. Professor Teodorović pioneered the applications of Artificial Intelligence Techniques (Fuzzy Systems, Swarm Intelligence) in traffic, transportation and logistics. He has published numerous articles in scientific journals and conference proceedings. These articles have been cited over 9,000 times in world literature. He is the author of the books "Transportation Networks", "Airline Operations Research", "Traffic Control and Transport Planning: A Fuzzy Sets and Neural Networks Approach", "Transportation Engineering: Theory, Practice and Modeling" and "Quantitative Methods in Transportation". From 2006 to 2009, Professor Teodorović was Vice-Rector of the University of Belgrade. He is the author of two collections of poems for children (in Serbian) published in 2019 and 2021.

CHAPTER FOUR

Situation-based genetic network programming to solve agent control problems

Mohamad Roshanzamir and Mahdi Roshanzamir
Department of Computer Engineering, Faculty of Engineering, Fasa University, Fasa, Iran

Contents

Abstract

Evolutionary algorithms are often used to generate the best solution based on a population of initial solutions and through successive generations. These algorithms have different types. One of them is Genetic Network Programming (GNP). This algorithm is one of the new evolutionary algorithms in which the structure of each individual is defined as a directed graph. This directed graph can be considered as a flowchart or strategy that can be used by the agent(s) to make decisions in the environment. So, this algorithm can be used for the automatic generation of solutions for agent control problems. Using GNP, researchers try to find the best individual (strategy) for an agent. However, if there is more than one agent in the environment, it is not easy to find a strategy that can optimally achieve the goal if all agents behave according to it. In this research, instead of looking for an optimal strategy for all agents, separate strategies are created for each agent based on its situation. This way, the goal can be achieved more easily and quickly because finding a strategy that can guide all the agents in final goal achievement is more difficult than finding different strategies that are created based on the situation of each agent. Generating different strategies gives more flexibility to the GNP algorithm for finding better solutions. For this purpose, the situation-based GNP

(SB-GNP) algorithm has been proposed, which generates a strategy for each agent based on the situation of that agent. The results of applying the proposed method on Tile-World as a benchmark show that this method can improve the performance of traditional GNP. An important advantage of this method is that it can be added to all versions of the GNP without additional overhead.

1. Introduction

Swarm intelligence [1–7] and evolutionary computation [8–10] algorithms are a set of optimization methods in the field of artificial intelligence that try to solve problems by imitating the behaviors of living organisms. In evolutionary algorithms, mechanisms such as reproduction, mutation, recombination, and replacement are used that usually imitate biological evolution. These types of algorithms are heuristic-based approaches for solving problems that are not easily solvable in polynomial time, such as classical NP-hard problems and other problems that take a too long time to fully be solved. However, these types of algorithms differ in the source from which they are inspired or the structures they use which are known as chromosomes. The most common types of evolutionary algorithms are genetic algorithms [11], evolutionary strategy [12], genetic programming [13], and evolutionary programming [14].

Different types of evolutionary algorithms are used to solve different problems and it cannot be guaranteed that one type of them can perform better than others in solving all problems. This feature is known as No-Free-Lunch (NFL) theorem [15]. Therefore, it is theoretically impossible to have a general optimization method that is better than other methods in solving all problems. Consequently, there has always been an interest to propose new types of algorithms to solve a particular problem. Following the NFL theorem, one of the successful paths is extending the tree structure of genetic programming to graph structure. In genetic programming, the structure of each individual in the population is a tree and it is mostly used to automatically generate computer programs. This structure can be extended to a graph [16–18]. This change allows easier modeling of complex systems. In this research, the genetic network programming (GNP) algorithm [17,19–21] has been used among the extensions that use graph structure. In this algorithm, a directed graph is used to represent each individual. This graph has three types of nodes. These three types of nodes are the start node, judgment node, and processing node. These nodes are used to create a flexible and efficient strategy (flowchart or program) based on which agents can determine their behavior in the environment.

The GNP algorithm has a very good performance in solving various problems such as stock trading [22–26], robot control [27–32], intrusion detection systems [33], social networks [34], data mining [35,36], and controlling the behavior of agents [19,20,37–40]. Different improvements have also been proposed to this algorithm [19,41]. However, in all proposed improvements, as well as in all applications, researchers have focused to find an optimal individual (strategy) in the population. In the agent or robot control problem, if there is only one agent or robot in the environment, this method is acceptable. However, if it is necessary to manage the behavior of several agents, it is better to find an optimal strategy for each agent according to its situation. In this research, a complete discussion about the features and advantages of this method which is called situation-based GNP (SB-GNP) will be provided. In another word, it will be explained why and how this method can improve the efficiency of the GNP algorithm. The important advantage of this innovation is that it can be applied to all improvements of the GNP algorithm if only there is more than one agent in the environment. In this research, an example of using this improvement has been applied to the Tile World environment [42] and the results have been compared and analyzed with the traditional GNP algorithm.

This research is organized as follows. First, in Section 2, the traditional GNP algorithm and its evolution mechanism are revisited. The proposed algorithm is comprehensively described in Section 3. In Section 4, the experimental results of applying this improvement of the GNP algorithm on the Tile World are explained, and finally, in Section 5 and Section 6, the discussion and conclusions are drawn respectively.

2. Related works

In this section, some of the previously published papers about different techniques developed for the agent control problems are investigated. In addition, different kinds of optimization techniques used for agent control problems are listed and explained how GNP is different from them, and why GNP is considered in this chapter.

In most real-world environments, there is more than one agent. In the fields such as intelligent transportation systems as well as the control of groups of robots, the cooperation between agents and making coordinated decisions to achieve a common goal is an integral part of agent control problems that will allow such systems to accomplish increasingly complex tasks. In some research such as [43], a good survey of the algorithms used to model the decision-making process to create cooperation between agents has been done. Algorithms like dynamic programming, reinforcement learning, neural networks, swarm intelligence, and evolutionary computing have been used for this purpose. These algorithms have been used to find optimal and suboptimal policies. In addition, in this research, the challenges in this field have also been investigated.

Onken et al. [44], a neural network approach was proposed for solving high-dimensional optimal control problems. The authors focused on multi-agent control problems with obstacle and collision avoidance. These problems immediately become high-dimensional. Their method fused the Pontryagin Maximum Principle and Hamilton-Jacobi-Bellman and parameterized the value function with a neural network. They showed their approach's effectiveness on a 150-dimensional multi-agent problem with obstacles. Rivière et al. [45], a provably safe, automated distributed policy generation was proposed to be used in multi-robot motion planning. In this method, the advantage of centralized planning of avoiding local minima was combined with the advantage of decentralized controllers of scalability and distributed computation. This method only requires relative state information of nearby neighbors and obstacles and computes a provably safe action. Hönig et al. [46], a new method was described for multi-robot trajectory planning. The authors tested their proposed method on a quadrotor swarm while navigating in a warehouse. They claimed that their approach can compute trajectories safely, smoothly, and quickly for hundreds of quadrotors in dense environments with obstacles.

An approximate algorithm for the decentralized control of multi-agent systems based on a genetic algorithm was proposed in Ref. [47]. As the first step, the problem was formalized using decentralized partially observable Markov decision processes. Then, a joint policy as a way of representing a solution in a chromosome is introduced. Finally, a genetic algorithm is proposed as a search mechanism. A multi-agent tiger problem is used as an experimental framework to show the effectiveness of the algorithm. Oliehoek et al. [48] new methods for decentralized partially observable Markov decision processes were proposed, which were general models for collaborative multi-agent planning under uncertainty. A generalized multi-agent A* algorithm was introduced which reduced the problem to a tree of one-shot collaborative Bayesian games. Additionally, new hybrid heuristic representations were introduced that were more compact. Consequently, they were able to find the solution for larger decentralized partially observable Markov decision processes. They also provided theoretical guarantees that, when a suitable heuristic is used, both incremental clustering and incremental expansion yield algorithms that are complete and search equivalent. Finally, extensive empirical results demonstrated that the algorithm which synthesizes these advances can optimally solve decentralized partially observable Markov decision processes of unprecedented size.

Kumar et al. [49] proposed a new class of approximation algorithms using connections between machine learning and multi-agent planning. They formulated a multi-agent planning problem as an inference in a mixture of dynamic Bayesian networks. This approach leads to efficient inference techniques in dynamic Bayesian networks for multi-agent decision-making problems. To improve its scalability, certain conditions were identified that are necessary to extend the proposed method to multi-agent systems. They showed that some existing multi-agent planning models can satisfy these conditions. Experimental results on large planning benchmarks confirmed the superiority of the proposed method in terms of scalability and runtime compared to existing algorithms.

In the above-mentioned articles, the main goal of solving the agent control problems is to control their swarm movement or develop cooperation between several agents. The GNP algorithm tries to generate an optimal strategy for the agent's behavior in the environment. This strategy is displayed in a flowchart-like structure. The structure of each individual in the GNP algorithm is similar to a finite state machine, with the difference that in this algorithm, the nodes do not represent the state of the environment. They are used to check the conditions and to determine the actions that the agent should perform. This structure is used to control the behavior of agents in partially observable Markov decision process problems. The reactive policy does not work well in partially observable Markov decision process problems. However, GNP works well in this type of environment because the network structure of GNP does not create reactive policy. Additionally, this algorithm can use its past information implicitly.

3. GNP algorithm

As mentioned, the structure of individuals in GNP is a directed graph in which there are three types of nodes: start, judgment, and processing nodes. In this structure, the transition between nodes begins from the start node. Judgment nodes act as the agent's sensors and determine which path

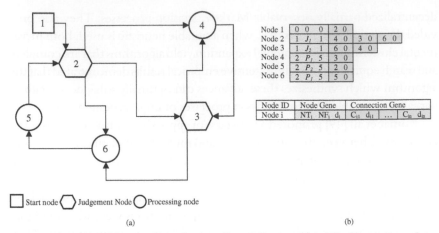

Fig. 1 The structure of an individual in the GNP algorithm (A) Phenotype of an Individual. (B) Genotype of an Individual.

the agent should follow in the graph according to the information from the environment. Processing nodes are defined according to the agent actors and specify what action should be taken. An example of phenotype (directed graph) and genotype (a hierarchical structural string) structure of an individual is shown in Fig. 1.

As can be seen in Fig. 1, node number 1, as the start node, specifies where the decision-making process of the agents begins. There is only one start node in each individual. In this figure, at first, the conditions specified in node number 2 are checked. This node has three outputs. According to the environmental conditions, one of the three output paths from this node will be selected. The agent continues its decision-making process according to the strategy specified in this structure to either reach the goal or pass a certain number of steps. In the genotype structure, each node has an identifier i and consists of two main parts: *Node Gene* and *Connection Gene*. *Node Gene* consists of three subsections:

- NT_i shows the node type. It may have three values 0, 1, and 2. The value 0 represents the start node, 1 represents the judgment nodes, and 2 shows the processing nodes.
- NF_i specifies the function to be executed.
- d_i specifies how much time is needed to execute this function.

The number of subsections of the *Connection Gene* depends on the type of node. In the judgment nodes, for each output, a subsection is created in the

Connection Gene. In the *Connection Gene* of processing nodes, there is only one subsection. Each subsection in *Connection Gene* consists of two parts:
- C_{ij} specifies which node the j_{th} branch of node i is connected to.
- d_{ij} specifies how much time is needed to transfer from the current node to the node C_{ij}.

Like other evolutionary algorithms, crossover and mutation operators perform the evolution process in successive generations. In the crossover, two individuals are selected by one of the selection algorithms, such as fitness proportionate or tournament selection [50]. Then, as shown in Fig. 2, the corresponding nodes in these two individuals are selected with the probability of p_c and their branches are exchanged. For example, node 3 is selected in two parents. In parent 1, the first and second branches of node 3 are connected to nodes 6 and 5, respectively. In parent 2, the first and second branches of node 3 are connected to nodes 2 and 4, respectively. In the crossover process, these connections are exchanged. It means that in parent 1, the first and second branches of node 3 will be connected to nodes 2 and 4,

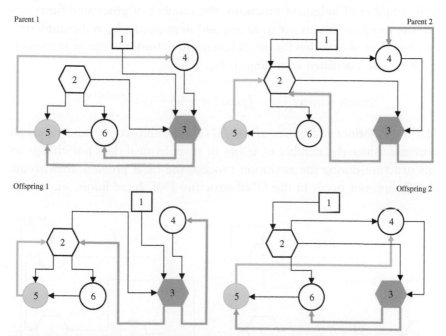

Fig. 2 The mechanism of crossover operator in GNP algorithm. Judgment nodes 3 (node in green) and processing nodes 5 (node in orange) are selected in parents. Then, they swapp their corresponding connections and generate two new offspring.

respectively, and in parent 2, these branches will be connected to nodes 6 and 5, respectively. The same process occurs for node number 5. This way, two new individuals will be generated.

In the mutation, like the crossover process, an individual is selected with one of the selection methods. Then, as shown in Fig. 3, each node of that individual is selected with a probability of p_m, and the branches of that node are changed randomly. For example, first node 1 is selected and its branch is changed from node 4 to 2. Then node 2 is selected and its branch is changed from node 5 to 6.

When the GNP algorithm is used to solve a problem, the number of decision and processing nodes is determined by the system designer. During the evolution process, the type of nodes, their functions, and the number of node branches do not change and only the branches of nodes (C_{ij}) are changed. The number of instances of each node is set manually and commonly equal for all nodes, which is called the program size. For example, if the program size is set to three, it means that from each judgment and processing node, there are three instances in each individual. Therefore, if the number of judgment functions, the number of processing functions and the program size are set to nj, np, and ps respectively, and considering an average of nb branches for the judgment functions, the size of the search space will be calculated according to Eq. (1).

$$\text{Search Dimension} = (ps \times (nj \times nb + np))^{(ps \times (nj+np))} \tag{1}$$

The most distinctive features of the GNP algorithm come from its graph structure. Since the number of nodes of an individual does not change in this structure during the evolution process, the bloat problem that occurs in GP does not occur in the GNP structure [50]. In addition, since only

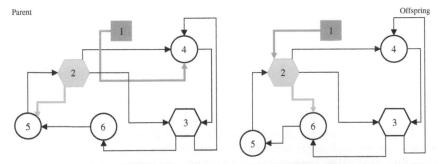

Fig. 3 The mechanism of mutation operator in GNP algorithm. Start node 1 (node in green) and judgment node 2 (node in orange) are selected in parent. Then, their corresponding connections are changed randomly and a new offspring is generated.

the connections of nodes, not their numbers are changed, the GNP algorithm produces the individuals (strategies) in a relatively compact form and tries to make optimal use of the limited number of nodes. The pseudo-code of the GNP algorithm is shown in Algorithm 1.

Algorithm 1 Pseudo code of GNP

1. Initialize the parameters, variables, and directed graph structure.
2. Generate the initial population.
3. While the termination condition does not meet
    3.1. For each individual
        3.1.1. Agents act one step according to that individual.
        3.1.2. If the agents have not reached the last step or the goal, go to 3.1.1.
        3.1.3. Calculate the fitness of that individual.
    3.2. Generate a predefined number of new individuals using the crossover operator.
    3.3. Generate a predefined number of new individuals using the mutation operator.
    3.4. Survival selection is applied.

---

4. Proposed algorithm

As mentioned before, in the traditional GNP algorithm as well as all its improvements, the algorithms look for an individual (strategy) that if the agents behave according to it, they can achieve the goal. In agent control problems, if there is only one agent in the environment, this solution is suitable. However, if there are several agents, this solution can be improved. It is necessary to notice the fact that instead of finding a solution for all agents, it is possible to find different strategies for different agents according to their situations. In this case, one of the advantages of the GNP algorithm, i.e. the optimal use of the minimum number of nodes, is better illustrated. For example, suppose that there is only one node of a specific function in the structure of an individual. If the GNP algorithm is looking for an optimal strategy for all agents, it may face resource limitations in creating this strategy. An agent may need to use this function in a specific position in an individual structure while another agent prefers to use it in another position of that individual. In this condition, generating different strategies for different agents can solve this problem. If this feature is not considered in the evolution process of this algorithm, searching for the optimal individual in the population will be faced with more challenges. In fact, by not considering this feature, the evolution process will be longer. The optimal individual that should be generated during this evolution process will have a more complex structure because this structure must be able to meet the needs of all agents in any situation they are in. Indeed, the main innovation of this research is proposing a new mechanism to improve the evolution process in the GNP algorithm when there are several agents in the environment. In Fig. 4, the difference between the mechanisms of traditional GNP and SB-GNP is illustrated. According to this figure, if the goal is arriving at the house and the agents use traditional GNP, they must learn to check the position of the house. Then, move toward it according to their locations. However, if SB-GNP is used the agent on the top of the house must learn

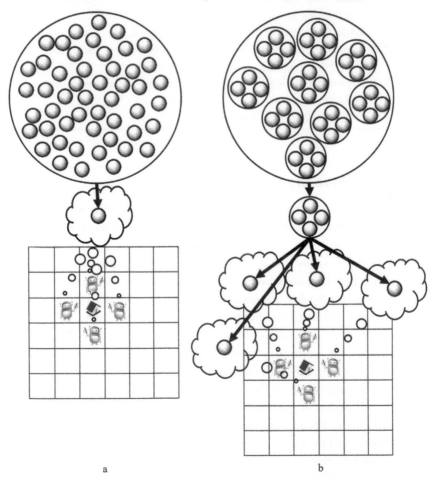

a   b

Fig. 4 The difference between the schema of (A) traditional GNP and (B) SB-GNP. The source of the robot icon in this figure is [51].

to move down, while the agent on the left side of the house must learn to move right and so on. Although the generated strategies using SB-GNP are not general and depend on the situation of the agents, it is not only faster and easier to learn but also leads to goal achievement. SB-GNP pseudo code is shown in Algorithm 2.

Algorithm 2 Pseudo code of SB-GNP

1. Initialize the parameters, variables, and directed graph structure.

2. Generate the initial population.

3. Group the individuals in n-ary groups. % n is the number of agents.

4. While the termination condition does not meet

    4.1. For each group

        4.1.1. Each agent acts one step according to its corresponding individual.

        4.1.2. If the agents have not reached the last step or the goal, go to 4.1.1.

        4.1.3. Calculate the fitness of the group.

    4.2. Generate a predefined number of new individuals using the crossover operator applied to the individuals corresponding to each agent.

    4.3. Generate a predefined number of new individuals using the mutation operator.

    4.4. Survival selection is applied to the groups.

---

5. Experimental results

The results of applying the SB-GNP algorithm to the Tile World are explained in this section. At first, this benchmark is introduced. Then, the results of this improvement to the GNP algorithm are explained in detail.

5.1 Tile world

As shown in Fig. 5, this environment is a grid world with agents, floors, obstacles, tiles, and holes. In this environment, some agents move simultaneously. The goal is to push the tiles into the holes as quickly as possible. After dropping a tile into a hole, that hole is converted into a floor.

In this environment, eight judgment functions are defined as agents' sensors. Using these functions, the agent can sense its surrounding environment. These functions are listed in Table 1. To better understand how these functions work, the results of using them in the conditions shown in Fig. 6A

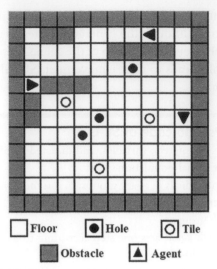

Fig. 5 An example of Tile World.

Table 1 Description of judgment and processing functions used in the structures of individuals.

Node Type	Symbol	Description	Outputs
Judgment Nodes	JF	Judge forward	Floor,
	JB	Judge backward	Obstacle,
	JL	Judge left	Tile, Hole,
	JR	Judge right	Agent
	JDNT	Judge the direction of nearest til	Forward,
	JDNH	Judge the direction of the nearest hole	Backward,
	JDNHNT	Judge the direction of the nearest hole	Left, Right
	JDSNT	from the nearest tile	
		Judge the direction of second nearest tile	
Processing Nodes	MF	Move forward	
	TL	Ninety-degree turn left	
	TR	Ninety-degree turn right	
	ST	Stay in its place	

is shown in Fig. 6B. Four processing functions are also defined as the actors of the agents and are listed in Table 1.

The fitness of each individual is calculated according to Eq. (2).

$$fitness = 100D + 3\left(SL - SU\right) + 20\left(\sum_{i=1}^{Total\ Number\ of\ Tiles}\left(ID_i - FD_i\right)\right) \quad (2)$$

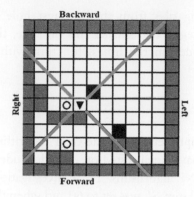

Judgment Functions	Outputs
JF	Obstacle
JB	Floor
JL	Floor
JR	Tile
JDNT	Right
JDNH	Backward
JDNHNT	Backward
JDSNT	Forward

a   b

Fig. 6 An example of the outputs of judgment functions.

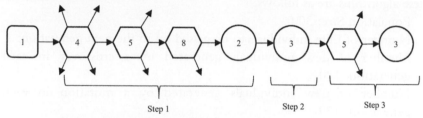

Step 1   Step 2   Step 3

Fig. 7 An example of the passing of three steps.

In this equation, D represents the number of tiles that the agents could drop into the holes. SL represents the number of steps that agents are allowed to take based on an individual's structure. In this research, each judgment node takes a one-time unit and each processing node takes five-time units. Each step is passed when the number of elapsed time units is greater than or equal to five. For example, if the agent has performed the judgment and processing functions shown in Fig. 7, three steps have been passed.

SU is the number of steps taken to reach the goal. ID_i is the initial distance of the i_{th} tile with its nearest hole, and FD_i is the final distance of the i_{th} tile with its nearest hole after passing the SU steps.

The three terms that are included in the fitness function are defined to efficiently achieve the goal of the problem. The main goal is to drop as many tiles as possible into the holes. This term is included in the fitness function with a factor of 100. The main goal must be achieved as soon as possible. Therefore, the lower the SU value, the lower the number of passed steps. This term is included in the fitness function with a factor of 3. Finally, if the agents cannot drop all the tiles into the holes during the SL steps, we consider the distance of the remaining tiles from their nearest hole as another criterion for evaluating a strategy. The smaller this distance, the better the strategy. This term is included in the fitness function with a factor of 20.

5.2 Experimental analysis

To fairly compare the impact of this improvement on the GNP algorithm, the results of the proposed algorithm have been tested on 30 independent runs of the traditional GNP and SB-GNP algorithms. These investigations were done when the maximum allowed step SL was set to 60 and when the program size was set to one, three, and five. These algorithms are stopped after 300,000 fitness function evaluations. The values of other parameters of these algorithms are as follows:

- Population Size: 300
- Number of Elite individuals to be passed on to the next generation: 1
- Number of new individuals generated by a crossover in each generation: 120
- Number of new individuals generated by a mutation in each generation: 179
- Tournament size: 2
- Crossover rate: 0.40
- Mutation rate: 0.01

The algorithms were implemented in MATLAB 2019.a under Windows 10 operating system with processor Intel(R) Core(TM) i7-3537U CPU 2.00 GHz–2.50 GHz and 6 GB of RAM.

The mean of fitness of traditional GNP and SB-GNP is shown in Figs. 8–10 for program sizes one, three, and five, respectively. According to Fig. 8, it is clear that SB-GNP is much better than traditional GNP when the program size is one. However, according to Fig. 9, this superiority is reduced when the program size is increased to three and is lost when the program size is increased to five as is clear in Fig. 10. In fact, these results show the advantage of the proposed method. It will be difficult to create an optimal strategy for all agents when the program size is small. In this situation, the proposed algorithm that defines a strategy for each agent separately will have a better performance. If the program size is increased, it will be possible to create an optimal strategy for all agents. Subsequently, the SB-GNP is less able to show its advantages than traditional GNP.

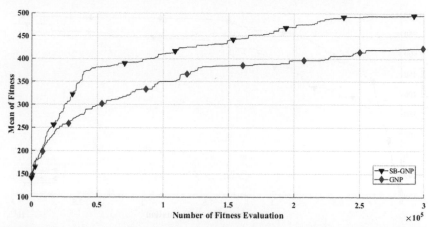

Fig. 8 Mean of fitness function during evolution process when the program size is set to one.

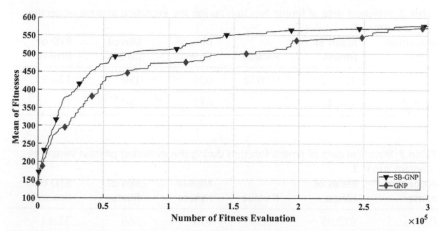

Fig. 9 Mean of fitness function during evolution process when the program size is set to three.

In Tables 2–4 more details of the obtained results are shown. In Table 2, the *P*-value [52] obtained from comparing the final results of these two algorithms is equal to 0.0072, which shows that the results of the SB–GNP algorithm are significantly better than GNP. The standard deviation value of SB–GNP is also lower than GNP, which indicates more stability of this algorithm. According to the *P*-value in Table 3, there is no significant difference between the results of the two algorithms while, in Table 4, the GNP

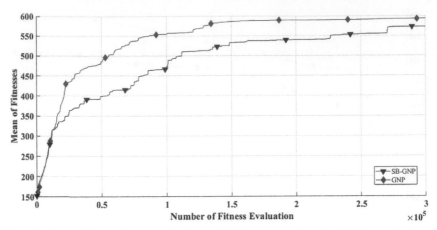

Fig. 10 Mean of fitness function during evolution process when the program size is set to five.

Table 2 Statistical data of fitness function during the evolution process when the program size is 1.

Algorithm name	Mean of fitness	P-value	Max of fitness	Min of fitness	STD of fitness
GNP	423.00	0.0072	626	180	138.2
SB-GNP	494.47	–	617	280	117.16

Table 3 Statistical data of fitness function during the evolution process when the program size is 3.

Algorithm name	Mean of fitness	P-value	Max of fitness	Min of fitness	STD of fitness
GNP	572.05	0.1591	614	360	52.43
SB-GNP	577.07	–	635	360	63.90

Table 4 Statistical data of fitness function during the evolution process when the program size is 5.

Algorithm name	Mean of fitness	P-value	Max of fitness	Min of fitness	STD of fitness
GNP	593.53	–	629	390	48.23
SB-GNP	572.71	0.021	620	509	32.79

algorithm has shown better performance. According to the results shown in these tables, the mean, maximum, and minimum fitness is improved while the program size is increased. In addition, the standard deviation of fitness is also improved. It clearly shows that having more nodes causes generating the optimal strategies more easily.

When the program size is one, according to the resource limitation, the SB–GNP algorithm has been able to achieve better efficiency. This achievement happens in a shorter time. However, when the program size is increased to three, according to the P-value, the efficiency of both algorithms becomes almost the same because now there is no resource limitation. Finally, when the program size is increased to five, according to the p-value, the traditional GNP algorithm performance is better than the SB–GNP algorithm because, in the SB–GNP, each agent has fewer individuals than the traditional GNP to find better strategies. In SB–GNP, the individuals in the population are divided among the agents. As a result, the diversity in the population and subsequently the efficiency of the program is reduced.

6. Discussion

The results show that when the program size is set to one, since there is more limitation in generating the optimal structure, the proposed algorithm clearly has better efficiency than the traditional GNP algorithm. The reason for this improvement is the greater flexibility provided by the SB–GNP algorithm. In the SB–GNP algorithm, as each agent has its own strategy, the nodes in the individual structure can be used in the best position according to the situation of the corresponding agent. Therefore, using SB–GNP can not only increase the efficiency of the generated strategies but also improve the speed of finding them. However, when the program size increases to three or five, this advantage is reduced because the traditional GNP can also generate solutions that are suitable for all agents.

Overall, generating a strategy based on the situation can increase the efficiency of the program. However, as it is clear from the name of the proposed algorithm, this method improves the efficiency of multi–agent systems when the positions of the agents are different. In situations where there is an agent in the environment or the agents are in a similar position, using this algorithm does not make a difference in the result.

Finally, the main advantage of the proposed method is that it can be used with all different versions of the GNP algorithm. In fact, this algorithm can be seen as an add-on to the GNP algorithm in solving problems in which different strategies must be generated for more than one agent.

7. Conclusion and future works

In this research, a new method is proposed to improve the performance of the GNP algorithm. The proposed method is effective when this algorithm is used in a multi-agent environment. In this condition, instead of generating an optimal strategy for all agents, each agent searches for its optimal strategy according to its situation in the environment. This leads to far better results. This improvement is much more obvious especially when there are limited resources. The most important strength of this algorithm is that it can be used with all the previous improvements. In future works, this subject will be investigated. In addition, to better determine the effectiveness of this method, it is better to investigate new environments with different positions for agents. Diversity reduction in the divided population is also another topic that must be investigated.

References

[1] A. Biswas, et al., Particle swarm optimisation with time varying cognitive avoidance component, Int. J. Comput. Sci. Eng. 16 (1) (2018) 27–41.
[2] A. Biswas, B. Biswas, K.K. Mishra, An atomic model based optimization algorithm, in: 2016 2nd International Conference on Computational Intelligence and Networks (CINE), 2016.
[3] A. Biswas, B. Biswas, Swarm intelligence techniques and their adaptive nature with applications, in: Q. Zhu, A.T. Azar (Eds.), Complex System Modelling and Control through Intelligent Soft Computations, Springer International Publishing, Cham, 2015, pp. 253–273.
[4] A. Biswas, et al., An improved random inertia weighted particle swarm optimization, in: 2013 International Symposium on Computational and Business Intelligence, 2013.
[5] A. Biswas, et al., Physics-inspired Optimization algorithms: a survey, J. Optimiz. 2013 (2013) 438152.
[6] A. Biswas, A. Kumar, K.K. Mishra, Particle swarm optimization with cognitive avoidance component, in: 2013 International Conference on Advances in Computing, Communications and Informatics (ICACCI), 2013.
[7] A. Biswas, Atom stabilization algorithm and its real life applications, J. Intell. Fuzzy Syst. 30 (2016) 2189–2201.
[8] A. Biswas, B. Biswas, Visual analysis of evolutionary optimization algorithms, in: 2014 2nd International Symposium on Computational and Business Intelligence, 2014.
[9] D. Sarkar, N. Mishra, A. Biswas, Genetic Algorithm-Based Deep Learning Models: A Design Perspective, Singapore, Springer Singapore, 2022.

[10] A. Biswas, B. Biswas, Regression line shifting mechanism for analyzing evolutionary optimization algorithms, Soft Comput. 21 (21) (2017) 6237–6252.

[11] A. Ghosh, S. Tsutsui, Advances in Evolutionary Computing: Theory and Applications, Springer Science & Business Media, 2012.

[12] H.-G. Beyer, H.-P. Schwefel, Evolution strategies – a comprehensive introduction, Nat. Comput. 1 (1) (2002) 3–52.

[13] J.R. Koza, R. Poli, Genetic programming, in: E.K. Burke, G. Kendall (Eds.), Search Methodologies: Introductory Tutorials in Optimization and Decision Support Techniques, Springer US, Boston, MA, 2005, pp. 127–164.

[14] V.W. Porto, Evolutionary programming, in: Evolutionary Computation 1, CRC Press, 2018, pp. 127–140.

[15] D.H. Wolpert, W.G. Macready, No free lunch theorems for optimization, IEEE Trans. Evol. Comput. 1 (1) (1997) 67–82.

[16] J.F. Miller, Cartesian genetic programming: its status and future, Genet. Program. Evolvable Mach. 21 (1) (2020) 129–168.

[17] K. Hirasawa, et al., Comparison between genetic network programming (GNP) and genetic programming (GP), in: Proceedings of the 2001 Congress on Evolutionary Computation (IEEE Cat. No.01TH8546), 2001.

[18] A. Teller, M. Veloso, PADO: Learning Tree Structured Algorithms for Orchestration into an Object Recognition System, Carnegie-Mellon Univ Pittsburgh PA Dept Of Computer Science, 1995.

[19] X. Li, H. Yang, M. Yang, Revisiting Genetic network Programming (GNP): towards the Simplified Genetic Operators, IEEE Access 6 (2018) 43274–43289.

[20] S. Mabu, K. Hirasawa, J. Hu, A graph-based evolutionary algorithm: Genetic network Programming (GNP) and its extension using reinforcement learning, Evol. Comput. 15 (3) (2007) 369–398.

[21] M. Roshanzamir, et al., Graph structure optimization for agent control problems using ACO, in: A. Biswas, C.B. Kalayci, S. Mirjalili (Eds.), Advances in Swarm Intelligence: Variations and Adaptations for Optimization Problems, Springer International Publishing, Cham, 2023, pp. 327–346.

[22] Y. Chen, et al., Trading rules on stock markets using genetic network programming with sarsa learning, in: Proceedings of the 9th Annual Conference on Genetic and Evolutionary Computation, ACM, London, England, 2007, p. 1503.

[23] Y. Chen, Z. Shi, Generating trading rules for stock markets using robust Genetic network Programming and portfolio Beta, J. Adv. Comput. Intell. Intell. Inform. 20 (3) (2016) 484–491.

[24] H.H. Bahar, M.H.F. Zarandi, A. Esfahanipour, Generating ternary stock trading signals using fuzzy genetic network programming, in: 2016 Annual Conference of the North American Fuzzy Information Processing Society (NAFIPS), 2016.

[25] Y. Chen, D. Mo, F. Zhang, Stock market prediction using weighted inter-transaction class association rule mining and evolutionary algorithm, Econ. Res.-Ekon. Istraz. 35 (1) (2022) 5971–5996.

[26] Y. Chen, Z. Xu, W. Yu, Agent-based artificial financial market with evolutionary algorithm, Econ. Res.-Ekon. Istraz. 35 (1) (2022) 5037–5057.

[27] P. Sung Gil, S. Mabu, K. Hirasawa, Robust Genetic Network Programming Using SARSA Learning for Autonomous Robots, ICCAS-SICE, 2009.

[28] S. Sendari, et al., Fuzzy Genetic network Programming with noises for Mobile robot navigation, JACIII 15 (7) (2011) 767–776.

[29] S. Sendari, S. Mabu, K. Hirasawa, Fuzzy genetic network programming with reinforcement learning for mobile robot navigation, in: IEEE International Conference on Systems, Man, and Cybernetics, 2011.

[30] W. Wang, et al., Multi-behaviour robot control using genetic network Programming with fuzzy reinforcement learning, in: J.-H. Kim, et al. (Eds.), Robot Intelligence

Technology and Applications 3: Results from the 3rd International Conference on Robot Intelligence Technology and Applications, Springer International Publishing, Cham, 2015, pp. 151–158.

[31] A.H. Findi, et al., Collision prediction based Genetic network Programming-reinforcement learning for Mobile robot navigation in unknown dynamic environments, J. Electr. Eng. Technol. 12 (2) (2017) 890–903.

[32] F. Foss, T. Stenrud, P.C. Haddow, Investigating genetic network programming for multiple nest foraging, in: 2021 IEEE Symposium Series on Computational Intelligence (SSCI), 2021.

[33] Y. Xu, et al., Attribute selection based Genetic network Programming for intrusion detection system, J. Adv. Comput. Intell. Intell. Inform. 26 (5) (2022) 671–683.

[34] S. Agarwal, S. Mehta, GNPA: a hybrid model for social influence maximization in dynamic networks, Multimed. Tools Appl. (2022).

[35] K. Taboada, et al., Genetic Network Programming based data mining method for extracting fuzzy association rules, in: IEEE Congress on Evolutionary Computation, 2008.

[36] H. Zhou, K. Hirasawa, Traffic density prediction with time-related data mining using genetic network programming, Comput. J. 57 (9) (2014) 1395–1414.

[37] M. Roshanzamir, M. Palhang, A. Mirzaei, Efficiency improvement of genetic network programming by tasks decomposition in different types of environments, Genet. Program. Evolvable Mach. 22 (2) (2021) 229–266.

[38] M. Roshanzamir, M. Palhang, A. Mirzaei, Tasks decomposition for improvement of genetic network programming, in: 2019 9th International Conference on Computer and Knowledge Engineering (ICCKE), 2019.

[39] M. Roshanzamir, M. Palhang, A. Mirzaei, Graph structure optimization of Genetic network Programming with ant colony mechanism in deterministic and stochastic environments, Swarm Evol. Comput. 51 (2019) 100581.

[40] X. Li, K. Hirasawa, A learning classifier system based on genetic network programming, in: 2013 IEEE International Conference on Systems, Man, and Cybernetics, 2013.

[41] X. Li, W. He, K. Hirasawa, Genetic Network Programming with Simplified Genetic Operators, Springer Berlin Heidelberg, Berlin, Heidelberg, 2013.

[42] M.E. Pollack, M. Ringuette, Introducing the Tileworld: Experimentally Evaluating Agent Architectures, AAAI, 1990.

[43] Y. Rizk, M. Awad, E.W. Tunstel, Decision making in multiagent systems: a survey, IEEE Trans. Cogn. Develop. Syst. 10 (3) (2018) 514–529.

[44] D. Onken, et al., A Neural Network Approach Applied to Multi-Agent Optimal Control, European Control Conference (ECC), 2021, p. 2021.

[45] B. Rivière, et al., GLAS: global-to-local safe autonomy synthesis for multi-robot motion planning with end-to-end learning, IEEE Robot. Autom. Lett. 5 (3) (2020) 4249–4256.

[46] W. Hönig, et al., Trajectory planning for quadrotor swarms, IEEE Trans. Robot. 34 (4) (2018) 856–869.

[47] M.A. Mazurowski, J.M. Zurada, Solving decentralized multi-agent control problems with genetic algorithms, in: 2007 IEEE Congress on Evolutionary Computation, 2007.

[48] F.A. Oliehoek, et al., Incremental clustering and expansion for faster optimal planning in Dec-POMDPs, J. Artif. Intell. Res. 46 (2013) 449–509.

[49] A. Kumar, S. Zilberstein, M. Toussaint, Probabilistic inference techniques for scalable multiagent decision making, J. Artif. Intell. Res. 53 (2015) 223–270.

[50] A.E. Eiben, J.E. Smith, Introduction to Evolutionary Computing, Vol. 53, Springer, 2003.

[51] NAHUweibkx, Thinking Robot, 2017, [cited 2022; Thinking Robot—Artificial Intelligence Robot Education PNG]. Available from: https://favpng.com/png_view/thinking-robot-artificial-intelligence-robot-education-png/E6qhBaeh.

[52] F. Wilcoxon, Individual comparisons by ranking methods, Biometrics 1 (6) (1945) 80–83.

About the authors

Mohamad Roshanzamir received a BSc in computer engineering-software from the Isfahan University of Technology, Iran 2003, an MSc in computer engineering-artificial intelligence From the University of Isfahan, Iran 2006, and a PhD in computer engineering-artificial intelligence From Isfahan University of Technology, Iran 2020. From 2021 till now, he has been a lecturer at Fasa University, Iran. His primary research interests include data mining, machine learning, and evolutionary optimization.

Mahdi Roshanzamir received a BSc in computer engineering-software from Iran University of Science and Technology, Iran 2010, an MSc in computer engineering-artificial intelligence From Iran University of Science and Technology, Iran 2013, and a PhD in computer engineering-artificial intelligence From the University of Tabriz, Iran 2020. His primary research interests include evolutionary optimization, multi-agent systems, machine learning, and data mining.

CHAPTER FIVE

Small signal stability enhancement of large interconnected power system using grasshopper optimization algorithm tuned power system stabilizer

Prasenjit Dey[a], Anulekha Saha[b], Aniruddha Bhattacharya[c], Priyanath Das[d], Boonruang Marungsri[e], Phumin Kirawanich[a,f], and Chaiyut Sumpavakup[g]

[a]Cluster of Logistics and Rail Engineering, Mahidol University, Nakhon Pathom, Thailand
[b]Department of Electrical Engineering, Chulalongkorn University, Bangkok, Thailand
[c]Department of Electrical Engineering, NIT, Durgapur, India
[d]Department of Electrical Engineering, NIT, Agartala, India
[e]School of Electrical Engineering, Suranaree University of Technology, Nakhon Ratchasima, Thailand
[f]Department of Electrical Engineering, Mahidol University, Nakhon Pathom, Thailand
[g]Research Centre for Combustion Technology and Alternative Energy—CTAE and College of Industrial Technology, King Mongkut's University of Technology North Bangkok, Bangkok, Thailand

Contents

Advances in Computers, Volume 135
ISSN 0065-2458
https://doi.org/10.1016/bs.adcom.2023.11.004

Abstract

This chapter reports a relatively new method to optimize the design of power system stabilizers. Small scale disturbances are a common occurrence in large-scale systems (interconnected), and presence of oscillations in the low-frequency (LFO) range present a major problem. Therefore, small signal stability study of a system is of utmost importance to assess its performance and stability. Power System Stabilizers (PSS) finds use in interconnected networks in damping LFOs. In this chapter, grasshopper optimization algorithm (GOA) has been applied to find the optimal control parameters in designing of PSS. Optimal tuned parameters are obtained to improve the objective function (OF) which comprises of eigenvalues as well as damping ratios of the lightly damped electromechanical modes. Comparison of the results obtained using GOA with those of symbiotic organisms search (SOS) and water evaporation algorithm (WEA), established superiority of GOA for designing of PSS under varied operating conditions like line outages, load variations etc. Damping ratio using GOA improved by 10.11% and 4.23% for light load condition, by 11.24% and 4.52% for normal load condition and 13.7% and 2.06% for high load condition as compared to WEA and SOS respectively.

1. Introduction

The electrical power network is non-linear and complex system. Oscillations, especially the low frequency ones ranging from 0.2 to 2 Hz, pose a very big problem to the system [1]. Disturbances in the system like line outages, load variations and faults etc., are the root causes behind the occurrence of LFOs. Since these oscillations are associated with synchronous machines' rotor angles, they continue to increase resulting in loss of synchronism in absence of adequate damping. PSSs mostly find their use in damping out system oscillations and enhancing the electromechanical modes (EMs). LFOs can be categorized into local mode (0.8–2 Hz) and inter-area mode (0.2–0.8 Hz) of oscillations. This phenomenon is realized using eigenvalue analysis and may solve by integrating PSS to the system. When PSS parameters are tuned properly, they present component of the electrical torque in phase with generator rotor angle deviations which helps damp out LFOs. PSS inputs are accelerating power, rotor speed deviation, rotor frequency, etc. Literature survey shows the use of conventional PSS (CPSS) [2] consisting of lead-lag compensators to overcome the problem of low-frequency oscillations. Designing such type of CPSS implicates linearized dynamic model, which performs poorly under variable operating environments. Moreover, it fails in maintaining electrical power system stability under heavy loaded conditions. Although classical approaches like pole placement [3], H-infinity [4], LMI [5], techniques etc. perform satisfactorily but these methods are not suited for application to non-smooth, complex, non-convex, non-differentiable OFs and constraints. Meta-heuristic techniques gained popularity with regard to their ease of application and lesser computational burden as compared to classical methods. Heuristic algorithms have the ability to solve non–linear problems and are hence, more sought after.

Presently, different Artificial Intelligent (AI) algorithms are being implemented to solve problems pertaining to LFOs. Artificial neural network (ANN) [6–10] has been broadly employed towards design of PSS among all the available AI techniques. But there are certain demerits of controllers based on ANN. They are restricted by longer training period as well as in the selection of number of neurons and layers. Fuzzy logic controller (FLC) belonging to AI techniques, has also found its use in controlling PSS signal [11–13]. FLCs are more advantageous as they are able to provide control signals based on language rules provided by the operator. Since FLCs may be designed with the help of linguistic rules provided by the control system, accuracy in plant modeling is not needed. Fine-tuning of the controllers to achieve adequate signal is very tedious and is a demerit of this process. Evolutionary based optimization techniques (EA) are attaining popularity in designing PSS. CPSS has been modeled with different EAs like Genetic algorithm (GA) [14,15], Particle swarm optimization (PSO) [16–18], differential evolution (DE) [19,20], firefly algorithm (FA) [21], cuckoo search (CS) [22], evolutionary programming [23], tabu search [24], simulated annealing [25], BAT [26] and bacteria foraging [27].

A recently developed evolutionary algorithm termed Grasshopper Optimization Algorithm (GOA) [28] is presented to design PSS for mitigating LFOs in the multi-machine power system. As a test system WSCC 3 machine 9 bus system is considered and state space representation is used for modeling purpose. The system behavior has been studied using eigenvalue analysis technique as it is simple and efficient as compared to other techniques such as: synchronizing and damping torque analysis, and frequency-response and residue - based analysis. It can identify the EMs easily as compared to other mentioned techniques. Results demonstrate ability of the proposed GOA in enhancing overall stability and mitigating problems related to LFOs effectively as compared to other optimization techniques like SOS [29] and WEA [30]. There is ample scope for applying newer optimization techniques [31–47] with the hope of obtaining better results. However, this chapter presents the applications of SOS, WEA, and GOA.

The rest part of the chapter has been organized as follows: Section 2 presents mathematical model of the power system; Section 3 presents description of PSS; Section 4 presents objective functions used; Section 5 presents description of GOA; Section 6 presents analysis of results and their discussion and Section 7 presents Conclusion.

2. Power system mathematical model

A small signal stability analysis needs to be conducted by considering adequate models for each component used in an interconnected power system. This section describes all the models used in the study. The dynamic and steady-state behaviors of the system were represented by differential and algebraic equations, respectively.

2.1 Differential equations for generator

Power plants frequently employ synchronous generators to transmit large amounts of electric power to load centers via high-voltage (HV) or extra-high-voltage (EHV) transmission lines. This chapter examines intricate generator models, such as machine models, excitation models, and prime mover controllers. By selecting the proper base quantities, it is usual practice to represent voltages, currents, and impedances in per-unit values. Rotor swings of the synchronous generators have an impact on the stability of power systems. d–q axis transformation was considered to represent power system dynamics. A large interconnected network is typically modeled in the form of a constant-impedance matrix along with loads. For modeling, 4th order generator model with four states have been used. Static exciter has been used for the excitation system. Modeling of state equations is done according to [48]. The subsequent differential-algebraic equations (DAE) from Eqs. (1)–(6) for the m machine, n bus system with the IEEE–Fast acting exciter [47]. All the abbreviations and symbols are listed in Table 1.

$$\dot{\delta} = \omega_i - \omega_s \tag{1}$$

$$\dot{\omega} = \frac{T_{Mi}}{M_i} - \frac{\left[E'_{qi} - X'_{di}I_{di}\right]I_{qi}}{M_i} - \frac{\left[E'_{di} - X'_{qi}I_{qi}\right]I_{di}}{M_i} - \frac{D_i(\omega_i - \omega_s)}{M_i} \tag{2}$$

$$\dot{E}'_{qi} = -\frac{E'_{qi}}{T'_{doi}} - \frac{\left(X_{di} - X'_{di}\right)I_{di}}{T'_{doi}} + \frac{E_{fdi}}{T'_{doi}} \tag{3}$$

Table 1 List of abbreviations and symbols.

LFO	Low frequency	AI	Artificial intelligent
PSS	Power system stabilizers	ANN	Artificial neural network
GOA	Grasshopper optimization algorithm	FLC	Fuzzy logic controller
OF	Objective function	GA	Genetic algorithm
SOS	Symbiotic organisms search	PSO	Particle swarm optimization
WEA	Water evaporation algorithm	FA	Firefly algorithm
EMs	Electromechanical modes	CS	Cuckoo search
CPSS	Conventional PSS	WECC	Western system coordinating council
T_{d0}'	Short circuit direct axis transient time constant	T_M	Mechanical torque to the shaft
T_{q0}'	Short circuit quadrature axis transient time constant	E_q'	Quadrature axis component of voltage behind X_d'
V_i	Generator terminal voltage	E_d'	Direct axis component of voltage behind X_q'
X_d	Direct axis synchronous reactance	K_A	Voltage regulator gain
X_q	Quadrature axis synchronous reactance	T_A	Voltage regulator time constant
X_d'	Direct axis transient reactance	X_q'	Quadrature axis transient reactance
DE	Differential evolution	E_{fd}	Field voltage
FE	Fitness evaluation	P_m	Input mechanical power
maxFE	Maximum fitness evaluation	K_{PSS}	Power system stabilizer gain
$\Delta\omega$	Angular frequency deviation	$T_1 \ and \ T_3$	Phase-lead time constants
T_w	Time constant of washout filter	T_2 and T_4	Phase-lag time constants
ω	Rotor electrical angular velocity	p.u.	Per unit
δ	Rotor electrical angular position	I	Objective function

$$\dot{E}'_{di} = -\frac{E'_{di}}{T'_{qoi}} - \frac{\left(X_{qi} - X'_{qi}\right)I_{qi}}{T'_{qoi}} \tag{4}$$

$$\dot{E}_{fdi} = -\frac{E_{fdi}}{T_{Ai}} + \frac{K_{Ai}}{T_{Ai}}\left(V_{refi} - V_i + V_{Si}\right) \tag{5}$$

Eqs. (1)–(5) can be represented in state space as follows:

$$\dot{X} = AX + BU \tag{6}$$

where $X = [\delta, \omega, E'_q, E'_d, E_{fd}]$ and δ denote the rotor angle, ω, the speed, E'_q, E'_d the internal voltage along quadrature axis and direct axis respectively and E_{fd}, the field voltage. $\dot{\delta}$, $\dot{\omega}$, $\dot{E}'_q$, $\dot{E}'_{di}$, $\dot{E}_{fdi}$ denotes first order derivatives of δ, ω, E'_q, E'_d, and E_{fd} respectively. i represents number of generators.

Matrix A has a dimension of $5p \times 5p$ and B matrix is of dimension $5 \times q$, where p denote the total machines, q denote the number of installed PSS. Fig. 1 represents the popular Western System Coordinating Council (WECC) 3-machine, 9-bus system [41]. This is also the system appearing in [42] and widely used in the literature. The base MVA is 100, and system frequency is 60 Hz.

The study was carried out for different operating conditions such as line tripping, varying loading conditions, etc. Table 2 presents the different loading conditions for which the study was carried out. System data of [49] have

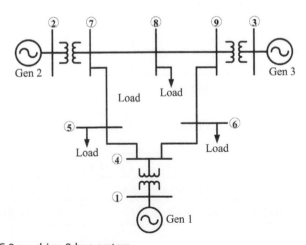

Fig. 1 WSCC 3 machine 9 bus system.

Table 2 Loading conditions (p.u.) studied.

	Light loading		Normal loading		Heavy loading	
	P	Q	P	Q	P	Q
Generator						
G1	0.9649	0.2330	1.7164	0.6205	3.5730	1.8143
G2	1.0000	−0.1933	1.6300	0.0665	2.2000	0.7127
G3	0.4500	−0.2668	0.8500	−0.1086	1.3500	0.4313
Load						
L5	0.7000	0.3500	1.2500	0.5000	2.0000	0.9000
L6	0.5000	0.3000	0.9000	0.3000	1.8000	0.6000
L8	0.6000	0.2000	1.0000	0.3500	1.6000	0.6500
Local load at G1	0.6000	0.2000	1.0000	0.3500	1.6000	0.6500

been considered. Small signal stability finds EMs and state variables that effectively participate in the system. Hence, participation factors can be used to identify various modes of oscillations [48].

3. Power system stabilizer

PSS provides supplementary torque to the exciter for dampen out LFOs. Speed based (CPSS) is the most commonly used PSS and is considered for designing purpose in the present study. CPSS is represented by the following functional block of Fig. 2.

The above figure demonstrates a double–stage PSS. It consists of PSS gain block, washout block and compensator. K_{PSS} represents gain of the PSS and ranges from 0.01 to 50 [29]. The gain of PSS provides adequate damping torque necessary to damp out the LFOs. This damping is proportional to the gain till it touches critical values, and then starts to decrease. The washout block serves as high–pass filter that allows the desired frequencies whereby removing steady-state signals delivered by the PSS, whose absence would result in modification of the terminal voltage of generator. T_w represents time- constant of washout filter. Literature review suggest that, for significant enhancement in system damping T_w should be

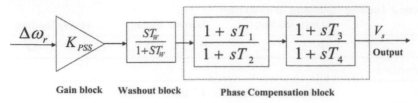

Fig. 2 Functional block diagram of two-staged PSS.

10 seconds (s) [29]. Phase lead–lag compensation block compensates the lag among electrical torque and PSS output. It also eliminates delay between electrical and excitation torque.

4. Objective functions (OF)

Whenever there is any disturbance in the system, there are oscillations which are reduced by the system's inherent damping factors and their amplitudes are determined by the damping ratio. An OF with two-part has been formulated for tuning parameters of PSS which are assessed by eigenvalue analysis. The first part of the function minimizes the real part of EMs and maximization of damping ratio is done by the second part, as shown in [49]. A stable system is ensured by eigenvalues having larger negative real parts and higher damping ratio. The damping coefficient is obtained from complex and real parts of eigenvalues. The chief objective is to improve the damping ratio and real part of eigenvalues. Mathematical representation of the objective is as follows:

$$\text{Minimize } I = I_1 + I_2 \tag{7}$$

$$\text{Where, } I_1 = \sum_{i=1}^{n} (\sigma_0 - \sigma_i)^2 \text{ and } I_2 = \sum_{i=1}^{n} (\zeta_0 - \zeta_i)^2; \tag{8}$$

Here, n denotes total eigenvalues related to the electromechanical (EM) modes. I_1 denotes part of the OF associated with the real part of EMs and responsible for shifting into the left half of S plane whereas I_2 concerns improving damping ratios. ζ is the damping ratio whereas σ represents the real part eigenvalues and. σ_0 and ζ_0 are considered correspondingly as -2.50 and 0.10 [49]. T_1 and T_3 represents phase-lead time constants between 0.10 and 1.50 s [48]. Phase-lag time constants are represented by

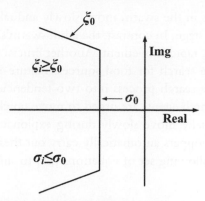

Fig. 3 Location of eigenvalues on the S-plane for l [29].

T_2 and T_4 which lies between 0.01 and $0.15\,\mathrm{s}$ [48]. K_{PSS}, T_1, T_2, T_3 and T_4 are improved with the help of various optimization methods whereas T_w is set to $10\,\mathrm{s}$. Fig. 3 shows the effect of OF subject to the inequality constraints shown below:

$$\left. \begin{array}{c} K_{PSS}^{\min} \leq K_{PSS} \leq K_{PSS}^{\max} \\ T_1^{\min} \leq T_1 \leq T_1^{\max} \\ T_2^{\min} \leq T_2 \leq T_2^{\max} \\ T_3^{\min} \leq T_3 \leq T_3^{\max} \\ T_4^{\min} \leq T_4 \leq T_4^{\max} \end{array} \right\} \quad (9)$$

This chapter mainly focuses on GOA algorithm for tuning PSS parameters for improving stability of the system under varied operating environments.

5. Grasshopper optimization algorithm (GOA)

Grasshoppers cause harm to agricultural and crop output, which is why they are regarded as a pest. Although they are typically observed singly in the wild, they join one of the biggest swarms of all living things. The swarm's size may be on a continental scale, which would be a nightmare for farmers. The peculiar feature of the grasshopper swarm is that both nymphal and adult stages exhibit swarming behavior. Nymph grasshoppers in their millions jump and move like rolling cylinders. They consume practically all of the vegetation along their path. When they reach adulthood and stop acting in this way, they gather in the air in a swarm. Grasshoppers travel great distances in this manner.

The grasshoppers in the swarm move slowly and take little steps when they are in the larval stage. In contrast, the adult swarm's primary characteristic is long-distance, rapid movement. Another crucial aspect of the grasshopper swarm is the search for food sources. Nature-inspired algorithms logically separate the search process into two tendencies: exploration and exploitation. The search agents are urged to move quickly during exploration, while they usually move slowly during exploitation. In addition to target finding, grasshoppers automatically carry out these two tasks.

GOA uses the following set of equations [28] to simulate its swarming behavior:

$$P_i = S_i + G_i + W_i \qquad (10)$$

where in the above equation, the position of the ith grasshopper is defined by P_i, while S_i represents social interaction, G_i denotes the gravity force on the ith grasshopper, and W_i indicates wind flow. Randomness in the motion of grasshoppers can be introduced as $P_i = r_1 * S_i + r_2 * G_i + r_3 * W_i$, where r_1, r_2 and r_3 are random numbers in [0,1].

The grasshoppers interact among themselves to decide their course of movement and this social interaction is defined as:

$$S_i = \sum_{\substack{k=1 \\ k \neq i}}^{N} s(d_{ik}) \hat{d}_{ik} \qquad (11)$$

where N the number of GHs, d_{ik} represents distance from ith to kth GH. $d_{ik} = |P_k - P_i|$ and $\hat{d}_{ik} = \frac{P_k - P_i}{d_{ik}}$ denotes unit vector from ith GH to kth GH.

The swarm movement is also dictated by social forces, which is the attraction/repulsion between grasshoppers, denoted by s, and is represented as:

$$s(r) = f e^{\frac{-r}{l}} - e^{-r} \qquad (12)$$

where the attraction intensity between the GHs is represented by f and the attraction length scale is represented by l. It is observed that the comfortable distance between grasshoppers for no attraction/repulsion is 2.079 units from each other [28]. The function s in Eq. (12) can divide the space between two grasshoppers into repulsion, comfort, and attraction regions. However, when the distance between grasshoppers is greater than 10 units, the function returns values close to zero, signifying repulsion. As a result, it cannot generate strong forces between grasshoppers that are far apart.

The different regions of comfort zone, attraction region, and repulsion region are also significantly dictated by the parameters l and f.

The gravitational force G is obtained as:

$$G_i = -g\hat{c}_g \tag{13}$$

where g denotes gravitational constant and $\hat{c}_g$ denotes a unit vector to the center of earth.

In nymph stage, grasshoppers have no wings, and hence, their movements are highly dependent on wind direction. The component of wind flow can be written as follows:

$$W_i = u\hat{b}_w \tag{14}$$

where u denotes drift constant and $\hat{b}_w$ represents unit vector in the direction of wind.

Substituting values of S, G, W in the parent Eq. (10), position of GH can be elaborated as follows:

$$P_i = \sum_{\substack{k=1 \\ k \neq i}}^{N} s(|P_k - P_i|)\frac{P_k - P_i}{d_{ik}} - g\hat{c}_g + u\hat{b}_w \tag{15}$$

Eq. (15) brings the initial random population closer to form a united, regulated swarm after some point in time.

But Eq. (15) cannot be directly used for solving optimization problems since the GHs move to comfort zone quickly, causing swarm divergence. The modified version of Eq. (15) is as follows:

$$P_i^d = c\left(\sum_{\substack{k=1 \\ k \neq i}}^{N} c\frac{ub^d - lb^d}{2}s(r)\left(|P_k^d - P_i^d|\right)\frac{P_k - P_i}{d_{ik}} \right) + \hat{T}^d \tag{16}$$

where ub^d denotes upper bound in the dth dimension, lb^d denotes lower bound in the dth dimension, T^d denotes the target—value (best solution so far) in dth dimension, c denotes the shrinking coefficient for decreasing comfort, attraction and repulsion zones of GHs.

In the modified equation, no G component is considered and it is assumed that wind direction (W component) is always towards a target (T^d).

The coefficient c is calculated as:

$$c = \max(c) - Iter\left(\frac{\max(c) - \min(c)}{\max Iter}\right) \tag{17}$$

where max (c) denotes maximum values of c and min (c) denotes minimum values of c. *Iter* signifies current iteration and *maxIter* denotes maximum iterations.

In Eq. (16), the next position of a grasshopper gets updated based on its present position, the target position, and the position of all other grasshoppers. The initial part of this equation takes into account the position of the current grasshopper relative to the other grasshoppers. Specifically, the status of all the grasshoppers have been evaluated to determine the placement of the search agents around the target. There is only one position vector for every search agent in GOA, unlike PSO. The position of a search agent in GOA is updated based on its present location, global best, and the position of all other search agents. It necessitates the involvement of all search agents in determining the subsequent position of each search agent.

In Eq. (16), the inner parameter c decreases the strength of repulsion/ attraction forces between grasshoppers as the number of iterations increases. On the other hand, the outer c factor reduces the search coverage around the target as the number of iterations increases.

To summarize, the first term in Eq. (16), which is the sum, takes into account the position of other grasshoppers and incorporates the interaction of grasshoppers in nature. The second term, T^d, models their inclination to move towards the food source. Furthermore, the parameter c represents the deceleration of grasshoppers as they approach the food source and eventually consume it. To introduce more randomness, both terms in Eq. (16) can be multiplied by random values. Additionally, individual terms can be multiplied by random values to introduce randomness in either the interaction of grasshoppers or the tendency towards the food source.

Fig. 4 represents the flowchart of GOA. The first block initiates the program. In the second block, random initialization of the swarm (SW) of GOA for a pre-specified swarm size is done. In the next block, specify the control parameters of GHs, which are nothing but Power System Stabilizer Gain (K_{PSS}), Phase-lead time constants(T_1 and T_3) and Phase-Lag Time Constants (T_2 and T_4) of PSS within their lower and upper limits. In the fourth block, perform load-flow analysis for analyzing small-signal stability. Obtain eigenvalues while the control variables are lying within bounds as given in Eq. (9) in the next block. In the 6th block, evaluate the

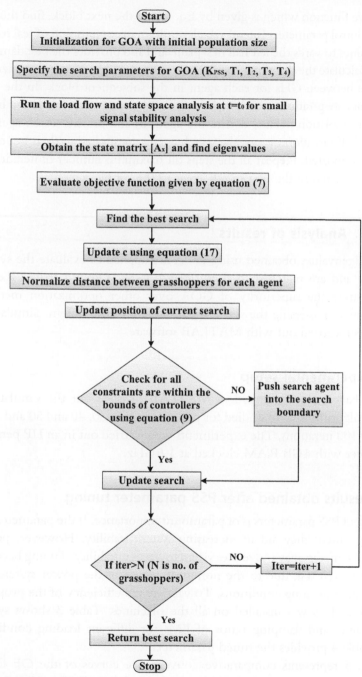

Fig. 4 Flowchart of GOA.

objective function which is given by Eq. (7). In the next block, find the best set of control parameters (agent) which results in maximum shift of real part of eigenvalues towards the left half of s-plane and also provides the best damping ratio. Calculate the shrinking coefficient c in the next block and normalize the distance between GHs for each agent in the subsequent block. In the next block, update position of GHs and check for any constraint violation in block 11. If no violation occurs, update the agent by the present search agent in block 12. If violation occurs, the search agents are pushed to their boundary and then updated. Repeat all the steps till maximum number of iterations is reached and return the best search agent in the last block.

6. Analysis of results

Eigenvalues obtained using GOA are applied to evaluate the system stability and are matched to those found using SOS and WEA. Results demonstrate the superiority of GOA over other optimization methods mentioned in assessing the small signal stability of the system. Simulations have been carried out with MATLAB software.

6.1 Experimental setup

Codes have been developed in MATLAB to carry out the simulations. All the algorithms were studied for population sizes 30, 40 and 50 and were run for 100 iterations. The experiments were carried out in an HP personal computer with 4 GB RAM clocked at 1.6 GHz.

6.2 Results obtained after PSS parameter tuning

Tuning of PSS parameters is of paramount importance. If the parameters are suitably tuned, they aid in increasing system stability. However, poorly tuned parameters may move the system towards instability. Tuning becomes a complicated task due to the nonlinear nature of the power system and its varying operating conditions. To validate the efficiency of the proposed algorithm, PSS was installed on all the machines. Table 3 shows system eigenvalues and damping ratios of EMs for different loading conditions and Table 4 provides the tuned parameters.

Fig. 5 represents comparative convergence curves of the OF for all the algorithms considered, signifying shifting of all the EM modes towards the D-space of the negative part of s-plane. GOA gives fastest converges (31 iterations) followed by SOS (45 iterations) and WEA (62 iterations).

Table 3 EMs and damping ratios for light-load, medium-load and high-load conditions.

	No stabilizer	WEA PSS	SOS PSS	GOA PSS
Light load	−1.436 ± 13.275i, 0.10755	−5.6796 ± 14.745i, 0.35944	−5.8176 ± 14.172i, 0.37975	−5.9328 ± 13.765i, 0.39581
	−0.37734 ± 9.131i, 0.04129	−2.2756 ± 9.1889i, 0.24039	−2.0746 ± 9.368i, 0.21622	−2.2506 ± 9.0169i, 0.24217
Normal load	−0.907 ± 13.57i, 0.066656	−6.2387 ± 15.008i, 0.38385	−6.3037 ± 14.084i, 0.40853	−6.4239 ± 13.604i, 0.42699
	−0.185 ± 9.0462i, 0.020433	−4.049 ± 9.3948i, 0.39579	−3.9049 ± 9.7611i, 0.37143	−3.8959 ± 9.0331i, 0.39603
High load	−0.79932 ± 13.633i, 0.058531	−6.2211 ± 15.17i, 0.37943	−6.365 ± 13.646i, 0.42271	−6.438 ± 13.463i, 0.43141
	−0.16828 ± 8.7924i, 0.019136	−3.9415 ± 8.3872, 0.42532	−3.8426 ± 8.6807i, 0.40478	−3.8251 ± 7.8673i, 0.43726

Table 4 Tuned PSS parameters for WEA, SOS and GOA.

		WEA	SOS	GOA
Generator1	K_{pss}	15.38000	28.33100	18.96000
	$T1(sec)$	1.500000	1.500000	1.500000
	$T2(sec)$	0.150000	0.126070	0.150000
	$T3(sec)$	1.380800	0.717250	1.136600
	$T4(sec)$	0.018270	0.028756	0.014153
Generator2	K_{pss}	3.090600	3.900000	3.564700
	$T1(sec)$	1.500000	1.500000	1.457100
	$T2(sec)$	0.010000	0.010000	0.010000
	$T3(sec)$	0.231120	0.165590	0.217620
	$T4(sec)$	0.010000	0.016922	0.010082
Generator3	K_{pss}	1.000000	1.988000	4.846500
	$T1(sec)$	0.750050	0.756900	0.745690
	$T2(sec)$	0.011114	0.011114	0.020197
	$T3(sec)$	1.500000	1.50000	0.317110
	$T4(sec)$	0.016095	0.018994	0.010000

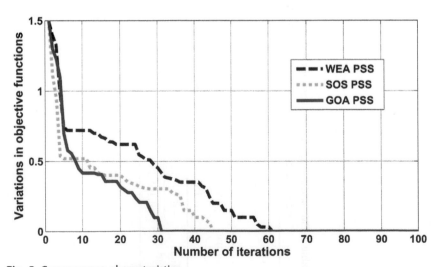

Fig. 5 Convergence characteristics.

Although PSS is primarily used for improving the damping torque, it is anticipated to influence synchronizing torque enhancement during disturbances to some extent. Also, after disturbance, contribution of synchronizing torque should be a bit positive other than the main and positive damping torque contribution. This signifies a slight increase in the imaginary part of EMs.

Table 3 shows that GOA shifted real parts of EMs to left of s-plane. Damping ratios are also enhanced when the system is subjected to different loading conditions in comparison to SOS and WEA. Parameters of PSS are tuned for a single operating point, for establishing robustness of the proposed design. This means, best group of that operating point is determined with respect to which all other operating points are assessed to obtain their responses. GOA-tuned PSS demonstrates improved performance and enhanced damping in comparison to SOS and WEA-tuned PSS for every operating condition. SOS, WEA and GOA-tuned PSS parameters are listed in Table 4.

6.3 System's response for different loading conditions

To illustrate superiority of GOA, a 3ϕ (phase) fault is considered near to bus 7 at $0.1\,\mathrm{s}$, which is cleared at $0.2\,\mathrm{s}$ without tripping any line.

Figs. 6 and 7 show responses of $\Delta\omega_{12}$ and $\Delta\omega_{13}$ obtained by each of SOS, WEA and GOA, when the system is operated under normal loaded conditions. It can be seen that GOA takes the least settling time of $2.2\,\mathrm{s}$ for $\Delta\omega_{12}$ and $1.8\,\mathrm{s}$ for $\Delta\omega_{13}$.

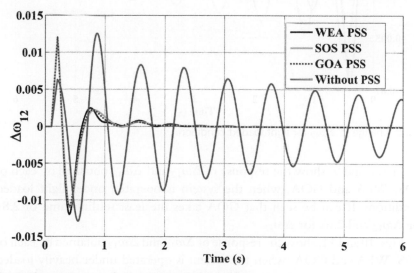

Fig. 6 Change in $\Delta\omega_{12}$ for normal loaded condition.

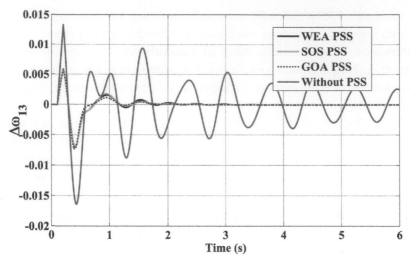

Fig. 7 Change in $\Delta\omega_{13}$ for normal loaded condition.

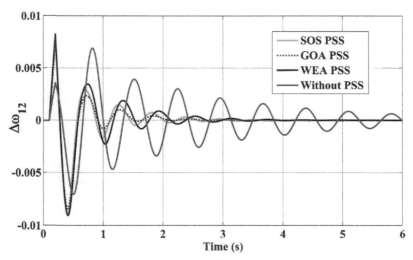

Fig. 8 Change in $\Delta\omega_{12}$ for light loaded condition.

Figs. 8 and 9, show the response of $\Delta\omega_{12}$ and $\Delta\omega_{13}$ obtained by each of SOS, WEA and GOA, when the system is operated under light loaded conditions. It can be seen that GOA takes the least settling time of 2.8 s for $\Delta\omega_{12}$ and 2.6 s for $\Delta\omega_{13}$.

Figs. 10 and 11, show the response of $\Delta\omega_{12}$ and $\Delta\omega_{13}$ obtained by each of SOS, WEA and GOA, when the system is operated under heavily loaded

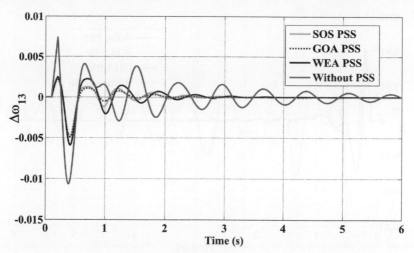

Fig. 9 Change in $\Delta\omega_{13}$ for light loaded condition.

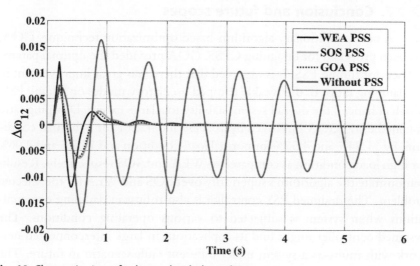

Fig. 10 Change in $\Delta\omega_{12}$ for heavy loaded condition.

conditions. It can be seen that for heavily loaded case also GOA takes the least time to settle (1.9 s for $\Delta\omega_{12}$ and 1.2 s for $\Delta\omega_{13}$) as compared to SOS and WEA. Hence, it can be concluded that GOA keeps the system more stabilized and needs lower settling time in mitigating system oscillations compared to SOS and WEA.

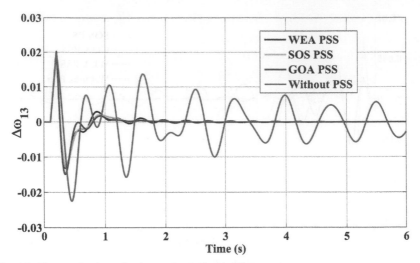

Fig. 11 Change in $\Delta\omega_{13}$ for heavy loaded condition.

7. Conclusion and future scopes

A new evolutionary algorithm-based optimization technique, GOA, has been proposed for designing CPSS. GOA provided the optimal parameter sets for tuning PSS. It is observed that there was great improvement in the damping ratios of the weakly damped oscillatory modes on adding PSS, which enhanced the overall system stability to a large extent. Damping ratio obtained using GOA improved by 10.11% and 4.23% for light load condition, by 11.24% and 4.52% for normal load condition and 13.7% and 2.06% for high load condition as compared to WEA and SOS respectively. Results demonstrate the algorithm's superiority over SOS and WEA for the selected problem. The designed PSS controller is also robust in damping out oscillations when system is subjected to various operating conditions. The designed controller might find its applications in large interconnected network with multi-area system under different fault scenario in future. The study can also be extended in future to include the following:

- Virtual inertia concept as well as energy storage system may be integrated to the system for improving overall system stability due to improved settling time and inertial response for wide range of loading conditions.
- If low inertia machines were employed with an HVDC link, they will be less vulnerable to faults on the AC side of the grid. Increase in the number of HVDC links in a large grid, will result in more dominant local modes which will be easier to monitor using local control mechanisms.

- Model predictive control and model reference adaptive control could be used to handle the non-linearity of the system for more accurate assessment of the system stability.
- LFOs appear in railway networks due to interaction of multiple electric units which leads to traction blockade as well as small signal stability issues [50,51], GOA based PSS can be used to effectively reduce these LFOs.

References

[1] G. Rogers, Power System Oscillations, Springer Science & Business Media, 2012.

[2] P.W. Sauer, M.A. Pai, J.H. Chow, Power System Dynamics and Stability: With Synchrophasor Measurement and Power System Toolbox, John Wiley & Sons, 2017.

[3] M.A. Abido, Pole placement technique for PSS and TCSC-based stabilizer design using simulated annealing, Int. J. Electr. Power Energy Syst. 22 (8) (2000) 543–554.

[4] G.N. Taranto, J.H. Chow, A robust frequency domain optimization technique for tuning series compensation damping controllers, IEEE Trans. Power Syst. 10 (3) (1995) 1219–1225.

[5] H. Werner, P. Korba, T.C. Yang, Robust tuning of power system stabilizers using LMI-techniques, IEEE Trans. Control Syst. Technol. 11 (1) (2003) 147–152.

[6] M. Eslami, H. Shareef, A. Mohamed, Application of artificial intelligent techniques in PSS design: a survey of the state-of-the-art methods, Przegląd Elektrotechniczny (Electrical Review) 87 (4) (2011) 188–197.

[7] Y. Zhang, G.P. Chen, O.P. Malik, G.S. Hope, An artificial neural network based adaptive power system stabilizer, IEEE Trans. Energy Convers. 8 (1) (1993) 71–77.

[8] A. Mahabuba, M.A. Khan, Small signal stability enhancement of a multi-machine power system using robust and adaptive fuzzy neural network-based power system stabilizer, Eur. Trans. Electr. Power 19 (7) (2009) 978–1001.

[9] R. Segal, A. Sharma, M.L. Kothari, A self-tuning power system stabilizer based on artificial neural network, Int. J. Electr. Power Energy Syst. 26 (6) (2004) 423–430.

[10] L.H. Hassan, M. Moghavvemi, H.A. Mohamed, Power system stabilization based on artificial intelligent techniques; a review, in: 2009 International Conference for Technical Postgraduates (TECHPOS), IEEE, 2009, pp. 1–6.

[11] M.A. Abido, Y.L. Abdel-Magid, A genetic-based fuzzy logic power system stabilizer for multimachine power systems, in: 1997 IEEE International Conference on Systems, Man, and Cybernetics. Computational Cybernetics and Simulation, 1, IEEE, 1997, pp. 329–334.

[12] M.J. Bosco, A.D.J. Raju, Power System Stabilizer Using Fuzzy Logic Controller in Multimachine Power Systems, 2007 IET-UK International Conference on Information and Communication Technology in Electrical Sciences (ICTES 2007).

[13] P. Hoang, K. Tomsovic, Design and analysis of an adaptive fuzzy power system stabilizer, IEEE Trans. Energy Convers. 11 (2) (1996) 455–461.

[14] Z.Y. Dong, Y.V. Makarov, D.J. Hill, Genetic Algorithms in Power System Small Signal Stability Analysis, International Conference on Advances in Power System Control, Operation and Management, APSCOM-97, 1997, pp. 342–347.

[15] P. Zhang, A.H. Coonick, Coordinated synthesis of PSS parameters in multi-machine power systems using the method of inequalities applied to genetic algorithms, IEEE Trans. Power Syst. 15 (2) (2000) 811–816.

[16] A. Stativă, M. Gavrilaş, V. Stahie, Optimal tuning and placement of power system sta-
     bilizer using particle swarm optimization algorithm, in: 2012 International Conference
     and Exposition on Electrical and Power Engineering, IEEE, 2012, pp. 242–247.
[17] A. Safari, A PSO procedure for a coordinated tuning of power system stabilizers for
     multiple operating conditions, J. Appl. Res. Technol. 11 (5) (2013) 665–673.
[18] H.E. Mostafa, M.A. El-Sharkawy, A.A. Emary, K. Yassin, Design and allocation of
     power system stabilizers using the particle swarm optimization technique for an
     interconnected power system, Int. J. Electr. Power Energy Syst. 34 (1) (2012) 57–65.
[19] S. Panda, Robust coordinated design of multiple and multi-type damping controller
     using differential evolution algorithm, Int. J. Electr. Power Energy Syst. 33 (4)
     (2011) 1018–1030.
[20] S. Panda, Differential evolutionary algorithm for TCSC-based controller design, Simul.
     Model. Pract. Theory 17 (10) (2009) 1618–1634.
[21] A. Ameli, M. Farrokhifard, A. Ahmadifar, A. Safari, H.A. Shayanfar, Optimal tuning of
     power system stabilizers in a multi-machine system using firefly algorithm, in: 2013 12th
     International Conference on Environment and Electrical Engineering, IEEE, 2013,
     pp. 461–466.
[22] S.M. Abd Elazim, E.S. Ali, Optimal power system stabilizers design via cuckoo search
     algorithm, Int. J. Electr. Power Energy Syst. 75 (2016) 99–107.
[23] M.A. Abido, Y.L. Abdel-Magid, Optimal design of power system stabilizers using evo-
     lutionary programming, IEEE Trans. Energy Convers. 17 (4) (2002) 429–436.
[24] M.A. Abido, A novel approach to conventional power system stabilizer design using
     tabu search, Int. J. Electr. Power Energy Syst. 21 (6) (1999) 443–454.
[25] M.A. Abido, Robust design of multimachine power system stabilizers using simulated
     annealing, IEEE Trans. Energy Convers. 15 (3) (2000) 297–304.
[26] D.K. Sambariya, R. Prasad, Robust tuning of power system stabilizer for small signal
     stability enhancement using metaheuristic bat algorithm, Int. J. Electr. Power Energy
     Syst. 61 (2014) 229–238.
[27] S. Mishra, M. Tripathy, J. Nanda, Multi-machine power system stabilizer design by rule
     based bacteria foraging, Electr. Power Syst. Res. 77 (12) (2007) 1595–1607.
[28] S. Saremi, S. Mirjalili, A. Lewis, Grasshopper optimisation algorithm: theory and
     application, Adv. Eng. Softw. 105 (2017) 30–47.
[29] P. Dey, A. Bhattacharya, J. Datta, P. Das, Small signal stability improvement of large inter-
     connected power systems using power system stabilizer, in: 2017 2nd International
     Conference for Convergence in Technology (I2CT), IEEE, 2017, pp. 753–760.
[30] A. Saha, P. Das, A.K. Chakraborty, Water evaporation algorithm: a new metaheuristic
     algorithm towards the solution of optimal power flow, Eng. Sci. Technol. Int. J. 20
     (6) (2017) 1540–1552.
[31] P. Dey, S. Mitra, A. Bhattacharya, P. Das, Comparative study of the effects of SVC and
     TCSC on the small signal stability of a power system with renewables, J. Renew.
     Sustain. Energy 11 (3) (2019) 033305.
[32] P. Dey, A. Saha, A. Bhattacharya, B. Marungsri, Analysis of the effects of PSS and
     renewable integration to an inter-area power network to improve small signal
     stability, J. Electr. Eng. Technol. 15 (5) (2020) 2057–2077.
[33] P. Dey, A. Saha, S. Mitra, B. Dey, A. Bhattacharya, B. Marungsri, Improvement of
     small-signal stability with the incorporation of FACTS and PSS, in: Control
     Applications in Modern Power System, Springer, Singapore, 2021, pp. 335–344.
[34] P. Dey, A. Saha, P. Srimannarayana, A. Bhattacharya, B. Marungsri, A realistic approach
     towards solution of load frequency control problem in interconnected power
     systems, J. Electr. Eng. Technol. 17 (2) (2022) 759–788.
[35] E.V. Fortes, L.F.B. Martins, E.L. Miotto, P.B. Araujo, L.H. Macedo, R. Romero,
     Bio-inspired metaheuristics applied to the parametrization of PI, PSS, and UPFC–POD

controllers for small-signal stability improvement in power systems, J. Control Automat. Electr. Syst. 34(1) (2022) 1–16.

[36] A. Biswas, Atom stabilization algorithm and its real life applications, J. Intell. Fuzzy Syst. 30 (4) (2016) 2189–2201.

[37] A. Biswas, B. Biswas, A. Kumar, K.K. Mishra, Particle swarm optimisation with time varying cognitive avoidance component, Int. J. Comput. Sci. Eng. 16 (1) (2018) 27–41.

[38] A. Biswas, K.K. Mishra, S. Tiwari, A.K. Misra, Physics-inspired optimization algorithms: a survey, J. Opt. 2013 (2013) 1–16.

[39] A. Biswas, B. Biswas, Regression line shifting mechanism for analyzing evolutionary optimization algorithms, Soft. Comput. 21 (21) (2017) 6237–6252.

[40] D. Sarkar, N. Mishra, A. Biswas, Genetic algorithm-based deep learning models: a design perspective, in: Proceedings of the Seventh International Conference on Mathematics and Computing, Springer, Singapore, 2022, pp. 361–372.

[41] A. Biswas, B. Biswas, Visual analysis of evolutionary optimization algorithms, in: 2014 2nd International Symposium on Computational and Business Intelligence, IEEE, 2014, pp. 81–84.

[42] A. Biswas, A. Kumar, K.K. Mishra, Particle swarm optimization with cognitive avoidance component, in: 2013 International Conference on Advances in Computing, Communications and Informatics (ICACCI), IEEE, 2013, pp. 149–154.

[43] A. Biswas, A.V. Lakra, S. Kumar, A. Singh, An improved random inertia weighted particle swarm optimization, in: 2013 International Symposium on Computational and Business Intelligence, IEEE, 2013, pp. 96–99.

[44] A. Biswas, B. Biswas, Swarm intelligence techniques and their adaptive nature with applications, in: Q. Zhu, A. Azar (Eds.), Complex System Modelling and Control through Intelligent Soft Computations, Springer, Cham, 2015, pp. 253–273.

[45] A. Biswas, B. Biswas, K.K. Mishra, An atomic model based optimization algorithm, in: 2016 2nd International Conference on Computational Intelligence and Networks (CINE), IEEE, 2016, January, pp. 63–68.

[46] B. Dembart, A.M. Erisman, E.G. Cate, M.A. Epton, H. Dommel, Power System Dynamic Analysis: Phase I, Final report (no. EPRI-EL-484), Boeing Computer Services, Inc., Seattle, Wash. (USA). Energy Technology Applications Div, 1977.

[47] P.M. Anderson, A.A. Fouad, Power System Control and Stability, John Wiley & Sons, 2008.

[48] P. Dey, A. Bhattacharya, J. Datta, P. Das, Parameter tuning of power system stabilizer using a meta-heuristic algorithm, in: 2017 Second International Conference on Electrical, Computer and Communication Technologies (ICECCT), IEEE, 2017, pp. 1–8.

[49] P. Dey, A. Bhattacharya, P. Das, Tuning of power system stabilizer for small signal stability improvement of interconnected power system, Appl. Comput. Inform. 16 (2) (2017) 3–28.

[50] P. Dey, P. Kirawanich, C. Sumpavakup, B. Aniruddha, Tuning of controller parameters for suppressing low frequency oscillations in electric railway traction networks using meta-heuristic algorithms, IET Electr. Syst. Transp. 13 (2) (2023) e12075.

[51] P. Dey, C. Sumpavakup, P. Kirawanich, Optimal control of grid connected electric railways to mitigate low frequency oscillations, in: 2022 Research, Invention, and Innovation Congress: Innovative Electricals and, Electronics, IEEE, 2022, pp. 70–75.

About the authors

Dr. Prasenjit Dey completed his B.E. degree in Electrical and Electronics Engineering from Anna University, M. Tech and Ph.D. degrees from National Institute of Technology, Agartala. He is presently a Foreign Expert in the CLARE division of Mahidol University, Salaya, Thailand. Earlier, Dr. Dey worked as an Assistant Professor in the Department of Electrical and Electronics Engineering, in the National Institute of Technology Sikkim. His research interests include Power System Stability, Optimization Techniques, and Railway Electrification.

Dr. Anulekha Saha received her B.Tech degree in Electrical Engineering from Birbhum Institute of Engineering and Technology, India, and M.Tech and Ph.D. from the Department of Electrical Engineering, NIT Agartala, India. She has worked as an Assistant Professor in the Department of Electrical Engineering, NIT Silchar, and the Department of Electrical and Electronics Engineering, NIT Sikkim, India. Presently, she is a Postdoctoral Research Fellow in the Department of Electrical Engineering, at Chulalongkorn University, Thailand. Her research area primarily focuses on power system optimization using soft-computing techniques, transient stability analysis and energy trading in electricity markets.

Dr. Aniruddha Bhattacharya did his BSc in Electrical Engineering from Regional Institute of Technology, Jamshedpur, India in 2000 as well as MEE and Ph.D. in Electrical Power System from Jadavpur University, Kolkata, India in 2008 and 2011, respectively. His employment experiences include Siemens Metering Limited, India; Jindal Steel & Power Limited, Raigarh, India; Bankura Unnayani Institute of Engineering, India; Dr. B.C. Roy Engineering College, India; NIT Agartala, India. He is currently an Assistant Professor at the Electrical Engineering Department, NIT Durgapur, India. His areas of interest include optimal power flow, hydrothermal scheduling, power system stability, power system optimization, and electric vehicle.

Dr. Priyanath Das obtained his B.Tech and M.Tech in Electrical Engineering in 1994 and 2002 respectively. He completed his PhD from Jadavpur University in 2013. He is presently working as a Professor in the Department of Electrical Engineering, NIT Agartala, India. His areas of interest include the Application of FACTS & HVDC, Deregulated Power Systems, and High Voltage Engineering.

Dr. Boonruang Marungsri (Member, IEEE) was born in Nakhon Ratchasima, Thailand, in 1973. He received the B.Eng. and M.Eng. degrees in electrical engineering from Chulalongkorn University, Thailand, in 1996 and 1999, respectively, and the D. Eng. degree in electrical engineering from Chubu University, Kasugai, Aichi, Japan, in 2006. He is an Assistant Professor at the School of Electrical Engineering, Suranaree University of Technology, Thailand. His research interests include electrical power technologies, smart grids, energy management systems, and high-voltage insulation technologies.

Dr. Phumin Kirawanich received the M.S. and Ph.D. degrees in electrical engineering from the University of Missouri, Columbia, in 1999 and 2002, respectively. From 2002 to 2007, he was with the Power Electronics Research Center and the High Power Electromagnetic Radiation Laboratory, University of Missouri, as a Postdoctoral Fellow. From 2007 to 2008, he was with the Department of Electrical and Computer Engineering, University of Missouri, as a Research Assistant Professor. Since 2008, he has been a Faculty Member with the Department of Electrical Engineering, Mahidol University, Salaya, Thailand, as an Associate Professor. His research interests include railway EMC, pulsed-power technology for biomedical and agricultural applications, terahertz-pulse generation, and electromagnetic and semiconductor device physics computation. For practical and professional experience, Dr. Kirawanich is an EMC specialist managing EMC compliance of the whole railway as a complete system.

Dr. Chaiyut Sumpavakup received the B.E., M.E. and Ph.D. degrees in Electrical Engineering from Suranaree University of Technology, Thailand in 2005, 2008, and 2017 respectively. He is currently an Assistant Professor in the Department of Power Engineering Technology, College of Industrial Technology, King Mongkut's University of Technology North Bangkok, Thailand. His research interests are energy management in electric railway system, electric vehicle, vehicle electrification and application of artificial intelligence techniques to power system operation, management, and control.

PART II

Ecological and economic systems

CHAPTER SIX

Air quality modeling for smart cities of India by nature inspired AI—A sustainable approach

Nishant Raj Kapoor[a,b], Ashok Kumar[a,b], Anuj Kumar[a,b,c],
Aman Kumar[a,b], and Harish Chandra Arora[a,b]
[a]CSIR-Central Building Research Institute, Roorkee, Uttarakhand, India
[b]Academy of Scientific and Innovative Research (AcSIR), Ghaziabad, Uttar Pradesh, India
[c]Haldaur Group of College, Bijnor, Uttar Pradesh, India

Contents

Abstract

Nature inspired artificial intelligence (AI) techniques are gaining traction in air quality modeling. A mathematical simulation of how air pollutants distribute and react in the atmosphere to impact ambient air quality is known as air quality modeling. Modeling aids in estimating the correlation between pollution levels and their impacts on air quality. Outdoor ambient air quality is important because it affects indoor air quality. Unprecedented growth of urban agglomeration, population blast, and unrestricted use of natural resources as well as nonrenewable fuels eventually results in higher pollution and driving the world away from the sustainable development goals. Rising air pollution levels around the globe is an alarming situation for all lifeforms. It also affects the nonliving things such as structures, metals, and water bodies. In this chapter, to predict air quality in seven Indian smart cities, a data-driven artificial neural network (ANN) is merged with particle swarm optimization (PSO), a nature-inspired optimization technique. The PSO technique presented in this chapter functions as an ANN

Advances in Computers, Volume 135
ISSN 0065-2458
https://doi.org/10.1016/bs.adcom.2023.11.012

129

optimization technique. The data were obtained from the online web repository called Kaggle. The data-sets for 12 pollutants were selected for 7 Indian smart cities. The datasets were recorded during the last 5 years from 2016 to 2020 and were included for computation. Models can be used by air quality managers to forecast the effects of prospective new emissions. Policy makers often use models to predict ambient air pollution concentrations under various scenarios as a decision-making tool. Because of the robustness and effectiveness of AI models, the public will be able to get early warning for a high concentration level of pollutants in big cities, potentially lowering cardiovascular and respiratory mortality. The findings proved PSO's outstanding performance while evaluating high-dimensional data.

1. Introduction

Clean air is an essential requirement for approximately all lifeforms on earth to survive and develop. Nowadays, with growing air pollution due to the expansion of industrialization, enhanced automobiles, and increased fossil fuels consumption the air quality is decreasing. Many pollutants such as NO_x, SO_x, CO, CO_2, and coarse and fine particulates exist in the atmosphere. A significant number of scientists and researchers worldwide have researched air pollution and air quality predictions, focusing on pollutant predictions. Pollutant sources may be broken down into two categories: (a) anthropogenic sources (man-made sources) and (b) natural sources. The primary causes of air pollution are anthropogenic sources such as discharges from construction activities, industrial operations, fuel combustion, and vehicle pollutants. Sulfur, metal compounds, nitrogen, hydrogen, oxygen, and particulate matter are only a few of the pollutants produced by man-made pollution sources. Natural causes of pollution are natural occurrences that emit hazardous chemicals as well as gases that have detrimental environmental consequences. Air pollutants, including NO_2, CO_2, CO, SO_2, and sulfate, are caused by the nature primarily due to volcanic eruptions or fires in forests. In many areas throughout the globe, air pollution is a major issue, with two major concerns: first, the influence on human health, such as cardiovascular illnesses as well as infections, and second, the impact on the environment, such as global warming, climate change, and acid rain.

Contaminated air has become one of the main contributors in the global share of growing infections. As indicated by World Health Organization (WHO) [1], 92% global population is living under poor air conditions and the pollutants concentration levels are much higher than prescribed limits that are not safe for well-being of both human and other life-forms.

Chronic exposure to poor air quality has been connected to an augmented risk of strokes, lung cancer, heart attacks, premature death, continuous wheezing, coughing and shortness of breath, and other major life-threatening diseases [2–4]. Poor quality of outdoor air also affects the indoor environmental quality. Buildings occupants consider air quality as an important parameter for their indoor comfort [5–7]. This represents the necessity for precise and reliable solutions to reduce the air pollution impacts on human health. Air quality impacts health and has an impact on economic development. According to an OECD estimate, outer air pollution may cost the world economy $2.6 trillion per year by 2060, including sick days, medical expenses, and lost agricultural production. Furthermore, by 2060, the costs of welfare linked with premature mortality will have risen to as much as $25 trillion [8]. Hence, it is a considerable issue for the global scientific community to find modern solutions involving techniques like artificial intelligence (AI) [9].

AI is the most frequently utilized technology instrument employed in recent years for managing and preventing the damaging effects of various air pollutants and it is gaining considerable interest in the fields of atmospheric research [10]. AI technology has advanced at a breakneck pace, with a plethora of new emerging AI techniques. Artificial neural networks (ANNs), support vector machines, genetic algorithms, fuzzy logic (FL), and with optimized natural inspired algorithms, have been used in agriculture, finance, engineering, security, education, medical, nanotechnology, and other fields [11–19]. AI positively affects the atmospheric research as well. The algorithms used to forecast the anomaly exploration in the quality of air have been focused on by Bai et al. [20] and Aggarwal et al. [21]. Deep learning (DL) applications have subsequently demonstrated a significant potential for exploring additional features [22–24]. Rahman et al. [25] studied applications of soft computing for air quality modeling by analyzing and discussing conventional as well as evolutionary ANNs, DL, FL, neuro-fuzzy systems, and a variety of hybrid models. Sayeed et al. [26] presented a deep convolutional ANNs-based AI algorithm to predict Texas' 24-hour ozone concentration in order to compare the results of several time periods in 2017. Several authors [27–30] looked for air quality models that might be used to evaluate air quality issues in the atmosphere. Using detailed area information with accurate and new forecasting techniques, these models can depict the health status of cities. These models have the benefit of being able to offer an early warning, as well as significantly reducing the amount of manual data gathering measures. AI will be an essential component for making cities smarter as it can sustainably provide reliable predictions without utilizing many resources [31]. Because of adaptability and well-generalized performance, advanced nature-inspired modeling techniques are gaining a lot of interest in air quality modeling.

This chapter uses ANN-PSO to predict the quality of air in Indian smart cities. The organization of this chapter contains six more sections apart from introduction: Section 2 contains the existing similar work, Section 3 explains the working methodology used in this study, and Section 4 deals with data processing part followed by results and discussion presented in Section 5. Finally, Section 6 contains the conclusion of the study.

2. Related work

2.1 Artificial neural network

Walter Pitts and Warren McCulloch pioneered the field by developing a model for neural networks (NN) in 1943 [32, 33]. Hebb developed a learning theory based on the process of brain functioning in the late 1940s, termed as Hebbian learning [34]. Clark and Farley were the first to utilize computational devices, then known as "calculators," to mimic a Hebbian network in 1954 [35]. F. Rosenblatt, a psychologist, built the perceptron, the first ANN, in 1958, with funding from the US Office of Naval Research [36, 37]. Ivakhnenko and Lapa presented first multiple levels functional network with the data handling group methods in 1965 [38]. Dynamic programming techniques were employed by Kelley and Bryson in the context of control theory to create the principles of continuous backpropagation (BP) in 1960 and 1961, respectively [39–42]. Dreyfus employed BP to alter controller settings with respect to the error gradients in 1973 [43]. Multilayer network training became feasible because of Werbos' BP approach. In 1982, he developed a method for applying Linnainmaa's AD approach to NN that have since become extensively utilized [44]. The particular function of the human marrow has inspired the development of ANNs, despite the fact that they are only loosely related. Even though ANNs do not reach the complication of the marrow, there are only two fundamental similarities between them. The network function is defined by the connectivity between the different neurons, and their building blocks are very interconnected and simple forms of computing devices. There are 10^{10} computational neurons in the brain of human beings that are connected by a single network with around 10^4 connections per element. ANNs can provide optimal solutions by tweaking their internal architecture using proper elementary and sufficient information. Information will be gained from the environment via suitable inputs to the ANN, simulating brain activity, and afterward users can be able to recollect this information. ANN has many variation and variety, but when it comes to the learning process, there are two basic groups that are immediately distinguishable: supervised and unsupervised [45]. Both categories of ANNs have been used in environmental and air quality investigations. The outcome is predicted by a supervised learning model. The unsupervised learning approach discovers hidden patterns in data. In supervised learning, input data, as well as output data, are presented to the model. Contrarily, only input data are presented to the algorithms in unsupervised learning.

2.1.1 Multilayer perception

The structure of multilayer perception (MLP) is a feedforward ANN architecture and consists of a minimum of minimum three-knot layers: (i) input, (ii) hidden, and (iii) output layers. The input nodes use a nonlinear activation function and the nonlinear activation function helps to identify the MLP from the linear perception. It can differentiate information that is not removable in a linear manner. The MLP neural model is the most often used a prediction model. The product of weights (w_{ij}) and input elements (a_i), as well as the bias (b_j), is added together at nodes in feedforward networks as expressed in Eq. (1) [46]:

$$x = \left(\sum_{i=1}^{n} w_{ij} a_i \right) + b_j \tag{1}$$

After imposing a transfer function F on X, an output is generated, expressed in Eq. (2):

$$F(X) = F\left[\left(\sum_{i=1}^{n} w_{ij} a_i \right) + b_j \right] \tag{2}$$

There are two roots in the aforementioned equations, which are bias and weight. Each neuron in the construction system can create connected outputs due to a set of weighted inputs. Weight is a function of technological and biological systems, according to professionals working in the area of AI and ML, where ANNs are frequently utilized. Fig. 1 illustrates the simple arrangement of the ANN, which is mainly made up of inputs, neurons, and a single output.

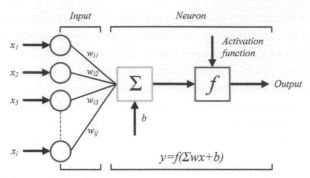

Fig. 1 Artificial neural network.

MLP-based algorithms can handle higher-dimensional input spaces in comparison with polynomial expansions, which is a useful feature in many real-world scenarios that require the modeling of several variables' interactions. Environmental study involves a wide range of variables. As a result, the MLP structure may be able to connect the air quality model to real-world conditions.

2.1.2 Feedforward neural network

An feedforward neural network (FFNN) differs from a common NN; the nodes do not create a cycle of connections. It was the first as well as the most basic form of ANN. The information in FFNN flows only in a single direction that is from input to output nodes, through hidden nodes [38]. When inputs enter the first layer, no computations are done. The hidden layer is where computations begin, by activating both linear and nonlinear activation functions as shown in Eq. (3):

$$Z_j = V_j \left(\sum_{i=1}^{p} w_{ij}^h x_i + p_j \right) \tag{3}$$

where V_j represents the hidden layer activation function, p_j is bias, number of hidden layers is h, and input is termed as p. The complex values between the input and output variables learn and recognize with the help of activation functions. They provide the network's nonlinear characteristics. Their primary function is to change a node's input signal to an output signal via ANN. In the next layer, the previous output signal is utilized as an input.

2.1.3 Backpropagation neural networks

BP is used in ANN to determine the gradient required to compute the weight employed in the architecture. This approach is useful for training deep NN with single or multiple hidden layers [38]. BP is an automated differentiation method. "During the learning phase, the gradient descent optimization technique employs backpropagation to determine the gradient of the loss function for modifying neuron weights" [38]. The BP algorithm is the most often employed in ANNs. In the feedforwarding process, the data are created by processing the information from the input-to-output layer [47]. Fig. 2 shows a typical back–propagation neural network design, which has one input layer and one output layer with a number of hidden layers in the middle of the network.

2.1.4 Levenberg–Marquardt algorithm

Levenberg–Marquardt (LM) algorithm is used to give a numerical explanation for decreasing a problem. It has consistent convergence, is quick, and can be used to train medium– as well as small-sized ANN problems [48]. The LM algorithm is used in mathematics and computers to solve nonlinear least-squares problems, commonly called the "damped least-squares" approach. The least-squares fit curve is used to reduce these issues particularly. ANN principles were typically used to train, validate, and test the ANN model, with 70%, 15%, and 15% of the dataset, respectively. Several architectures were assessed using testing data from the network to identify the best-anticipated architecture. "A layer of input neurons, one or more hidden layers that connect the input and output layers, and a layer of output neurons are common

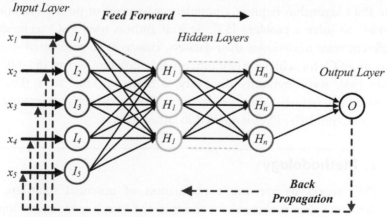

Fig. 2 Back-propagation neural network.

components of an ANN architecture" [49]. The number of neurons was established based on the number of inputs and outputs. Furthermore, for the ANN model, both one and two hidden layers were tried. After considering the finest findings among them, one of them is then recommended for modeling. The LM method is often used for learning and optimization algorithm in training as expressed in Eq. (4) [50]:

$$O = f\left(T + \sum w_i x_i\right) \qquad (4)$$

where f represents the nonlinear sigmoid function and the value of threshold bias is T. The error was estimated after training and testing periods resulted on the difference between computed and targeted outputs. The error function listed below is utilized in this case as follows in Eq. (5) [51]:

$$E = \frac{1}{p} \sum_p \sum_k \left(t_{pk} - z_{pk}\right)^2 \qquad (5)$$

where z_{pk} is the kth element of the outcome vector for the pth pattern input, t_{pk} is the kth element of the pth desired pattern vector, k is the output vector element index, and p is the index of the p training pairs of vectors.

2.2 Particle swarm optimization

Kennedy and Eberhart are credited for first proposing PSO in 1995, which was first used to simulate social behavior as an idealized depiction of the movement of creatures in a fish school or bird flock [52]. The PSO was discovered as a new heuristic method. Shi and Eberhart proposed the most well-known form of PSO among the different versions. A simple version of the PSO algorithm requires a population (swarm) of possible solutions (particles) to solve a problem [53]. Several authors used PSO techniques in different areas to optimize their datasets. Biswas et al. performed several studies on PSOs for achieving best outcomes in different sectors [54–60]. In a study [54], for community discovery, a fuzzy agglomerative (FuzAg) method that iteratively updates node membership degree is suggested by the authors. Similarly, PSO is applied in other studies.

3. Methodology

PSO works by creating a population of potential solutions, or "particles," and then moving them about in the search space according to a straightforward mathematical formula depending on their position and velocity. Each particle has a local best-known position (BKP) that guides its travel, but it also moves in the direction of the BKP in the search space, which are updated when other particles find better locations. The swarm is predicted to migrate toward the best options as a result of this. PSO is a method for tuning the initial weights as well as biases in NN that are inspired by the social behavior of fish schooling or birds. Fig. 3 shows the workflow of using the PSO-ANN hybrid algorithm for predicting air quality. Particles are used to represent possible solutions. Particles with randomly initialized locations (Eq. 6) refer to a set of possible weights as well as biases from ANN models and velocities in Eq. (7):

$$X^0 = \left[x_1^0, x_2^0, x_3^0, ..., x_i^0, ..., x_M^0\right] \tag{6}$$

$$V^0 = \left[v_1^0, v_2^0, v_3^0, ...v_i^0, ..., v_M^0\right] \tag{7}$$

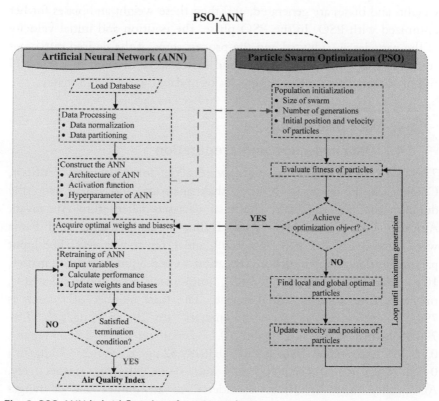

Fig. 3 PSO-ANN hybrid flowchart for AQI prediction.

where X represents the initial location of particles, v is the velocity, in x_M^0 M is the swarm's specified size, whereas 0 is the initial phase. Particles are drawn to their global best position Pg and the personal best position in history P_i for each repetition [42]. Eqs. (8) and (9) show the updated velocity of ith particle and position of (x_i), respectively.

$$v_i^{t+1} = \omega v_i^t + c_1 \varepsilon_1 \left(P_i^t - x_i^t \right) + c_2 \varepsilon_2 \left(P_g^t - x_i^t \right) \tag{8}$$

$$x_i^{t+1} = x_i^t + v_i^{t+1} \tag{9}$$

where t is the current generation, inertia weight is ω, ϵ_1 and ϵ_2 are the independent arbitrary values that lie between 0 and 1, and c_1 and c_2 are the acceleration constants. To increase search efficiency, velocity and particle location are constrained in ranges $[v_{min}, v_{max}]$ and $[x_{min}, x_{max}]$, respectively:

The selected normalized dataset is loaded in the ANN algorithm and splitted into training and testing datasets. The activation functions between the input-to-hidden layer and the hidden-to-output layer are applied. The weights and biases are generated, and then these weight and biases further optimized with PSO. In the PSO phase, the position and initial velocity of the particles are adjusted and the process stops, if the desired outcomes are achieved. These new weights and biases were updated in the ANN and predictions are made on the bases of updated weights and biases.

4. Data processing

4.1 Data collection and preparation

The data considered in this work were taken for Amaravati, Amritsar, Chandigarh, Hyderabad, Delhi, Vishakhapatnam, and Patna from the duration of 2016 to 2020. The total data-sets taken from Amaravati, Amritsar, Chandigarh, Hyderabad, Delhi, Vishakhapatnam, and Patna were 680, 634, 299, 1764, 1460, 1899, and 1166, respectively. The input parameters considered in this investigation were 2.5 μm particulate matter, 10 μm particulate, NO, NO_2, NO_x, ammonia, carbon monoxide, SO_2, ozone, xylene, toluene, and benzene. The total number of datasets is 7902. The ranges of these parameters are 2–685.36 μg/m³, 7.8–796.88 μg/m³, 0.25–221.41 μg/m³, 0.17–266.46 μg/m³, 0–259.54 μg/m³, 0.11–166.70 μg/m³, 0–30.44 μg/m³, 0.36–89.52 μg/m³, 0.1–257.73 μg/m³, 0–170.36 μg/m³, 0–123.36 μg/m³, and 0–64.33 μg/m³, respectively. Air quality index (AQI) is considered as an output and it ranges between 22 and 737. Tables 1 and 2 show the statistical parameters of

Table 1 Collection of AQI parameters data.

Cities	$PM_{2.5}$	PM_{10}	NO	NO_2	NO_x	NH_3	CO
Amaravati	4.65–139.38	7.8–230.27	0.25–43.76	1.52–140.17	0.86–103.49	1.28–33.72	0–1.81
Amritsar	8.8–249.1	18.03–486.99	0.4–103.44	3.08–102.85	7.46–150.96	0.5–129.46	0–3.83
Hyderabad	4.83–571.02	10.54–485.88	1.22–38.06	0.62–92.33	1.58–63.29	2.5–78.57	0–8.83
Chandigarh	6.86–154.85	15.78–242.22	0.44–70.94	2.47–52.22	3.66–70.53	0.65–100.35	0.2–1.3
Delhi	10.24–685.36	19.51–796.88	3.57–221.03	10.63–162.5	1.87–254.8	6.78–166.7	0–30.44
Vishakhapatnam	2–203.05	8.16–326.4	0.44–90.84	0.17–130.26	0–117.99	0.11–68.42	0.11–2.11
Patna	10.78–645.5	30.03–276.34	0.7–221.41	0.66–266.46	1.11–259.54	1.14–49.83	0.07–8.06

Table 2 Collection of AQI parameters data.

Cities	SO_2	O_3	B_{en}	T_{ol}	X_{yl}	AQI
Amaravati	4.65–66.39	12.29–138.18	0–53.89	0–26.79	0.1–125.18	24–312
Amritsar	0.71–67.26	3.71–66.7	0.52–18.73	0.28–14.6	0.65–29.17	39–478
Hyderabad	0.71–70.39	0.1–98.75	0–8.06	0–60.02	0–28.24	22–737
Chandigarh	4.27–18.14	3.2–62.28	0–64.33	0–25.74	0–116.62	26–335
Delhi	2.65–71.56	6.94–257.73	0.22–20.64	0–103	0–23.3	51–716
Vishakhapatnam	1.23–61.41	1.55–162.33	0–15.97	0–34.23	0–23.31	23–387
Patna	0.36–89.52	0.6–142.43	0–28.55	0–123.36	0–170.37	53–619

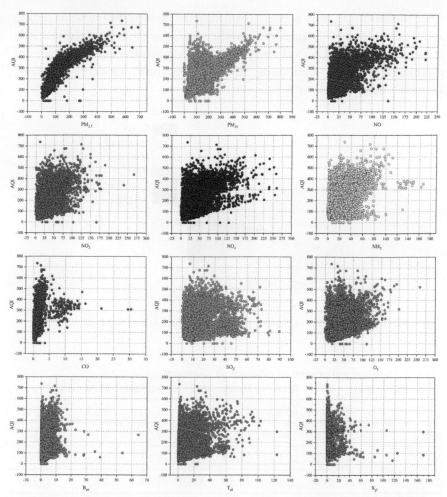

Fig. 4 Distribution of AQI output with respect to input variables (particulate matter, ammonia, nitro-oxygen compounds, CO, SO₂, ozone, xylene, toluene, and benzene).

different air pollutants. The relationship of AQI with all input parameters is shown in Fig. 4.

Fig. 5 depicts the air quality levels distribution according to the AQI on the basis of the study dataset. Data standardization was used previously in the installation of ML models to assure the learning speed and prediction accuracy of the models. Data were standardized between −1 and +1 using Eq. (10):

$$Y = \left(2 \times \frac{y - y_{\min}}{y_{\max} - y_{\min}}\right) - 1 \tag{10}$$

After the standardization process, the data were divided into two sets testing and training. The 80% data contain to the training set and other residual 20% of the data were used as the testing set. Table 3 shows the statistical values of parameters such as minimum, mean, maximum, standard deviation, kurtosis, and skewness in detail.

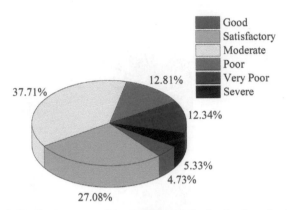

Fig. 5 AQI level distribution according to NAQI on the basis of study dataset.

Table 3 Statistical parameters of different air pollutants.

Parameters	Unit	Min.	Max.	Mean	Std.	Kurtosis	Skewness	Type
$PM_{2.5}$	$\mu g/m^3$	2	685.36	79.32	73.36	10.39	2.33	Input
PM_{10}	$\mu g/m^3$	7.8	796.88	135.33	94.4	7.36	1.79	Input
NO	$\mu g/m^3$	0.25	221.41	21.49	26.17	12.57	2.75	Input
NO_2	$\mu g/m^3$	0.17	266.46	35.58	23.97	7.48	1.49	Input
NO_x	$\mu g/m^3$	0	259.54	36.73	32.34	9.06	2.13	Input
NH_3	$\mu g/m^3$	0.11	166.70	22.11	17.04	9.85	1.94	Input
CO	$\mu g/m^3$	0	30.44	1.19	1.49	67.24	6.37	Input
SO_2	$\mu g/m^3$	0.36	89.52	14.62	11.71	6.56	1.84	Input
O_3	$\mu g/m^3$	0.1	257.73	39.50	24.78	6.57	1.48	Input
B_{en}	$\mu g/m^3$	0	64.33	2.75	2.89	51.81	3.92	Input
T_{ol}	$\mu g/m^3$	0	123.36	10.32	12.62	11.94	2.51	Input
X_{yl}	$\mu g/m^3$	0	170.37	3.17	6.73	154.40	9.06	Input
AQI	–	22	737	172.96	114.93	3.62	1.15	Output

As per the dataset, $PM_{2.5}$ is maximum in Delhi and least in Vishakhapatnam, and PM_{10} is maximum in Delhi and least in Amravati. NO, NO_2, and NO_x are maximum at Patna. NH_3 and CO are maximum at Delhi. SO_2 was maximum at Patna. Ozone was maximum at Delhi. Benzene was maximum at Chandigarh. Toluene and xylene were maximum at Patna. AQI was comparably high in Delhi, Hyderabad, and Patna. AQI was comparably lower in Amravati and Chandigarh.

4.2 Performance indices

To compare forecasted and actual results, the Pearson correlation coefficient (R), root mean square error (RMSE), mean absolute error (MAE), and mean absolute percentage error (MAPE) were used (as presented in Eqs. 11–14). R-values closer to 1 suggest the best correlation results. The smaller values of RMSE, MAPE, and MAE show the superior results of ANN and PSO–ANN models:

$$R = \frac{\sum_{i=1}^{N}(t_i - \mu_T)(p_i - \mu_P)}{\sqrt{\sum_{i=1}^{N}(t_i - \mu_T)^2}\sqrt{\sum_{i=1}^{N}(p_i - \mu_P)^2}} \tag{11}$$

$$MAE = \frac{1}{N}\sum_{i=1}^{N}|t_i - p_i| \tag{12}$$

$$MAPE = \frac{100\%}{N}\sum_{i=1}^{N}\left|\frac{t_i - p_i}{t_i}\right| \tag{13}$$

$$RMSE = \sqrt{\frac{\sum_{i=1}^{N}(t_i - p_i)^2}{N}} \tag{14}$$

where p_i is the forecasted value, t_i is the measured or target value, μ_P is the average values of predicted data and μ_T is the average values of the target data, and N is the number of values in the recorded dataset.

5. Results and discussion

To establish the unit dimensions of data, all datasets were standardized in the range of -1 to $+1$ as mentioned earlier. The calculation process is carried out in a MATLAB environment on a laptop computer (Dell Inspiron 15R, i5 CPU @160GHz 2.30GHz, 6GB RAM). The ANN model was trained from 3 neurons to 30 neurons, to identify the best-fitted neuron. The best-fitting model for both training and testing data was found to be 13 neurons based on the ranking systems of performance indicators like R^2, RMSE, and MAPE. The R-value of the ANN model for both training and testing data is 0.9697 and 0.9653, sequentially. Fig. 6 represents the plot of experimental and predicted values with respect to the number of sample datasets, error frequency distribution, and errors distribution with respect to the number of sample datasets for ANN model.

To enhance the capacity of the proposed model, a hybrid natural inspired algorithm was applied on the ANN model called PSO-ANN. To achieve this aim, different particle sizes were used from 30 to 600 with 13 neurons in the hidden layer. The hit and trial technique was used to achieve the optimum number of swarm particles. Again by using ranking system on the performance indices, the particle size 120 number shows the best results.

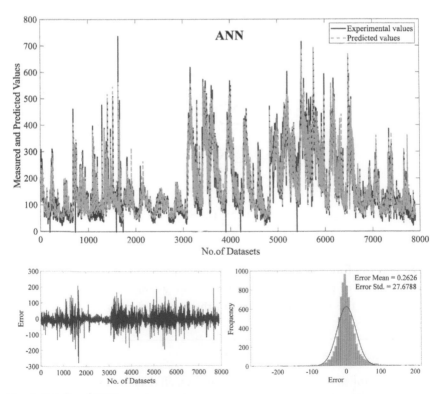

Fig. 6 Results of ANN.

The R-value of the PSO-ANN model for training and testing data is 0.9725 and 0.9637, sequentially as shown in Table 4. The other values of performance indices are shown in Table 5. Fig. 7 represents the plot of

Table 4 Comparison of training and testing results with ANN and PSO-ANN.

ML models	R			R^2		
	Training	Testing	All	Training	Testing	All
ANN	0.9697	0.9653	0.9612	0.9403	0.9318	0.9239
PSO-ANN	0.9725	0.9637	0.9722	0.9458	0.9287	0.9452

Table 5 Statistical analysis results of ANN and PSO-ANN.

Method	Statistical parameters for AQI					
	R	R^2	MSE	RMSE	MAE	Std.
ANN	0.9612	0.9239	834.2708	28.83	19.36	111.30
PSO-ANN	0.9722	0.9452	795.15	28.19	19.67	112.85

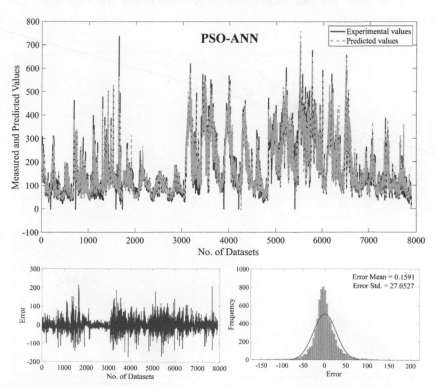

Fig. 7 Results of PSO-ANN.

experimental and predicted values with respect to the number of sample datasets, error frequency distribution, and error distribution with respect to the number of sample datasets for PSO-ANN model. The R-value of training PSO-ANN model data is 0.2887% higher than the conventional ANN model. For testing data, R-values are very closer to each other. In the whole dataset, the R-values of ANN as well as PSO-ANN models are 0.9612 and 0.9722, sequentially, which is 1.14% higher than the ANN model.

Fig. 8 presents the graph between measured and estimated values of training, and testing data for both ANN and PSO-ANN models. As a result, based on the AQI values of the data-sets included in this work, the suggested model has the potential to be used effectively to prediction-related problems. The hybrid PSO-ANN model may be regarded as a model capable of estimating AQI values to a higher degree of accuracy. Overall, the study's findings are interesting for environmental and environmental health engineers since they can measure the AQI using AI without spending a lot of money or time (in situ studies for AQI estimations are expensive and time-consuming).

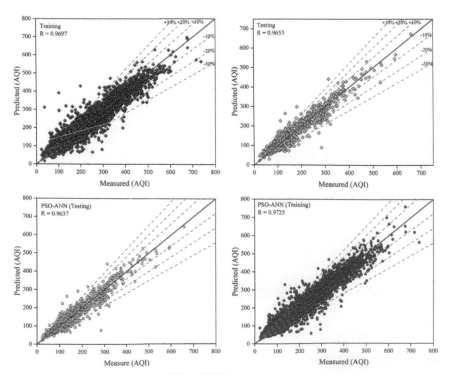

Fig. 8 Prediction of AQI with ANN and PSO-ANN.

6. Conclusion

Statistical methods and soft computing are the two main types of prediction techniques. The aim of this study is to develop an ML-based model (PSO and ANN) to predict the quality of air. The approaches described in this paper are realistic, rapid, and capable of assessing pollutants' combined effect in city regions in order to minimize cardiovascular and respiratory illness as well as deaths. As presented in Fig. 5, more than 50% of the air quality of Indian cities lies in the category range of satisfactory to severe. In this present study, the influence of particulate matter and pollutant gases on the AQI was investigated using ANN and PSO-ANN. These air quality model predictions for high-dimensional data evaluation performed exceptionally well, according to the findings. Among these models, PSO-ANN was quite accurate with a good correlation coefficient (R^2) of 0.9452. This proposed forecasting equation is only applicable for the AQI data that lie in the range of 22–737, a limitation of this study. According to the findings of this study, PSO-ANN has a high accuracy, applicability, and practicality in the prediction of air quality, and it can be successfully used in relevant projects with similar circumstances and input parameter ranges. In future work, some of the other nature-inspired optimized algorithms such as plant propagation and Social Spider Algorithm, artificial bee colony, etc. can be used to reach the R^2 value equal to one.

References

[1] Inheriting a sustainable world: Atlas on children's health and the environment. https://www.who.int/publications/i/item/9789241511773, (accessed 09.04.21).

[2] X.-Q. Jiang, X.-D. Mei, D. Feng, Air pollution and chronic airway diseases: what should people know and do? J. Thorac. Dis. 8 (1) (2016) E31.

[3] N.R. Kapoor, A. Kumar, C.S. Meena, A. Kumar, T. Alam, N.B. Balam, A. Ghosh, A systematic review on indoor environmental quality in naturally ventilated school classrooms: a way forward, Adv. Civ. Eng. 2021 (2021) 1–19. 8851685.

[4] N. Raj, A. Kumar, A. Kumar, S. Goyal, Indoor environmental quality: impact on productivity, comfort, and health of Indian occupants, in: Proceedings of the Abstract Proceedings of International Conference on Building Energy Demand Reduction in Global South (BUILDER'19), New Delhi, India, 2019, pp. 13–14.

[5] N.R. Kapoor, A. Kumar, A. Kumar, A. Kumar, M.A. Mohammed, K. Kumar, S. Kadry, S. Lim, Machine learning-based CO2 prediction for office room: a pilot study, Wirel. Commun. Mob. Comput. 2022 (2022) 1–16. 9404807.

[6] N.R. Kapoor, A. Kumar, T. Alam, A. Kumar, K.S. Kulkarni, P. Blecich, A review on indoor environment quality of Indian school classrooms, Sustainability 13 (21) (2021) 11855.

[7] N.R. Kapoor, J.P. Tegar, Human comfort indicators pertaining to indoor environmental quality parameters of residential buildings in Bhopal, Int. Res. J. Eng. Technol 5 (2018). 2395–0056.

[8] A. Hunt, J. Ferguson, F. Hurley, A. Searl, Social Costs of Morbidity Impacts of Air Pollution, 2016.

[9] F.M. Bublitz, A. Oetomo, K.S. Sahu, A. Kuang, L.X. Fadrique, P.E. Velmovitsky, R.M. Nobrega, P.P. Morita, Disruptive technologies for environment and health research: an overview of artificial intelligence, blockchain, and internet of things, Int. J. Environ. Res. Public Health 16 (20) (2019) 3847.

[10] L. Bai, J. Wang, X. Ma, H. Lu, Air pollution forecasts: an overview, Int. J. Environ. Res. Public Health 15 (4) (2018) 780.

[11] P. Wang, Y. Liu, Z. Qin, G. Zhang, A novel hybrid forecasting model for $PM_{1}0$ and SO_2 daily concentrations, Sci. Total Environ. 505 (2015) 1202–1212.

[12] H. Zhang, P. Wu, A. Yin, X. Yang, M. Zhang, C. Gao, Prediction of soil organic carbon in an intensively managed reclamation zone of eastern China: a comparison of multiple linear regressions and the random forest model, Sci. Total Environ. 592 (2017) 704–713.

[13] D. Wang, S. Wei, H. Luo, C. Yue, O. Grunder, A novel hybrid model for air quality index forecasting based on two-phase decomposition technique and modified extreme learning machine, Sci. Total Environ. 580 (2017) 719–733.

[14] D.M. Chambers, C.M. Reese, L.G. Thornburg, E. Sanchez, J.P. Rafson, B.C. Blount, J.R.E. Ruhl III, V.R. De Jesus, Distinguishing petroleum (crude oil and fuel) from smoke exposure within populations based on the relative blood levels of benzene, toluene, ethylbenzene, and xylenes (BTEX), styrene and 2, 5-dimethylfuran by pattern recognition using artificial neural networks, Environ. Sci. Technol. 52 (1) (2018) 308–316.

[15] A. Kumar, H.C. Arora, M.A. Mohammed, K. Kumar, J. Nedoma, An optimized neuro-bee algorithm approach to predict the FRP-concrete bond strength of rc Beams, IEEE Access 10 (2021) 3790–3806.

[16] A. Kumar, H.C. Arora, N.R. Kapoor, M.A. Mohammed, K. Kumar, A. Majumdar, O. Thinnukool, Compressive strength prediction of lightweight concrete: machine learning models, Sustainability 14 (4) (2022) 2404.

[17] A. Kumar, J.S. Rattan, N.R. Kapoor, A. Kumar, R. Kumar, Structural health monitoring of existing reinforced cement concrete buildings and bridge using nondestructive evaluation with repair methodology, in: Advances and Technologies in Building Construction and Structural Analysis, BoD-Books on Demand, 2021, p. 87.

[18] A. Kumar, N. Mor, Prediction of accuracy of high-strength concrete using data mining technique: a review, in: Proceedings of International Conference on IoT Inclusive Life (ICIIL 2019), NITTTR Chandigarh, India, Springer, 2020, pp. 259–267.

[19] A. Kumar, J.S. Rattan, A journey from conventional cities to smart cities, in: Smart Cities and Construction Technologies, IntechOpen, 2020.

[20] X.X. Bai, J. Dong, X.G. Rui, H.F. Wang, W.J. Yin, Very short-term air pollution forecasting, 2019. US Patent 10,438,125, (October 8).

[21] A. Aggarwal, D. Toshniwal, Detection of anomalous nitrogen dioxide (NO_2) concentration in urban air of India using proximity and clustering methods, J. Air Waste Manage. Assoc. 69 (7) (2019) 805–822.

[22] C.J. Torney, D.J. Lloyd-Jones, M. Chevallier, D.C. Moyer, H.T. Maliti, M. Mwita, E.M. Kohi, G.C. Hopcraft, A comparison of deep learning and citizen science techniques for counting wildlife in aerial survey images, Methods Ecol. Evol. 10 (6) (2019) 779–787.

[23] A.J. Fairbrass, M. Firman, C. Williams, G.J. Brostow, H. Titheridge, K.E. Jones, CityNet—deep learning tools for urban ecoacoustic assessment, Methods Ecol. Evol. 10 (2) (2019) 186–197.

[24] S. Christin, É. Hervet, N. Lecomte, Applications for deep learning in ecology, Methods Ecol. Evol. 10 (10) (2019) 1632–1644.

[25] M.M. Rahman, M. Shafiullah, S.M. Rahman, A.N. Khondaker, A. Amao, M. Zahir, et al., Soft computing applications in air quality modeling: past, present, and future, Sustainability 12 (10) (2020) 4045.

[26] A. Sayeed, Y. Choi, E. Eslami, Y. Lops, A. Roy, J. Jung, Using a deep convolutional neural network to predict 2017 ozone concentrations, 24 hours in advance, Neural Netw. 121 (2020) 396–408.

[27] A. Alimissis, K. Philippopoulos, C.G. Tzanis, D. Deligiorgi, Spatial estimation of urban air pollution with the use of artificial neural network models, Atmos. Environ. 191 (2018) 205–213.

[28] N. Agarwal, C.S. Meena, B.P. Raj, L. Saini, A. Kumar, N. Gopalakrishnan, A. Kumar, N.B. Balam, T. Alam, N.R. Kapoor, et al., Indoor air quality improvement in COVID-19 pandemic, Sustain. Cities Soc. 70 (2021) 102942.

[29] M. El Raey, Air Quality and Atmospheric Pollution in the Arab Region Economic and Social League of Arab States, Report by Commission for Western Asia Joint Technical Secretariat of the Council of Arab Ministers Responsible for the Environment, 2007.

[30] S.M. Cabaneros, J.K. Calautit, B.R. Hughes, A review of artificial neural network models for ambient air pollution prediction, Environ. Model. Softw. 119 (2019) 285–304.

[31] A. Kumar, N.R. Kapoor, H.C. Arora, A. Kumar, Smart cities: a step toward sustainable development, in: Smart Cities, CRC Press, 2022, pp. 1–43.

[32] W.S. McCulloch, W. Pitts, A logical calculus of the ideas immanent in nervous activity, Bull. Math. Biol. 52 (1) (1990) 99–115.

[33] S.C. Kleene, et al., Representation of events in nerve nets and finite automata, Autom. Stud. 34 (1956) 3–41.

[34] D.O. Hebb, The Organization of Behavior: A Neuropsychological Theory, Psychology Press, 2005.

[35] B.W.A.C. Farley, W. Clark, Simulation of self-organizing systems by digital computer, Trans. IRE Prof. Group Inf. Theory 4 (4) (1954) 76–84.

[36] F. Rosenblatt, The perceptron: a probabilistic model for information storage and organization in the brain, Psychol. Rev. 65 (6) (1958) 386.

[37] M. Olazaran, A sociological study of the official history of the perceptrons controversy, Soc. Stud. Sci. 26 (3) (1996) 611–659.

[38] J. Schmidhuber, Deep learning in neural networks: an overview, Neural Netw. 61 (2015) 85–117.

[39] J. Schmidhuber, Deep learning, Scholarpedia 10 (11) (2015) 32832, https://doi.org/10.4249/scholarpedia.32832. (Revision #184887.

[40] S.E. Dreyfus, Artificial neural networks, back propagation, and the Kelley-Bryson gradient procedure, J. Guid. Control Dynam. 13 (5) (1990) 926–928.

[41] E. Mizutani, S.E. Dreyfus, K. Nishio, On derivation of MLP backpropagation from the Kelley-Bryson optimal-control gradient formula and its application, in: Proceedings of the IEEE-INNS-ENNS International Joint Conference on Neural Networks. IJCNN 2000. Neural Computing: New Challenges and Perspectives for the New Millennium, vol. 2, IEEE, 2000, pp. 167–172.

[42] H.J. Kelley, Gradient theory of optimal flight paths, Ars J. 30 (10) (1960) 947–954.

[43] S. Dreyfus, The computational solution of optimal control problems with time lag, IEEE Trans. Autom. Control 18 (4) (1973) 383–385.

[44] A.V. Balakrishna, M. Thoma, Lecture Notes in Control and Information Sciences, Springer-Verlag, 1978.

[45] Y.-S. Park, S. Lek, Artificial neural networks: multilayer perceptron for ecological modeling, in: Developments in Environmental Modelling, vol. 28, Elsevier, 2016, pp. 123–140.

[46] H.K. Ghritlahre, R.K. Prasad, Exergetic performance prediction of solar air heater using MLP, GRNN and RBF models of artificial neural network technique, J. Environ. Manage. 223 (2018) 566–575.

[47] H. Oğuz, I. Sarıtas, H.E. Baydan, Prediction of diesel engine performance using biofuels with artificial neural network, Expert Syst. Appl. 37 (9) (2010) 6579–6586.

[48] B.M. Wilamowski, H. Yu, Improved computation for Levenberg-Marquardt training, IEEE Trans. Neural Netw. 21 (6) (2010) 930–937.

[49] B. Khoshnevisan, S. Rafiee, M. Omid, H. Mousazadeh, M.A. Rajaeifar, Application of artificial neural networks for prediction of output energy and GHG emissions in potato production in Iran, Agr. Syst. 123 (2014) 120–127.

[50] Z. Zhao, T.L. Chow, H.W. Rees, Q. Yang, Z. Xing, F.-R. Meng, Predict soil texture distributions using an artificial neural network model, Comput. Electron. Agric. 65 (1) (2009) 36–48.

[51] M.K.D. Kiani, B. Ghobadian, T. Tavakoli, A.M. Nikbakht, G. Najafi, Application of artificial neural networks for the prediction of performance and exhaust emissions in SI engine using ethanol-gasoline blends, Energy 35 (1) (2010) 65–69.

[52] D. Wang, D. Tan, L. Liu, Particle swarm optimization algorithm: an overview, Soft Comput. 22 (2) (2018) 387–408.

[53] S. Cheng, H. Lu, X. Lei, Y. Shi, A quarter century of particle swarm optimization, Complex Intell. Syst. 4 (3) (2018) 227–239.

[54] A. Biswas, B. Biswas, A. Kumar, K.K. Mishra, Particle swarm optimisation with time varying cognitive avoidance component, Int. J. Comput. Sci. Eng. 16 (1) (2018) 27–41.

[55] A. Biswas, B. Biswas, Swarm intelligence techniques and their adaptive nature with applications, in: Complex System Modelling and Control Through Intelligent Soft Computations, Springer, 2015, pp. 253–273.

[56] A. Biswas, A. Kumar, K.K. Mishra, Particle swarm optimization with cognitive avoidance component, in: 2013 International Conference on Advances in Computing, Communications and Informatics (ICACCI), IEEE, 2013, pp. 149–154.

[57] A. Biswas, A.V. Lakra, S. Kumar, A. Singh, An improved random inertia weighted particle swarm optimization, in: 2013 International Symposium on Computational and Business Intelligence, IEEE, 2013, pp. 96–99.

[58] D. Sarkar, N. Mishra, A. Biswas, Genetic algorithm-based deep learning models: a design perspective, in: Proceedings of the Seventh International Conference on Mathematics and Computing, Springer, 2022, pp. 361–372.

[59] A. Biswas, K.K. Mishra, S. Tiwari, A.K. Misra, Physics-inspired optimization algorithms: a survey, J. Optim. 2013 (2013) 1–16. 438152.

[60] A. Biswas, B. Biswas, K.K. Mishra, An atomic model based optimization algorithm, in: 2016 2nd International Conference on Computational Intelligence and Networks (CINE), IEEE, 2016, pp. 63–68.

About the authors

Dr. Nishant Raj Kapoor hails from Kota, Rajasthan, India. Dr. Kapoor is an accomplished alumnus, having earned degrees from AcSIR (CSIR-CBRI, Roorkee), NITTTR in Bhopal, and RTU in Kota, India. His diverse research interests encompass real-time challenges in the built environment, with a focus on areas such as COVID-19, indoor human comfort, indoor environmental quality, comfort perceptions, building energy efficiency, artificial intelligence, and environmental engineering. With a prolific scholarly record, Dr. Kapoor has authored over 45 research papers, review articles, patents, conference articles, and book chapters. These contributions have been published in esteemed peer-reviewed international journals and books, reflecting his commitment to advancing scientific knowledge. Dr. Kapoor played a vital role in formulating ventilation-related guidelines to curb the spread of SARS-CoV-2 in Indian office and residential buildings, showcasing his practical impact on public health. Additionally, he actively engages with the global scientific community by serving as an editor for international scientific book projects and contributing as a reviewer for SCI/Scopus indexed international journals. Recognizing his outstanding contributions, Dr. Kapoor was honored with the Diamond Jubilee Best Technology Award in 2021, a testament to his excellence in the field. Further attesting to his prowess, he received the Best Paper Award in September 2022, underscoring the quality and impact of his research.

Dr. Ashok Kumar is a distinguished professional with a rich and diverse experience of more than 33 years in research, academia and industry. He superannuated as Outstanding Scientist and Professor from the prestigious building research lab of India (CSIR-CBRI, Roorkee). His educational journey includes earning a B.Sc., B. Arch. (Gold Medal), M.U.R.P. (Hons.), and a Ph.D. from the Indian Institute of Technology Roorkee (IITR). He has made significant contributions to research and academia, showcasing expertise in green and energy-efficient buildings, green retrofits, and affordable housing. Dr. Kumar has a proven track record of handling national and international collaborative R&D projects in areas such as building energy efficiency, school and healthcare buildings, and green affordable housing. Having led and participated in over 95 international and national projects, Dr. Kumar has left an indelible mark on the field. Notable projects include the design of the Medical College Complex in Haldwani, Uttarakhand, the concept design for prefab hospitals during the Covid-19 pandemic, and the development of guidelines on ventilation for residential and office buildings in the context of the SARS-CoV-2 virus. Dr. Kumar and his team received the Director's Diamond Jubilee Best Technology Award in 2021 for their work on HVAC Ducting System for Integration of Covid-19 Disinfection Solutions. He has also been honored with the National Design Award for Architectural Engineering in 2019 and various Best Paper Awards for his contributions to the field. He serves on the Governing Council of the Bureau of Energy Efficiency, Ministry of Power, representing architects of India. He is also involved in expert committees related to energy conservation, awards, and sustainable habitat. With over 120 publications in international and national journals, conferences, and book chapters, as well as holding two patents and three copyrights, Dr. Kumar's impact on the field is both broad and profound. His commitment to education is evident through his role as a visiting professor at the National University of Singapore. Dr. Kumar's commitment to research, education, and innovation sets a high standard in the industry.

Dr. Anuj Kumar assumed the role of Director at Haldaur Group of Colleges, Bijnor, UP, India, in September 2023. Dr. Kumar's educational background includes a B.Sc. and M.Sc. in Physics (Electronics), an M. Phil. in Instrumentation from the Indian Institute of Technology Roorkee, an M. Tech. in Instrumentation from the National Institute of Technology Kurukshetra, and a Ph.D. in Electrical Engineering from the Indian Institute of Technology Delhi. Dr. Kumar's academic journey is marked by diverse roles, starting as a Research Associate at the Instrument Design and Development Center (IDDC), Indian Institute of Technology Delhi. Subsequently, he served as an Assistant Professor at Amity University Noida and held a visiting Post-doctoral Fellowship at the

University of Seoul, Seoul, Korea. He further contributed as a Vice-Chancellor Postdoctoral Fellow at the University of Pretoria's Advanced Sensor Network (ASN) group and as a Research Fellow at the Electrical Machines and Drives Laboratory, National University of Singapore. His roles also extended to being a Ramanujan Fellow at CSIR-Central Building Research Institute Roorkee and a Consultant specializing in building efficiency at CSIR-CBRI Roorkee. Additionally, he served as an Assistant Professor in AcSIR, India. Dr. Kumar is a prolific author, with contributions including a book, three book chapters, and over 80 papers in international refereed journals and conferences. Over the last three years, he has secured two copyrights and filed three patents, showcasing his breadth of research in intelligent systems, IoT, smart buildings, and health monitoring. His editorial role as an Associate Editor of IEEE Access Journal and reviewer for prominent journals such as IEEE Transactions, Elsevier, and Springer underscores his recognition in the field. As a Senior Member of IEEE Societies of IES, Sensors, and Instrumentation & Measurement, he received the Ramanujan Fellow award in Engineering Sciences in 2015 and the IEEE Sensors Council President's most downloaded Research Paper award in 2014, 2015 & 2016. Acknowledged in Marquis "Who's Who in the World" in 2013 and "Who's Who in America" in 2015, Dr. Kumar is actively leading regional and international research projects, notably spearheading initiatives in smart buildings and multi-agent systems with a total funding of approximately 250 Lakhs INR. In addition to his academic endeavors, Dr. Kumar plays a crucial role as a consultant for both industrial and institutional sectors, providing valuable insights and guidance to businesses, as well as assisting in the establishment of new institutes or universities.

Er. Aman Kumar hails from Bilaspur, Himachal Pradesh, India. He holds a Master of Engineering degree in construction technology and management from the prestigious National Institute of Technical Teacher's Training and Research Institute in Chandigarh, India. Currently, he is fervently pursuing a Ph.D. in engineering sciences, specializing in structural engineering, at the renowned CSIR-Central Building Research Institute in Roorkee, India. His academic journey is underscored by a deep passion for various facets of civil engineering, including sustainability development, non-destructive testing, concrete technology, and strengthening techniques such as fiber reinforced polymer and fiber reinforced cementitious matrix. He is also deeply engaged in exploring corrosion protection techniques for structural design, as well as the cutting-edge domains of artificial intelligence and the Internet of Things. Now his focus to solve the complex structural engineering problems with machine learning algorithms. Aman Kumar's dedication to the field is evident through his extensive research endeavours and comprehensive technical surveys. His scholarly pursuits have culminated in numerous research papers and book chapters, which have been featured in esteemed international scientific publications.

Dr. Harish Chandra Arora currently holds the esteemed position of Principal Scientist in Structural Engineering Group at CSIR-Central Building Research Institute in Roorkee, India. With a distinguished career spanning more than 29 years, Dr. Arora is a renowned figure in the field of structural engineering. Dr. Arora is also functioning as an Associate Professor in Academy of Scientific and Innovative Research (AcSIR), Ghaziabad, India. His contemporary research areas include structural composites, structural corrosion, distress diagnosis, seismic evaluation and repair & retrofitting of structures and machine learning applications in structural engineering, etc. Dr. Arora's exceptional contributions to the field have garnered recognition in both national and international academic journals. Beyond his scholarly achievements, he has made a significant impact on the education and development of future engineers, having supervised and guided over 100^+ students in their pursuit of Bachelor of Technology and Masters of Technology degrees. Additionally, he continues to mentor and support research scholars pursuing doctoral programs at the Central Building Research Institute in Roorkee, India. Furthermore, Dr. Arora actively contributes to the scholarly community as a reviewer for journals published by Springer Nature and Elsevier. His commitment to maintain the quality and rigor of academic publications is highly regarded. Beyond his academic pursuits, Dr. Arora has undertaken numerous consultancy and Research & Development projects within the field of structural engineering, further showing his dedication to advancing the science and practice of sustainable construction.

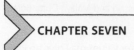

CHAPTER SEVEN

Genetic algorithm for the optimization of infectiological parameter values under different nutritional status

Zakir Hussain and Malaya Dutta Borah
Department of Computer Science and Engineering, National Institute of Technology Silchar, Silchar, Assam, India

Contents

Abstract

Each infectious disease has three infectiological parameters called susceptibility, infection, and recovery. These parameters are affected by a lot of environmental and biological/clinical factors associated with the pathogen. These parameters may determine the severity of disease, and also these parameters react differently under different levels of

Advances in Computers, Volume 135
ISSN 0065-2458
https://doi.org/10.1016/bs.adcom.2023.12.001

the nutritional status. Optimization of infectiological parameter values may give an idea about the possible precautionary measures to be taken for mitigation of the losses in adverse situations. Genetic algorithm has potential to optimize the values within a predefined range. In this chapter, we present the working process of our formulated mathematical equations for susceptibility, infection, and recovery under normal nutritional status and malnutrition. We integrate nutritional status levels with the genetic algorithm and use it to solve the formulated equations. We perform the experiments using two sets of values. Our experimental results depict that susceptibility is best optimized under normal nutritional status considering heuristic crossover for both the sets of values, and infection is best optimized under normal nutritional status using heuristic crossover for the first set of values and under normal nutritional status using arithmetic crossover for the second set of values, while recovery is best optimized under normal nutritional status considering heuristic crossover for both the sets of values. Malnutrition is found to be negatively influencing the optimization of infectiological parameters.

1. Introduction

A disease is a condition that disturbs normal functioning of cells, tissues, and organs. Disease is broadly classified into infectious disease and noninfectious disease. Infectious disease, often called as communicable disease, spreads person to person by transfer of pathogens like bacteria, virus, fungi, parasite, etc. [1]. Noninfectious disease, often called as non-communicable disease, does not spread person to person and it restrains itself within a person. Noninfectious diseases are also caused by pathogens, but these diseases are mostly influenced by genetics, age, nutritional status, gender, environment, lifestyle, etc. The pathogens that cause diseases are very small in size or simply microscopic in size and are often referred to as microbes or microorganisms. We are always surrounded by microbes living in air, water, and soil. Though microbes can be an agent of infection, most of the microbes do not cause any disease in human. In fact, a collection of microbes called microbiome inhabit in our body and are very important and beneficial. Microbes cause diseases when they enters into a part of the body where they should not. Based on the microorganisms, the infections can be classified as follows:

- Viral infections: Virus are often called dead unless they are in contact with living cells. The living cells where they find feasible environment to reproduce and spread are called as host [2]. Since virus need living cells for survival, they do not do well on surfaces outside the living cells and can survive only for few hours. Virus enter into our body through natural or unnatural openings and also spread through exchange of body fluids. There are not much cures for the diseases caused by the virus. Antibiotics do not fight virus. Some antiviral medications are there but those do not fully cure the disease, rather it reduces the symptoms and shortens the time of being sick. In serious viral infections, vaccines are used to prevent.

- Bacterial infections: These are single-cell microorganisms that can reproduce and mutate themselves. These organisms can survive on the surfaces for days or weak or even more [3]. They enter into our body through the natural or unnatural openings. Antibiotics are the main treatment for bacterial infections. Antibiotics either kill the bacteria or stop them from reproducing. Not all bacterial infections need an antibiotic to get cured. In some cases our body can fight them off.

- Fungal infections: Fungal infections are not that much serious but annoying. These infections mostly spread by air or touch [4]. It is suggested to maintain good hygiene and keep the skin neat, clean, and dry specially in the folds. Also, it is suggested to wear a mask when working in an area having fungal spores such as chicken coops. Antifungal drugs are used to manage fungal infections. Also, over-the-counter medications are used to manage the symptoms.

- Parasitic infections: These infections are spread by parasites, often by bugs or smaller microorganisms. Parasites spread through contact with contaminated soil, water, food, waste, blood, insects, etc. [5]. It is suggested to wear globes when working in an area with animal feces. Also, it is suggested not to drink water from areas without having adequate processing facilities. Often antiparasitic drugs are used as treatment of parasitic diseases. Also, over-the-counter medications are used to manage the symptoms.

There are a number of ways by which a person can get infected with an infectious agent. The medium or route of transmission of infectious agents is a very important factor that determines how quickly an infectious agent spreads through a population. In some cases, humans have to come into direct contact with the agent or source of the infection like contaminated food, water, dirty material, body fluids, or animal products. An agent that can transmit through air has greater potential to infect a large number of individuals than that of spreading through direct contact. Another important factor in transmission of infectious agent is the survival time of the agent in the environment. An agent that is capable of surviving a few seconds cannot infect as many people as an agent that can survive for hours, days, or even longer.

All the infectious diseases have some similar kind of attributes. Most of the infectious diseases spread too fast. The spread of any infectious disease is determined by the basic reproduction number of that pathogen [6]. The other factors that are used to predict different scenarios or estimate possible outcomes of the infectious diseases are as follows:

- Susceptibility: Susceptibility refers to a state of being predisposed to or sensitive to developing a certain disease [7]. That means this set of population has a chance of being infected. In general, this is the set of population excluding the population who are vaccinated and immune to that particular disease. From a susceptible stage a person can move to an infected stage.
- Infection: Infection refers to a state of suffering from the disease. That means the individual came into contact with the pathogen and has developed the disease already [8]. From this stage, an individual can die or recover. There is even a possibility of being in susceptible stage, but the possibility is less as most of the recovered cases develop immunity against that pathogen.
- Recovery: Recovery refers to a stage of being cured of the infection. That means the set of population who could win the fight against the pathogen [9]. This set comes from the set of infection. The set of recovered individuals can eventually be in the susceptible stage but may not be immediately after getting cured.

All these infectiological parameters are greatly influenced by the nutritional status of our body. Nutritional status is the physical state of our body. Nutritional status is measured by macronutrients and micronutrients present in our body [10, 11]. Low amount of nutrients and high amount of nutrients develop malnutrition [12, 13]. Both undernutrition and overnutrition are together called malnutrition [14]. Nutritional status is related to all types of diseases, be it infectious or noninfectious. Out of the three nutritional status levels, undernutritional status is considered to be the worst form in the context of infectious disease. The overnutritional status is considered to be the mediocre form, while the normal nutritional status is considered to be the best form. In the context of infectious disease, our main expectation remains on maximizing the susceptibility, minimizing the infection, and maximizing the recovery.

Genetic algorithms are very popular for performing optimization [15]. Genetic algorithms are capable of working with both maximization and minimization problems. These algorithms have an interesting way of solving problems. These algorithms solve the problems from a set of solutions [16].

Out of the set of solutions, these algorithms find the best solution. The solution space that is considered is generally called as population [17]. From the population, a parent population is selected using fitness function. The fitness function varies based on the type of problems (maximization or minimization). The parent population then undergoes genetic operations like crossover and mutation. These genetic operations find the best child population for the next generation. This process continues till the termination condition is satisfied [18]. In the case of infectious diseases, we do not have any exact value for the infectiological parameters. Unavailability of the numbers related to infectiological parameters hampers our preparation for the situation that is supposed to be created by infectious diseases. So, we can assume a minimum limit and a maximum limit of these parameters. Then the values of these parameters can be optimized using genetic algorithm for better decision-making and preparing for the forthcoming situations. The case of susceptibility is a maximization problem, infection is a minimization problem, while recovery is a maximization problem. So, genetic algorithm is a good fit for finding optimal values of infectiological parameters of any infectious disease.

1.1 Motivation

The current pandemic and our unpreparedness to tackle the unprecedented situation taught us a lot. Due to that, we are bearing huge loss in every aspect of our life. We as well as the concerned authorities tried to handle the situation, but the disease spreads too fast that our strategy to handle the situation became a failure. We could not estimate any figure for susceptibility, infection, and recovery. We did not have any clue about the limit of the spread. We realized that if we could have a hint about the numbers of susceptibility, infection, and recovery, we could have prepared better and tackle the situation with diligence. So, one of the solutions is to consider a lower limit and an upper limit of susceptibility, infection and recovery, and optimize the values using evolutionary algorithms that are well known for optimization purpose. Hence, we thought of working on finding optimal values for infectiological parameters with a predefined lower limit and upper limit.

1.2 Related studies

There are many works that analyze different aspects of infectious diseases. The disease dynamics of infectious disease are basically analyzed from clinical perspectives or computational perspectives. In computational perspectives, disease dynamics are analyzed using various soft computing techniques and mathematical models. We perform the literature survey to find the applications of a very popular optimization algorithm called genetic algorithm in the optimization of infectiological parameters or close enough studies. We find that a recent study by Qiu et al. [19] combines genetic algorithm with an improved susceptible-exposed-infectious-removed (SEIR) model for predicting epidemic trends in China. This study uses genetic algorithm for optimizing the parameters to be used for predicting. This study claims that the improved SEIR model is capable of predicting the epidemic situation better and obtaining more accurate parameters related to epidemic. Another study by Rouabah et al. [20] proposes an optimized dynamical epidemic model using genetic algorithm and cross-validation method to overcome over-fitting problem. This study reveals an inverse relationship between the size of training sample and required number of generations in genetic algorithm. The presented enhanced compartmental model has proven to be a reliable tool to estimate important epidemic parameters. One more study by Bansal et al. [21] proposes a new semisupervised feature learning technique. The three-step architecture of the proposed technique consists of a feature extractor based on the convolutional autoencoder, a feature selector based on multiobjective genetic algorithm, and a binary classifier based on bagging ensemble of support vector machine. One more study by Yarsky [22] uses genetic algorithm for fitting the parameters of SEIR model in the United States. Genetic algorithm was used to adjust population-dependent parameters to fit the SEIR model to the data of different states. This study claims that genetic algorithm produces very good agreement between available data and the model.

1.3 Scope for contribution

From the literature survey, we find that there are many studies that design mathematical models for infectiological parameters like susceptibility, infection, and recovery, but the work on optimizing the values of these parameters is limited. So, the main aim of this study is to optimize the infectiological parameters called susceptibility, infection, and recovery. To achieve this aim we incorporate the following:

- Formulate differential equations for susceptibility, infection, and recovery integrating three levels of the nutritional status like undernutritional status, normal nutritional status, and overnutritional status.

- Solve the formulated differential equations using genetic algorithm under predefined lower limits and upper limits of susceptible individuals, infected individuals, and recovered individuals for finding optimal values.
- Find how do different levels of nutritional status impact the optimization of susceptibility, infection, and recovery.

2. Methods and materials
2.1 Preliminary concepts

Almost every infectious disease dynamics is analyzed using three infectiological parameters called susceptibility, infection, and recovery. The nutritional status of our body is broadly classified into normal nutritional status and malnutrition [12, 23]. As per the clinical facts, malnutrition impacts our body in a negative way. That means, under the condition of malnutrition the susceptibility gets reduced, infection becomes more, and recovery becomes less. The susceptibility, infection, and recovery get impacted by other factors like pathogen type, feasibility of spread, basic reproduction number, environmental factors, etc. Considering the facts, affecting factors, and making some assumptions, we formulate the differential equations as shown in subsequent sections.

2.2 Formulation of differential equations

During the formulation of differential equations, we refer to a model called NICOV available in a recent study by Hussain et al. [12]. Considering the assumptions made in this model, we make some assumptions about the infectious disease as follows:

- The population of a particular area at time t can be classified into:
  1. Susceptible individuals with normal nutritional status $SNN(t)$.
  2. Susceptible individuals with malnutrition $SMN(t)$.
  3. Infected individuals with normal nutritional status $INN(t)$.
  4. Infected individuals with malnutrition $IMN(t)$.
  5. Recovered individuals with normal nutritional status $RNN(t)$.
  6. Recovered individuals with malnutrition $RMN(t)$.
- The pathogen population at time t is $P(t)$.
- Rate of recruitment of population is Λ.
- Population with normal nutritional status may move to the category of malnutrition at a rate of M_r.

- Population with malnutrition may move to the category of population with normal nutritional status at a rate of N_r.
- The infection rate of the pathogen is $\lambda = \frac{eP}{C+P}$, where e is the exposure and C is the concentration of pathogen P in a medium.
- Susceptible individuals SNN and SMN acquire the infection at the rates λ and $\alpha\lambda$ and turns into INN and IMN, respectively, where $\alpha > 1$.
- INN and IMN recover at the rates r_n and r_m, respectively.
- INN and IMN die at the rates η and $\beta\eta$, respectively, where $\beta > 1$.
- Natural death rate of individual d.
- Natural death rate of pathogen μ.
- The pathogen grows itself at a rate of g.
- Pathogen grows in INN and IMN at the rates $\sigma1$ and $\sigma2$, where $\sigma2 > \sigma1$. By this, each infected individual contributes to the pathogen population in a medium.

Combining the above facts and assumptions, we formulate the following differential equations:

The differential equation for susceptibility under the condition of normal nutritional status is:

$$SNN'(t) = \Lambda \cdot SNN + N_r \cdot SMN - \frac{e \cdot (P - \mu) \cdot g}{C + (P - \mu) \cdot g} \cdot SNN$$
$$- d \cdot SNN \tag{1}$$

and the differential equation for susceptibility under the condition of malnutrition is:

$$SMN'(t) = \Lambda \cdot SMN + M_r \cdot SNN - \alpha \cdot \frac{e \cdot (P - \mu) \cdot g}{C + (P - \mu) \cdot g} \cdot SMN$$
$$- d \cdot SMN \tag{2}$$

For the case of infection, the differential equation under normal nutritional status is:

$$INN'(t) = \frac{e \cdot (P + \sigma1 - \mu) \cdot g}{C + (P + \sigma1 - \mu) \cdot g} \cdot SNN - (r_n + \eta + d) \cdot INN \tag{3}$$

and the differential equation under malnutrition is:

$$IMN'(t) = \alpha \cdot \frac{e \cdot (P + \sigma2 - \mu) \cdot g}{C + (P + \sigma2 - \mu) \cdot g} \cdot SMN - (r_m + \beta\eta + d) \cdot IMN \tag{4}$$

For recovery, the differential equation under the situation of normal nutritional status is:

$$RNN'(t) = r_n \cdot INN - d \cdot RNN \tag{5}$$

and the differential equation under the situation of malnutrition is:

$$RMN'(t) = r_m \cdot IMN - d \cdot RMN \tag{6}$$

2.3 Genetic algorithm with infectiological parameters

Like all other evolutionary algorithms, genetic algorithm also passes through some stages. The different stages and the incorporated modifications are shown below:

- Population initialization: We need to randomly initialize the population strictly following the boundary set of a variable. The size of the population is equal to the number of variables in the objective function. All the variables are independent if the constraints are only the inequality constraints. However, one variable is considered as dependent on the rest of the variables if there is an equality constraint.
- Fitness calculation: The fitness of the parent population with integration of nutritional status are shown in Algorithm 1.

  The relative fitness and expected count is calculated using:

$$Fit_{Relative} = \frac{f_r}{\sum f_r} \tag{7}$$

$$Expected\_count = \frac{Fit_{Relative}}{AVG(Fit_{Relative})} \tag{8}$$

where f_r is the fitness of rth population and $r = 1, 2, ..., Pop\_size$. $Expected\_count$ determines the actual number of parents of a type that will move to the selected population for genetic operations.

- Population selection: In most of the cases, we use a Roulette wheel selection method for selection of the parent population based on the fitness and the associated $Expected\_count$.
- Genetic operations: The genetic operations are performed using Algorithm 2.
- Population selection for the next generation: After mutation, the child population is sent for the next generation.
- Termination: Printing results and exit once the termination condition is satisfied.

ALGORITHM 1 Fitness calculation.

Require: Function value $f(x)$, penalty $= \lambda \times (violation)^2$ {λ is the penalty multiplier and $violation =$ difference between actual value and the limit.}

Require: Promotion constant X, Reduction constant Y, Susceptibility, Infection, Recovery

Ensure: Fitness for Susceptibility, Infection, Recovery

1: **if** Parameter $=$ SUSCEPTIBILITY **then**
2: **if** Nutritional_Status $=$ NORMAL **then**
3:     Fitness_max $= \frac{f(x) \times X}{1 + penalty}$
4: **else if** Nutritional_Status $=$ MALNUTRITION **then**
5:     Fitness_max $= \frac{f(x)}{(1 + penalty) \times Y}$
6: **end if**
7: **end if**
8: **if** Parameter $=$ INFECTION **then**
9: **if** Nutritional_Status $=$ NORMAL **then**
10:    Fitness_min $= \frac{1 \times X}{1 + f(x) + penalty}$
11: **else if** Nutritional_Status $=$ MALNUTRITION **then**
12:    Fitness_min $= \frac{1}{(1 + f(x) + penalty) \times Y}$
13: **end if**
14: **end if**
15: **if** Parameter $=$ RECOVERY **then**
16: **if** Nutritional_Status $=$ NORMAL **then**
17:    Fitness_max $= \frac{f(x) \times X}{1 + penalty}$
18: **else if** Nutritional_Status $=$ MALNUTRITION **then**
19:    Fitness_max $= \frac{f(x)}{(1 + penalty) \times Y}$
20: **end if**
21: **end if**

2.4 Experimentation

The ranges of the susceptibility, infection, and recovery for which the experiments have been carried out are shown in Table 1. We consider big ranges for susceptibility because we expect a greater number of people be deployed in the susceptibility. The range of infection has been kept small because we expect a smaller number of infections. Likewise, we keep comparatively a big number for recovery because we expect more recovery. We keep the set of the ranges significantly different to clearly see the difference in optimal values of the susceptibility, infection, and recovery. It is worth mentioning that these ranges are for representational purpose only. Users can

ALGORITHM 2 Genetic operations.

Require: Parent, Child

Ensure: Mutated child

1: **for** CROSSOVER **do**
2:    **if** CROSSOVER = ARITHMETIC **then**
3:       $B = r \times F + (1 - r) \times M$
4:       $G = (1 - r) \times F + r \times M$ {Here, B, G, F, M are child1, child2, parent1 and parent2 respectively. r is a uniform random number.}
5:    **else if** CROSSOVER = HEURISTIC **then**
6:       **if** $Fit_F > Fit_M$ **then**
7:          $Fit_F = best$
8:          $Fit_M = worst$
9:          $B = best + random \times (best - worst)$
10:         $G = best$ {Here, Fit_F, Fit_M, B, G are fitness of parent1, fitness of parent2, child1 and child2 respectively.}
11:    **end if**
12:   **end if**
13: **end for**
14: **for** MUTATION **do**
15:   $C'_{pq} = X_k^{min} + random \times (X_k^{max} - X_k^{min})$ {Here, C' is the mutated child produced by changing $(p \times q)$th chromosome of kth genome, X^{min} is the lowest boundary, X^{max} is the highest boundary, and $random$ is the uniform random number}
16: **end for**

Table 1 Range of working space.

Range	Symbol	Set-1	Set-2
Range of susceptibility under normal nutritional status	SNN	(40, 60)	(70, 90)
Range of susceptibility under malnutrition status	SMN	(40, 60)	(70, 90)
Range of infection under normal nutritional status	INN	(10, 20)	(30, 40)
Range of infection under malnutrition status	IMN	(10, 20)	(30, 40)
Range of recovery under normal nutritional status	RNN	(20, 30)	(40, 50)
Range of recovery under malnutrition status	RMN	(20, 30)	(40, 50)

change the limits based on the working space. The use of big ranges will increase computation time.

Other parameter values that are used in the experimentation are shown in Table 2. In Table 2, we keep some values in such a way that it matches

Table 2 Values used in the experiment.

Parameter	Symbol	Set-1	Set-2
Rate of recruitment	Λ	80%	90%
Percentage of population moving from malnutrition to normal nutritional status	N_r	2%	4%
Percentage of population moving from normal nutritional status to malnutrition	M_r	4%	6%
Percentage of exposure	e	10%	15%
Pathogen population	P	4	8
Natural death of pathogen	μ	2%	4%
Pathogen growth rate	g	3%	6%
Concentration of pathogen	C	2	3
Natural death of population	d	3%	5%
Pathogen growth in infected population under normal nutritional status	$\sigma 1$	0.5%	1%
Pathogen growth in infected population under malnutrition	$\sigma 2$	1.5%	2%
Recovery rate under normal nutritional status	r_n	15%	20%
Recovery rate under malnutrition	r_m	5%	10%
Infection rate under normal nutritional status	$\lambda(=\frac{eP}{C+P})$	0.067	0.109
Infection rate under malnutrition	$(\alpha=2.5)\lambda$	0.167	0.273
Death rate of infected population under normal nutritional status	η	10%	15%
Death rate of infected population under malnutrition	$(\beta=1.5)\eta$	15%	22.5%

with global figure or clinical facts. The percentage of population moving from normal nutritional status to malnutrition is much higher than that of population moving from malnutrition to normal nutritional status. That is why we keep $M_r > N_r$ in both the sets. Pathogen growth in infected population with malnutrition is higher than that of pathogen growth in infected population with normal nutritional status. So, we keep $\sigma 2 > \sigma 1$ in both the sets. The recovery rate in population with malnutrition is always lower than that of the recovery rate in population with normal nutritional status. That is why, we keep $r_m < r_n$. Other parameter values are kept in such a way that it shows clear differences on the results obtained for both the sets of values.

3. Results

We perform the experiments for both the situations of nutritional status separately. Also, under each nutritional status, we perform experiments for susceptibility, infection, and recovery separately. The obtained results are shown in subsequent sections. All the experiments are performed considering population size (Pop_size) of 5 and 10. The considered numbers of generations (Generation) are 20 and 40. The population size and the number of generations can be changed as per the requirement. However, greater population size and higher number of generations will increase the computation time. The values of the promotion constant (X) and the reduction constant (Y) used in the fitness function of genetic algorithm are considered to be 1.05 and 1.25, respectively.

3.1 Results under normal nutritional status

Under the situation of normal nutritional status, we perform the experiments for an optimal change in susceptibility, infection, and recovery. Table 3 shows the results obtained for susceptibility. Table 3 lists out the optimal susceptibility in the given range, optimal deployed population in the given range, and optimal change in susceptibility for arithmetic crossover as well as heuristic crossover under the situation of normal nutritional status. Table 4 lists out the optimal number of susceptibility in the given range, optimal number of infection in the given range, and optimal change in

Table 3 Susceptibility under normal nutritional status.

Value	Pop_size	Generation	Opt_SNN	Opt_PNN	Opt_SNNdt AC	HC
Set-1	5	20	51.0395	53.3150	40.0337	40.1073
	5	40	53.5066	53.2237	41.9688	47.0620
	10	20	50.8870	46.6792	39.9140	47.0620
	10	40	51.4618	49.8326	40.3649	47.0620
Set-2	5	20	76.3898	75.1775	66.4133	78.2460
	5	40	72.1347	83.5072	62.7138	77.2953
	10	20	78.6251	82.7297	68.3566	78.2460
	10	40	76.4061	77.5080	66.4274	78.2460

Table 4 Infection under normal nutritional status.

					Opt_INNdt	
Value	Pop_size	Generation	Opt_SNN	Opt_INN	AC	HC
Set–1	5	20	50.7747	15.4582	−4.0419	−3.3234
	5	40	58.0107	13.0867	−3.3371	−3.5420
	10	20	51.0894	14.1847	−3.6835	−4.1845
	10	40	48.5673	14.0582	−3.6624	−4.0664
Set–2	5	20	87.2416	35.9833	−12.5941	−11.3142
	5	40	77.9303	37.4858	−13.3872	−13.8002
	10	20	80.5575	38.3200	−13.6667	−12.5227
	10	40	79.8658	35.5598	−12.5769	−14.1469

Table 5 Recovery under normal nutritional status.

					Opt_RNNdt	
Value	Pop_size	Generation	Opt_INN	Opt_RNN	AC	HC
Set–1	5	20	14.8267	26.9316	1.4161	1.9990
	5	40	14.7159	23.9040	1.4903	2.3837
	10	20	17.9664	24.0482	1.9735	2.3429
	10	40	15.8741	24.7632	1.6382	2.4000
Set–2	5	20	36.2764	43.1800	5.0963	5.9770
	5	40	37.7014	49.1213	5.0842	5.9974
	10	20	37.4447	45.8928	5.1943	6.0000
	10	40	33.5532	44.6544	4.4779	6.0000

infection for arithmetic crossover and heuristic crossover. This table shows negative values for optimal change in infection because it is a minimization problem and our expectation is to reduce the number of infections. The negative values signify that the number of infections will get reduced by that amount. Table 5 shows the results obtained for recovery under the situation of normal nutritional status. This table lists out the optimal number of infection within the given range, optimal number of recovery in the given range and change in recovery with application of arithmetic crossover and heuristic crossover.

3.2 Results under malnutrition

We separately perform the experiments for susceptibility, infection, and recovery under the condition of malnutrition. Table 6 shows the results obtained for susceptibility under malnutrition. This table lists optimal values for susceptibility within the given range, optimal deployment of population within the range, and optimal change in susceptibility for arithmetic crossover and heuristic crossover. Table 7 shows the results obtained for infection under the situation of malnutrition. This table lists out optimal number of

Table 6 Susceptibility under malnutrition.

Value	Pop_size	Generation	Opt_SMN	Opt_PMN	Opt_SMNdt AC	HC
Set-1	5	20	50.1788	44.4231	39.9381	41.3411
	5	40	47.8304	58.4557	38.0690	44.2932
	10	20	46.4726	47.7726	36.9883	38.9451
	10	40	49.5778	52.9488	39.4597	38.5673
Set-2	5	20	76.9399	70.0286	66.0528	73.3193
	5	40	79.1847	72.6959	67.9800	71.0571
	10	20	89.5367	79.0281	76.8672	75.4139
	10	40	76.5882	78.1318	65.7509	69.8880

Table 7 Infection under malnutrition.

Value	Pop_size	Generation	Opt_SMN	Opt_IMN	Opt_IMNdt AC	HC
Set-1	5	20	45.2020	15.2847	−2.8766	−2.2260
	5	40	40.0160	19.9747	−4.0286	−2.2676
	10	20	41.9000	17.8674	−3.5173	−2.7789
	10	40	46.0061	19.5662	−3.8500	−2.2851
Set-2	5	20	77.9124	37.1862	−9.9235	−10.9636
	5	40	73.4546	36.0540	−9.7291	−10.3723
	10	20	76.6564	36.6452	−9.7855	−9.2267
	10	40	80.2471	39.9588	−10.8428	−8.5826

Table 8 Recovery under malnutrition.

Value	Pop_size	Generation	Opt_IMN	Opt_RMN	Opt_RMNdt	
					AC	HC
Set-1	5	20	13.4625	28.4045	−0.1790	0.2505
	5	40	16.9396	24.7487	0.1045	0.4000
	10	20	16.0702	23.1819	0.1081	0.3967
	10	40	15.9371	21.3628	0.1560	0.1581
Set-2	5	20	35.1557	45.9414	1.2185	1.1609
	5	40	36.9866	40.9698	1.6502	1.3805
	10	20	30.4908	48.9456	0.6018	1.1877
	10	40	33.2689	46.2767	1.0131	1.5336

susceptibility, optimal number of infection, and optimal change in infection under arithmetic crossover and heuristic crossover. Here also the optimal change in infection shows negative values that signify a reduction in infection by that amount. Table 8 shows the results obtained for recovery under the condition of malnutrition. This table lists out the optimal number of infection, optimal number of recovery, and optimal change in recovery with arithmetic crossover as well as heuristic crossover.

4. Discussion

The experimental results show the optimized values of the susceptibility, infection, and recovery under different situations of nutritional status and different genetic operations. The obtained results are within a predefined range of infectiological parameter values as shown in Table 1. The ranges can be changed depending on the area/infection space, situation, and requirement. Other parameter values as shown in Table 2 can also be changed based on the pathogen type, medium of spread, geographical location, environmental condition, feasibility of spread, etc. The optimization of different infectiological parameters is discussed in subsequent sections.

4.1 Optimization of susceptibility

Fig. 1 shows the optimal change in susceptibility under normal nutritional status as well as malnutrition. Fig. 1A shows the optimal changes under the

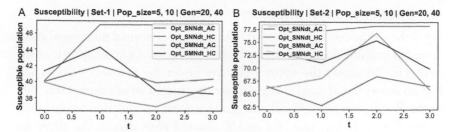

Fig. 1 Optimal changes in susceptibility under different nutritional status and different crossovers: (A) with Set-1 of values; (B) with Set-2 of values.

first set of values, while Fig. 1B shows the optimal changes under the second set of values. In the Y-axis of the plots, we put optimal changes in susceptible population against time in the X-axis. The BLUE-colored plots denote the curve for optimal changes in susceptibility under normal nutritional status considering arithmetic crossover. The YELLOW-colored plots denote the curve for optimal changes in susceptibility under normal nutritional status considering heuristic crossover. The GREEN-colored plots denote the curve for optimal changes in susceptibility under malnutrition considering arithmetic crossover, while the RED-colored plots denote the curve for optimal changes in susceptibility under malnutrition considering heuristic crossover. The plots for the first set of values show that, under the situation of normal nutritional status, the results obtained using heuristic crossover surpass the results obtained using arithmetic crossover. Under the situation of malnutrition, the results obtained using heuristic crossover surpass the results obtained using arithmetic crossover. Also, the results obtained under malnutrition considering arithmetic crossover is the worst while the results under normal nutritional status considering heuristic crossover are the best. Overall, the results under heuristic crossover are better than the results obtained using arithmetic crossover. In the plots of the second set of values also, under the situation of normal nutritional status, the results obtained using heuristic crossover surpass the results obtained using arithmetic crossover. Under the situation of malnutrition, the results obtained using heuristic crossover surpass the results obtained using arithmetic crossover. Here, the results obtained under normal nutritional status considering arithmetic crossover are the worst, while the results under normal nutritional status considering heuristic crossover are the best. As a whole the results under heuristic crossover are better than the results obtained using arithmetic crossover.

4.2 Optimization of infection

Fig. 2 shows the optimal change in infection under normal nutritional status as well as malnutrition considering arithmetic crossover and heuristic crossover. Fig. 2A shows the optimal changes under the first set of values and Fig. 2B shows the optimal changes under the second set of values. Here also, in the Y-axis, we put optimal changes in infected population against time in the X-axis. In this case, the BLUE-colored plots denote the curve for optimal changes in infection under normal nutritional status considering arithmetic crossover. The YELLOW-colored plots denote the curve for optimal changes in infection under normal nutritional status considering heuristic crossover. The GREEN-colored plots denote the curve for optimal changes in infection under malnutrition considering arithmetic crossover, while the RED-colored plots denote the curve for optimal changes in infection under malnutrition considering heuristic crossover. In the case of infection, it is a minimization problem and we expect less number of infections. That means the lesser value is better here. The plots for the first set of values show that, under the situation of normal nutritional status, the results obtained using heuristic crossover fall behind results obtained using arithmetic crossover. Under the situation of malnutrition, the results obtained using arithmetic crossover fall behind the results obtained using heuristic crossover. In this case, the results obtained under normal nutritional status considering heuristic crossover are the best and the results obtained under malnutrition considering heuristic crossover are the worst. In the plots for the second set of values, under the situation of normal nutritional status, the results obtained using arithmetic crossover fall behind the results obtained using heuristic crossover. Under the situation of malnutrition, the results obtained using arithmetic crossover fall behind results obtained using heuristic crossover. Here, the results obtained under normal

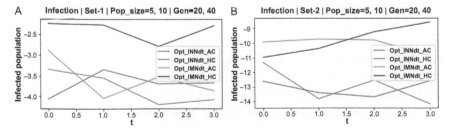

Fig. 2 Optimal changes in infection under different nutritional status and different crossovers: (A) with Set-1 of values; (B) with Set-2 of values.

nutritional status considering arithmetic crossover are the best, while the results under malnutrition considering heuristic crossover are worst. As a whole, the results under arithmetic crossover are better compared to the results obtained using heuristic crossover.

4.3 Optimization of recovery

Fig. 3 shows the plots for optimal changes in recovery under malnutrition and normal nutritional status considering arithmetic crossover and heuristic crossover. Fig. 3A is for the first set of values, while Fig. 3B is for the second set of values. Here also, in the Y-axis of the plots, we put optimal changes in recovered population against time in the X-axis. Like previous two cases, here also, the BLUE-colored plots denote the curve for optimal changes in recovery under normal nutritional status considering arithmetic crossover. The YELLOW-colored plots denote the curve for optimal changes in recovery under normal nutritional status considering heuristic crossover. The GREEN-colored plots denote the curve for optimal changes in recovery under malnutrition considering arithmetic crossover, while the RED-colored plots denote the curve for optimal changes in recovery under malnutrition considering heuristic crossover. The plots for the first set of values show that, under the situation of normal nutritional status, the results obtained using both the crossovers show very good results, and to be more specific, the results obtained using heuristic crossover surpass the results obtained using arithmetic crossover. Under the situation of malnutrition, the results obtained using heuristic crossover surpass the results obtained using arithmetic crossover. In this case, the results under normal nutritional status are very good compared to the results under malnutrition. Also, the

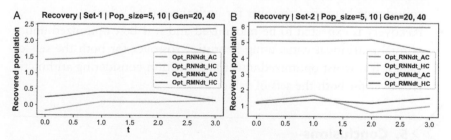

Fig. 3 Optimal changes in recovery under different nutritional status and different crossovers: (A) with Set-1 of values; (B) with Set-2 of values.

results obtained under malnutrition considering arithmetic crossover are the worst, while the results under normal nutritional status considering heuristic crossover are the best. Overall the results under heuristic crossover are better compared to the results obtained using arithmetic crossover. In the plots for the second set of values also, under the situation of normal nutritional status, the results under both the crossovers are better, and to be more specific, the results obtained using heuristic crossover surpass the results obtained using arithmetic crossover. Under the situation of malnutrition, the results obtained using heuristic crossover surpass the results obtained using arithmetic crossover. In this set also, the results under normal nutritional status are better compared to the results obtained under malnutrition. The results obtained under malnutrition considering arithmetic crossover are the worst, while the results under normal nutritional status considering heuristic crossover are the best.

4.4 Overall observations

From the experimental results and the associated discussion, we observe the following:

- Susceptibility should be maximized and it is best optimized under normal nutritional status considering heuristic crossover for both the sets of values. But, susceptibility is worst optimized under malnutrition considering arithmetic crossover for the first set of values and under normal-nutritional status considering arithmetic crossover for the second set of values.
- Infection should be minimized and it is best optimized under normal nutritional status using heuristic crossover for the first set of values and under normal nutritional status using arithmetic crossover for the second set of values. Infection is worst optimized under malnutrition considering heuristic crossover for both the sets of values.
- Recovery is expected to be maximized and it is best optimized under normal nutritional status using heuristic crossover for both the sets of values. It is worst optimized under malnutrition considering arithmetic crossover for both the sets of values.

5. Conclusions

Infectious diseases spread too fast that we need to bear huge loss within a short period of time. The severity of infectious disease is measured by infectiological parameters like susceptibility, infection, and recovery. These parameters are mostly affected by a lot of environmental and clinical factors. Nutritional status of our body greatly impacts these infectiological parameters. Optimization of these parameters helps us developing strategy to mitigate losses in adverse situation that is supposed to be created by infectious diseases. The well-known genetic algorithm has immense potential to optimize both maximization and minimization problems within a predefined range of values. In our case, susceptibility needs to be maximized, infection has to be minimized, while recovery should be maximized. This chapter presents our formulated equations for susceptibility, infection, and recovery under normal nutritional status and malnutrition. It also integrates nutritional status levels with genetic algorithm and use the same for solving the formulated equations. The experimental results reveal that, normal nutritional status in association with heuristic crossover best optimizes susceptibility, infection, and recovery. Malnutrition is proved to be negatively influencing the optimization of the infectiological parameters.

Conflict of interest

The authors declare no conflict of interest.

References

[1] S.L. Roberts, Emerging and re-emerging diseases, in: S. Romaniuk, M. Thapa, P. Marton (Eds.), The Palgrave Encyclopedia of Global Security Studies, Springer International Publishing, Cham, 2019, pp. 1–9, ISBN: 978-3-319-74336-3, https://doi.org/10.1007/978-3-319-74336-3_531-1.
[2] P.C. Calder, A.C. Carr, A.F. Gombart, M. Eggersdorfer, Optimal nutritional status for a well-functioning immune system is an important factor to protect against viral infections, Nutrients 12 (4) (2020) 1181, https://doi.org/10.3390/nu12041181.
[3] K. Boone, C. Wisdom, K. Camarda, P. Spencer, C. Tamerler, Combining genetic algorithm with machine learning strategies for designing potent antimicrobial peptides, BMC Bioinformatics 22 (1) (2021) 239, https://doi.org/10.1186/s12859-021-04156-x.
[4] S. Ochoa, G.M. Constantine, M.S. Lionakis, Genetic susceptibility to fungal infection in children, Curr. Opin. Pediatr. 32 (6) (2020), https://doi.org/10.1097/MOP.0000000000000948.
[5] J.K. Davis, T. Gebrehiwot, M. Worku, W. Awoke, A. Mihretie, D. Nekorchuk, M.C. Wimberly, A genetic algorithm for identifying spatially-varying environmental drivers in a malaria time series model, Environ. Model. Softw. 119 (2019) 275–284, https://doi.org/10.1016/j.envsoft.2019.06.010.
[6] Z. Hussain, M. Dutta Borah, Forecasting probable spread estimation of COVID-19 using exponential smoothing technique and basic reproduction number in indian context, in: R. Patgiri, A. Biswas, P. Roy (Eds.), Health Informatics: A Computational

Perspective in Healthcare, Springer Singapore, Singapore, 2021, pp. 183–196, ISBN: 978-981-15-9735-0, https://doi.org/10.1007/978-981-15-9735-0_10.

[7] I. Trejo, N.W. Hengartner, A modified susceptible-infected-recovered model for observed under-reported incidence data, PLOS ONE 17 (2) (2022) 1–23, https://doi.org/10.1371/journal.pone.0263047.

[8] J. Tolles, T. Luong, Modeling epidemics with compartmental models, JAMA 323 (24) (2020) 2515–2516, https://doi.org/10.1001/jama.2020.8420.

[9] O.N. Bjørnstad, K. Shea, M. Krzywinski, N. Altman, The SEIRS model for infectious disease dynamics, Nat. Methods 17 (6) (2020) 557–558, https://doi.org/10.1038/s41592-020-0856-2.

[10] Z. Hussain, R.K. Ahmed, M.D. Borah, A computational aspect to analyse impact of nutritional status on the performance of anaesthesia on surgical patients, Procedia Comput. Sci. 218 (2023) 514–523, https://doi.org/10.1016/j.procs.2023.01.033.

[11] Z. Hussain, M.D. Borah, A computational aspect to analyse impact of nutritional status on drug resistance, in: 2022 IEEE Silchar Subsection Conference (SILCON), 2022, pp. 1–6, https://doi.org/10.1109/SILCON55242.2022.10028912.

[12] Z. Hussain, M.D. Borah, NICOV: a model to analyse impact of nutritional status and immunity on COVID-19, Med. Biol. Eng. Comput. 60 (5) (2022) 1481–1496, https://doi.org/10.1007/s11517-022-02545-9.

[13] Z. Hussain, M.D. Borah, Nutritional status prediction in neonate using machine learning techniques: a comparative study, in: A. Bhattacharjee, S.K. Borgohain, B. Soni, G. Verma, X.-Z. Gao (Eds.), Machine Learning, Image Processing, Network Security and Data Sciences, Springer Singapore, Singapore, 2020, pp. 69–83, ISBN: 978-981-15-6318-8, https://doi.org/10.1007/978-981-15-6318-8_7.

[14] Z. Hussain, M.D. Borah, Predicting mental health and nutritional status from social media profile using deep learning, in: T.-P. Hong, L. Serrano-Estrada, A. Saxena, A. Biswas (Eds.), Deep Learning for Social Media Data Analytics, Springer International Publishing, Cham, 2022, pp. 177–193, ISBN: 978-3-031-10869-3, https://doi.org/10.1007/978-3-031-10869-3_10.

[15] S. Katoch, S.S. Chauhan, V. Kumar, A review on genetic algorithm: past, present, and future, Multimed. Tools Appl. 80 (5) (2021) 8091–8126, https://doi.org/10.1007/s11042-020-10139-6.

[16] M. Kaur, P.K. Shukla, J.K. Sandhu, A. Ahirwar, D. Ghai, P. Maheshwary, P.K. Shukla, Multiobjective genetic algorithm and convolutional neural network based COVID-19 identification in chest X-ray images, Math. Probl. Eng. 2021 (2021) 7804540, https://doi.org/10.1155/2021/7804540.

[17] N. Andelic, S.B. Segota, I. Lorencin, V. Mrzljak, Z. Car, Estimation of COVID-19 epidemic curves using genetic programming algorithm, Health Informatics J. 27 (1) (2021) 1460458220976728, https://doi.org/10.1177/1460458220976728.

[18] X. Lin, X. Wang, Y. Wang, X. Du, L. Jin, M. Wan, H. Ge, X. Yang, Optimized neural network based on genetic algorithm to construct hand-foot-and-mouth disease prediction and early-warning model, Int. J. Environ. Res. Public Health 18 (6) (2021) 2959, https://doi.org/10.3390/ijerph18062959.

[19] Z. Qiu, Y. Sun, X. He, J. Wei, R. Zhou, J. Bai, S. Du, Application of genetic algorithm combined with improved SEIR model in predicting the epidemic trend of COVID-19, China, Sci. Rep. 12 (1) (2022) 8910, https://doi.org/10.1038/s41598-022-12958-z.

[20] M.T. Rouabah, A. Tounsi, N.E. Belaloui, Genetic algorithm with cross-validation-based epidemic model and application to the early diffusion of COVID-19 in Algeria, Sci. Afr. 14 (2021) e01050, https://doi.org/10.1016/j.sciaf.2021.e01050.

[21] S. Bansal, M. Singh, R.K. Dubey, B.K. Panigrahi, Multi-objective genetic algorithm based deep learning model for automated COVID-19 detection using medical image data, J. Med. Biol. Eng. 41 (5) (2021) 678–689, https://doi.org/10.1007/s40846-021-00653-9.

[22] P. Yarsky, Using a genetic algorithm to fit parameters of a COVID-19 SEIR model for US states, Math. Comput. Simul. 185 (2021) 687–695, https://doi.org/10.1016/j.matcom.2021.01.022.
[23] Z. Hussain, M.D. Borah, Birth weight prediction of new born baby with application of machine learning techniques on features of mother, J. Stat. Manag. Syst. 23 (6) (2020) 1079–1091, https://doi.org/10.1080/09720510.2020.1814499.

About the authors

Zakir Hussain is rendering his services as a Guest Lecturer in the department of Computer Application, North-Eastern Hill University (NEHU), Tura Campus, Meghalaya, India. He is also a research scholar in the department of Computer Science and Engineering at National Institute of Technology Silchar. He has submitted his Ph.D. thesis on 31 July 2023. He has completed his Master of Technology in Information Technology from Gauhati University in the year 2015. He has completed his Bachelor of Engineering in Computer Science and Engineering in the year 2012 with Honors and bagged 4th position in the university. He was the state topper in the subject of General Science in board examination conducted by Assam state board in the year 2005. He was a Guest/ Part time faculty in the level of Assistant Professor at Barak Valley Engineering College; a government of Assam institution in the district of Karimganj. His research interest is in the field of Mathematical Biology, Machine Learning for Healthcare, Soft Computing, and Natural Language Processing. He has his credit to 20+ articles in Journal/Conference/Edited books. He has created a dataset available in IEEE dataport and holds a Germany patent.

Malaya Dutta Borah received her B. Tech degree in Computer Science and Engineering from the North Eastern Regional Institute of Science and Technology, Govt. of India, M.E. in Computer Technology and Applications from Delhi College of Engineering, and Ph.D. in Computer Science and Engineering from Delhi Technological University. Dr. Malaya Dutta Borah is currently working as an Assistant Professor Grade-I in the Department of Computer Science and Engineering, National Institute of Technology

Silchar, Assam, India. Her research areas include Data Mining, BlockChain Technology, Cloud Computing and e-Governance. She is the author of more than 70 papers including in reputed journals, Book Chapters, various National and International Conferences. She is the Editor in three multi-authored edited books published by IGI Global, Springer and CRC Press. She has three patent granted namely Patent Number: 202022101981, Germany patent, Patent Number: 2021105739, Australian Patent and Patent number: 2021107080, Australian Patent. She is a member of professional societies like IEEE, CSI and ACM.

CHAPTER EIGHT

A novel influencer mutation strategy for nature-inspired optimization algorithms to solve electricity price forecasting problem

Priyanka Singh[a] and Rahul Kottath[b,c]

[a]Department of Computer Science and Engineering, Indian Institute of Information Technology, Raichur, India
[b]School of Electrical and Electronics Engineering, VIT Bhopal University, Bhopal, India
[c]Digital Tower, Bentley Systems India Private Limited, Pune, India

Contents

Advances in Computers, Volume 135
ISSN 0065-2458
https://doi.org/10.1016/bs.adcom.2023.12.002

Abstract

Optimization is an active area of research in the field of soft computing. Over a period, various nature-inspired optimization algorithms have been proposed to solve complex engineering problems. This chapter discusses a new mutation strategy that can be integrated with Harris hawks optimization, moth flame optimization, particle swarm optimization, and whale optimization algorithms to improve the convergence rate and avoid local stagnation problems. Often, nature-inspired algorithms update the population by using the best particle's position. Here, we have modeled a new mutation strategy based on a group of particles labeled as an influencer group. The influencer group consists of top-performing particles, guiding the remaining particle toward the optimal point. This concept is inspired by the general tendency of humans who get influenced by a group of individuals while making decisions. Influencer mutation shows significant improvement in the performance of base optimization algorithms. The efficacy of improved algorithms is tested on standard benchmark functions. Further, an artificial neural network is integrated with proposed metaheuristic algorithms for solving electricity price forecasting problems. The statistical results show the efficacy of the proposed mutation strategy.

1. Introduction

Nature has always been a problem solver, which is why various algorithms are inspired by nature, known as nature-inspired algorithms (NIA). Meta-heuristic algorithms come under NIA, which can be broadly classified as evolution-based and swarm-based algorithms. The former is based on the natural evolution of species, whereas the latter deals with the intelligence behavior shown among a group of animals, birds, fish, etc. Genetic algorithms, genetic programming, differential evolution, and evolutionary strategies are some examples of evolution-based optimization algorithms. Swarm-based optimization methods include particle swarm optimization (PSO), cuckoo optimization algorithm (COA), ant colony optimization, follow the leader, and others. These algorithms try to minimize/maximize an objective function based on some given constraints. The objective of any optimization is to reach the global solution in a search space by avoiding multiple local solutions [1, 2]. Exploration and exploitation are two essential strategies of any optimization algorithm where exploration helps to search the space globally, while exploitation exploits the solutions locally using the best solution [3].

David E. Goldberg introduced mutation in genetic algorithm [4], which is inspired by human evolution. Mutation adds exploration to the population, and to date, many mutation strategies have been added to the literature

[5]. The traditional evolutionary algorithms introduced mutation operators to avoid local stagnation problems. The mutation operators can be categorized into artificial and biologically inspired mutation [6]. Mutation operators such as bit flipping and Gaussian mutation operator fall under the artificial mutation operator as their origin is independent of biological aspects. Frameshift, translocation, and homeostasis mutation are a few examples of biologically inspired mutation operators. In a different bio-inspired algorithm, A. Biswas proposed a new optimization by using Bohr's atomic model [7].

Recently, several swarm-based optimizations with the concept of mutation operator have been proposed by researchers. Zhang and Ming have added a mutation operator to improve the search strategy of grey wolf optimizer (GWO) and eliminated the restructuring mechanism for the poor search wolves [8]. In another work, improved GWO is proposed by incorporating ranking-based mutation operators and applied to system identification problem [9]. Singh and Dwivedi proposed a hybrid model by combining PSO with homeostasis mutation to predict short-term electricity load for NEPOOL, England power market [10]. Gupta et al. introduced a mutation operator in the sine cosine algorithm to add exploration and avoid local stagnation problems [11]. Panda et al. modified the spotted hyena optimizer algorithm by adding a mutation operator to solve optimization problems and train the neural network [12]. This shows the usability of mutation operators in various situations.

The area of optimization is very popular due to complex real-world problems, and the search for an appropriate algorithm is still going on. As mentioned earlier, various optimization algorithms have been proposed in the literature, and the search for the best algorithm continues. There are two possible approaches to finding a better algorithm: first is to develop a new algorithm, and second to upgrade the existing one. With this motivation, a new addition to the literature on mutation has been made in this work. The mutation operator is combined to improve the performance of existing optimization algorithms.

This chapter proposes a novel mutation operator inspired by human behavior termed influencer mutation. The performance of the proposed mutation strategy has been tested by implementing it in multiple well-known optimization algorithms such as ant lion optimization (ALO) [13], GWO [14], Harris hawks optimizer (HHO) [15], moth flame optimization (MFO) [16], PSO [17, 18], and whale optimization algorithm (WOA) [19]. The improved algorithms are evaluated on 20 benchmark functions.

Further, improved algorithms are combined with neural networks to find optimal weight parameters for electricity price prediction.

The rest of the chapter is organized as follows. Section 2 discusses the related works and mathematical background required to understand the paper. Section 3 details the influencer mutation operator in detail with its mathematical representation. Section 4 presents the detailed procedure of the influencer mutation-based optimization algorithm. Section 5 describes the two experiments used to test the efficacy of the proposed approach using benchmark functions and electricity price forecasting. Finally, Section 6 concludes the chapter by providing some possible future directions.

2. Related works

This section details the theoretical background of the discussed algorithms, along with the basics of artificial neural networks (ANN). The individual algorithms are discussed briefly with their particle update equations. More details can be found in the corresponding reference.

2.1 Artificial neural networks

ANN architecture imitates the working of biological neural networks. A general neural network has several weight connections between the input-hidden and hidden-output layers. The output (Z_k) for two-layered ANN with n input, h hidden, and m output neurons can be obtained by,

$$Z_k = \sum_{j=1}^{h} \left(W_{k,j} * \frac{1}{1 + exp\left(-\sum_{i=1}^{n} W_{j,i} X_i + b_i\right)} + b_k \right) \quad (1)$$

where X_i is the input given to the neural network with i input features, $W_{j,i}$ represents weights connecting input layer to hidden layer, and $W_{k,j}$ represents weights connecting hidden layer to output layer. b represents the bias in the hidden and output layer. Neural networks have been widely used in solving real-time problems. One of the commonly used applications is time series problems. Hyperparameter tuning has been one of the challenging tasks for neural networks due to the trial-and-error method [20]. Moreover, high variance and high bias are other problems of ANN models.

Let us assume that X_c is the current population and X_{new} represents the next position of the population. Then, the mathematical aspects of the optimization algorithm used in this work are explained here in this section.

2.2 Ant lion optimizer

ALO algorithms imitate the behavior of ant around antlions in the trap [13]. ALO algorithm follows a random walk, and thus, walk by ants is given as:

$$X_c = [0, C_{sum}(2R_1(T_1) - 1), C_{sum}(2R_1(T_2) - 1), \ldots, C_{sum}(2R_1(T_n) - 1)]$$

$$(2)$$

where C_{sum} represents cumulative sum and $R_1(T) = \begin{cases} 1 & random > 0.5 \\ 0 & otherwise \end{cases}$.

The search space defined for the ants is given as

$$X_c^t = \frac{(X_c^t - A_c)(D_c^t - C_c^t)}{B_c - A_c} + C_c^t \qquad (3)$$

Also, the random walk chosen by ants is affected by the antlion trap, i.e.,

$$C_c^t = AL_c^t + C^t, \quad D_c^t = AL_c^t + D^t, \quad C^t = \frac{C^t}{I_r}, \quad and \quad D^t = \frac{D^t}{I_r} \qquad (4)$$

where AL_c^t is antlion in tth generation. A_c and B_c represent minimum and maximum random walk of tth generation. C_c^t and D_c^t represent minimum and maximum of all variables in tth. I_r is the sliding ratio.

Each ant's position is updated using the average position of the best ant (R_b) and a random ant (R_r). Update AL_c by X_c if fitness of X_c is better than AL_c.

$$X_{new} = \frac{R_r^t + R_b^t}{2} \qquad (5)$$

2.3 Grey wolf optimizer

The hierarchical leadership and hunting behavior of the wolves are modeled in GWO [21]. The essence of the algorithm revolves around α, β, δ, and ω wolves. The three main steps of hunting strategies followed by GWO are:
- Encircling:

$$X_{new} = X_{prey,c} - A.|C.X_{prey,c} - X_c| \qquad (6)$$

$$C = 2.r_2 \qquad (7)$$

$$A = 2.ar_1 - a, \qquad a = \left(1 - \frac{I}{I_{max}}\right) * 2 \qquad (8)$$

where r_1 and r_2 are random numbers. I is generation and I_{max} is maximum generation. $X_{prey,c}$ is current position of prey.

- Hunting:

$$X_{c1} = X_\alpha - A_1 \cdot |C_1 X_\alpha - X_c|$$
$$X_{c2} = X_\beta - A_1 \cdot |C_2 X_\beta - X_c| \tag{9}$$
$$X_{c3} = X_\delta - A_1 \cdot |C_3 X_\delta - X_c|$$

where, X_α, X_β, and X_δ are positions of α, β, and δ wolves. Values of C_1, C_2, and C_3 are taken from Eq. (7). ω wolves update their position using:

$$X_c = \frac{X_{c1} + X_{c1} + X_{c3}}{3} \tag{10}$$

- Attack: Reach termination condition.

2.4 Harris hawks optimizer

HHO imitates the cooperative behavior and prey-catching manner of Harris hawks [22]. Two phases of the HHO algorithm can be represented as:

$$X_{new} = \begin{cases} X_r - r_1 |X_r - 2 \cdot r_2 \cdot X_c| & p \geq 0.5 \\ (X_{prey} - X_{avg}) - r_3(lb + r_4(ub - lb)) & p < 0.5 \end{cases} \tag{11}$$

where X_{prey} is best position of prey and X_{avg} is average location of hawks. p, r_1, r_2, r_3, and $r_4 \in [0, 1]$. The decreasing energy of prey during its escaping behavior can be modeled as follows:

$$E_g = E_0 \left(2 - \frac{2t}{T}\right) \tag{12}$$

$$X_{new} = \begin{cases} X_{prey} - X_c - E_g |2(1 - r_5) \cdot X_{prey} - X_c| & r_5 \geq 0.5 \; \& \; |E_g| \geq 0.5 \; \; Soft \; besiege \\ X_{prey} - E_g |X_{prey} - X_c| & r_5 \geq 0.5 \; \& \; |E_g| < 0.5 \; \; Hard \; besiege \end{cases} \tag{13}$$

Soft besiege with progressive behavior:

$$\left. \begin{array}{l} Y_1 = X_{prey} - E_g |2(1 - r_5) \cdot X_{prey} - X_c| \\ Z = Y + S \times LF(d) \end{array} \right|_{r_5 < 0.5 \; and \; |E_g| \geq 0.5} \tag{14}$$

where LF is Levy flight function and S is random vector. The final position of hawks is represented by:

$$X_{new} = \left\{ \begin{array}{ll} Y_1 & if \; F(Y_1) < F(X_c) \\ Z_1 & if \; F(Z_1) < F(X_c) \end{array} \right\} \tag{15}$$

Hard besiege with progressive behavior:

$$\left. \begin{aligned} Y_1 &= X_{prey} - E_g \left| 2(1 - r_5).X_{prey} - X_{avg} \right| \\ Z_1 &= Y + S \times LF(d) \end{aligned} \right| r_5 < 0.5 \, \& \, \left| E_g \right| < 0.5 \quad (16)$$

The final strategy to update the position of hawks can be mathematically represented as:

$$X_{new} = \left\{ \begin{aligned} Y_1 & \quad if \quad F(Y_1) < F(X_c) \\ Z_1 & \quad if \quad F(Z_1) < F(X_c) \end{aligned} \right\} \quad (17)$$

2.5 Moth-flame optimization

MFO algorithm is inspired by the navigation of moth fly at night [16]. The moth position is updated as follows:

$$X_{new} = \left| F_c - X_c \right| \cdot e^{bt} \cdot \cos(2\pi t) + F_c \quad (18)$$

where F_c is cth flame, b is constant, and $t \in [-1, 1]$.

2.6 Particle swarm optimization

PSO models the group intelligence of bird flocks for finding food [17]. Each bird tends to move toward its personal best and the group's best position. Birds update their velocity and position as follows:

$$V_{new} = w * V_c + c_1 * r_1 * (X_{pbest} - X_c) + c_2 * r_2 * (G_{best} - X_c) \quad (19)$$

$$X_{new} = dt * V_{new} + X_c \quad (20)$$

where V_c and V_{new} represent old and new velocity of bird, respectively, X_{pbest} is personal best position of bird, G_{best} is global best position, w is inertial weight, r_1 & $r_2 \in [0, 1]$, c_1 & c_2 are constants, and dt is retarding parameter.

2.7 Whale optimization algorithm

The bubble-net feeding behavior of the humpback whales during foraging is modeled in the WOA algorithm [23]. The exploitation phase of the algorithm can depict the process of encircling the prey, and the exploration phase randomly searches for prey. They can be mathematically represented as:

- Exploitation phase:

$$X_{new} = \left\{ \begin{aligned} & X_{best} - U.\left| C_1.X_{best} - X_c \right| & \quad if \quad r_2 < 0.5 \\ & \left| X_{best} - X_c \right|.e^{bl}.\cos(2\pi l) + X_{best} & \quad otherwise \end{aligned} \right. \quad (21)$$

where $U = 2.a.r_1 - a$, $C_1 = 2.r_1$, a is a linearly decreasing parameter from 2 to 0, b is constant, $l \in [-1, 1]$, and r_1 and $r_2 \in [0, 1]$.

- Exploration phase:

$$X_{new} = X_r - U \cdot |C_1.X_r - X_c| \tag{22}$$

where X_r is random whale. U is a random vector.

Local stagnation is one of the major issues faced by optimization algorithms. Even though researchers tried to minimize this by using various parameters, it either costs reduced exploitation or exploration in the algorithm. In this chapter, the authors tried to balance these two aspects of the optimization algorithm by introducing a mutation parameter.

3. Influencer mutation

Recently, several NIA such as salp swarm optimization [24], environmental adaptation method [25], grasshopper optimization algorithm (GOA) [26], and crow search algorithm [27] have been proposed. Most of these algorithms suffer from local stagnation problems because of which they are unable to perform well on a few benchmark functions and complex optimization problems [28]. For this purpose, modifications have been proposed to achieve global optima [29, 30]. To overcome the problem of local stagnation, we propose a novel mutation strategy named influencer mutation, which depicts the impact of a group of solutions on other solutions to solve problems.

The aim of the proposed mutation strategy is to generate random-based solutions to increase diversity among different algorithms and keep them away from the local optimum. In this work, a novel biologically inspired influencer mutation operator has been presented. Influencer mutation works on the concept that a group of best-performing solutions exists that affect the position of other solutions. Also, every solution tries to improve its performance by getting far away from the worst-performing solution. Moreover, a minor impact can also be seen on the performance of solutions because of some random solutions. This paradigm can be correlated to human behavior while making any decision. We often get influenced by a group of people we feel are the best performers for that characteristic. So, the concept of getting influenced by multiple individuals in a population has been modeled here. This allows the particles to move through the search space and fix the local stagnation problem. The mathematical representation is as follows:

$$X_{mut}^k = X_{new}^k + F * r_1 * (X_{inf} - X_{worst}) + F * r_2 * (X_{S_1} - X_{S_2}) \tag{23}$$

Here, k represents the solution on which mutation has been performed, X_{mut}^k is the position of kth solution generated from mutation operation, X_{new}^k is the position of kth solution, X_{S_1} and X_{S_2} are positions of two random solutions from the given population, r_1 & r_2 are random values, and F is scaling factor similar to differential equation algorithm whose value lie between 0 and 1 [31]. X_{inf} is a group of solutions with good performance, while X_{worst} is a nonperforming solution. For the implementation, 0.2 has been considered the mutation rate, and the influencer group consists of 10% top-performing solutions in the current generation.

4. Influencer mutation-based optimization algorithm

In this section, we describe the conceptual and mathematical details of influencer mutation and how it can be combined with various existing optimization algorithms. For this purpose, we have utilized six well-known nature-inspired optimization algorithms: ALO, GWO, HHO, PSO, MFO, and WOA. The basic idea is to amplify the efficacy of these algorithms by combining the proposed mutation strategy and profoundly improve their ability to explore and exploit to reach the optimal value. The flowchart to develop an influencer mutation-based algorithm is demonstrated in Fig. 1.

The process begins with the random initial population X_{old} of size Ps generated with the range of lb and ub. Set the maximum function evaluations (Max_fe), mutation rate (m_rate), and n_inf with the number of influencers. Next, find the fitness of X_{old} and set $fe = 1$. Now, if $fe \leq Max\_fe$, find influencer group X_{inf} and worst particle X_{worst}; else, particle with the best fitness value is the final solution. Find new position X_{new} of all the particles using the position update equation given in the previous section for a given algorithm. Now, choose two random particles X_{S_1} and X_{S_2} from X_{new} and apply the influencer mutation operator on X_{new} by using Eq. (1) to find X_{mut} when mutation condition is followed. Update X_{mut} to X_{old} and continue this process until termination condition satisfies.

The steps required for implementing the influencer mutation-based optimization algorithm are as follows:

4.1 Initialization

Set initial parameters of the algorithms along with general parameters such as population size (ps), search space boundary (lb, ub), maximum function evaluation (Max_{fe}), and influencer number (n_i, nf).

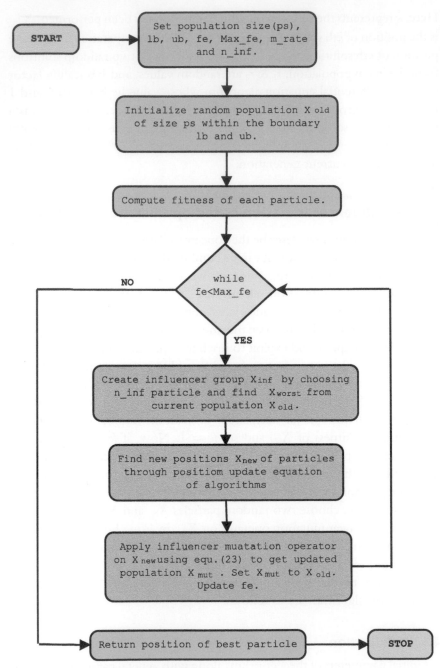

Fig. 1 Flowchart of proposed algorithm.

4.2 Compute influencer group

The initialized population goes through fitness evaluation for computing the influencer group. Then, the top n_i and nf number of particles form the influencer group. This will be used for the mutation operation step.

4.3 Update positions

The position of the individual particles is updated using the update equation of the respective algorithm. The corresponding equations of the algorithms can be referred to in Section 2.

4.4 Apply mutation

Once the particle updates its position, some percentage of the population undergoes mutation to improve the exploration of the particle. Mutation percentage is taken as a hyperparameter for the algorithm but generally set to 20–30%.

4.5 Selection

The mutation generates new particles from the current population. In order to maintain the population to ps, top-performing particles are selected from the entire set of particles, and the rest will be discarded. This ensures retaining the top performers of the population over the iterations.

4.6 Termination

In every iteration, the termination condition is checked, which can be maximum function evaluations or fitness of the best particle.

The flowchart shows that, at the end of every generation, X_{old} of any optimization algorithm is updated by passing through its population update procedure followed by the influencer mutation operator. This whole process of population update continues until the termination condition is reached, and finally, a particle with a near-optimal value is found. Following six improved optimization algorithms have been proposed by combining influencer mutation with ALO, GWO, HHO, MFO, PSO, and WOA algorithms:

- IALO (Influencer ALO)
- IGWO (Influencer GWO)
- IHHO (Influencer HHO)
- IMFO (Influencer MFO)
- IPSO (Influencer PSO)
- IWOA (Influencer WOA)

Here, two experiments are performed to prove the efficacy of the proposed mutation strategy. First, the influencer mutation operator is combined with six popular optimization algorithms. Thereafter, improved influencer mutation-based algorithms are integrated with ANN to solve real problems. The simulations of the first experiment were performed on MATLAB R2020b software and the second on Python 3.8 software.

5. Experimental results and analysis

This section discusses the implementation details and results of the proposed mutation operator over benchmark functions and the electricity price.

5.1 Dataset description

ISO, England electricity price dataset considers fourteen features, which are taken as input to ANN with improved influencer mutation-based algorithms. These input parameters include temperature, seasonal dependency, load, previous price, and holiday/weekend indicator [32]. The input parameter from $P_1 \ldots P_{14}$ used in our implementation is mentioned in Fig. 4. The dataset is normalized and divided into training and test data. The former includes data from 2004 to 2007 and the latter from the year 2008. This splitting ensures that the model is tested on out-of-sample data.

5.2 Performance evaluation metrics

The following five evaluation metrics have been utilized in this work for price prediction.
- *Mean absolute error*

$$\text{MAE} = \frac{1}{N} \sum_{j=1}^{n} |Y_j - Y_j'| \tag{24}$$

- *Normalized mean squared error*

$$\text{NMSE} = \frac{1}{\Delta^2 N} \sum_{j=1}^{n} (Y_j - Y_j')^2 \tag{25}$$

- *Root of mean squared error*

$$\text{RMSE} = \sqrt{\frac{1}{N} \sum_{j=1}^{n} \left(Y_j - Y_j'\right)^2} \tag{26}$$

- *Mean absolute percent error*

$$\text{MAPE} = \frac{1}{N} \sum_{j=1}^{n} \frac{|Y_j - Y_j'|}{Y_j} * 100 \tag{27}$$

- *Pearson's correlation coefficient*

$$r = \frac{\sum_{j=1}^{n} (Y_j - \bar{Y})(Y_j' - \bar{Y}')}{\sqrt{\sum_{j=1}^{n} (Y_j - \bar{Y})^2 \sum_{j=1}^{n} (Y_j' - \bar{Y}')^2}} \tag{28}$$

where Y_j is target value, Y_j' is obtained value, $\bar{Y}$ is mean of Y_j, $\bar{Y}'$ is mean of Y_j', and N is number of training data.

5.3 Experiment I: Benchmark function

Twenty standard benchmark functions have been utilized in this work to validate the performance of improved algorithms. Out of these, F1–F6 are fixed unimodal, F7–F10 are fixed multimodal, F11–F15 are unimodal, and F16–F20 are multimodal functions. The mathematical equations and their details are shown in Table 1. For this experiment, *Ps* is 100, and *Max_fe* = 10, 000 has been considered.

Table 2 shows the best and average optimal values generated after 25 epochs for F1–F10 benchmarks in 2D. The proposed algorithms, IALO, IGWO, IHHO, IMFO, IPSO, and IWOA, have been compared with their base algorithms to show improvement in their performance. The best value out of the base and improved optimization algorithm has been shown in bold, which implies that the results are upgraded for most of the fixed unimodal and fixed multimodal functions, or a very close value has been generated. The one-to-one comparison is used to evaluate the performance of the proposed algorithm [33, 34]. Tables 3 and 4 show the results of unimodal and multimodal functions on 2, 3, 5, and 10 dimensions. The bold values show that IALO, IGWO, IHHO, IMFO, IPSO, and IWOA achieved better results than their base algorithms in different dimensions. The table also reveals that IMFO and IWOA perform well for 2 dimensions,

Table 1 Benchmark functions.

F_{ID}	Mathematical equation	Dim	Domain
F1	$f(y,z) = y^2 + 2z^2 - 0.3\cos(3\pi y) - 0.4\cos(4\pi z) + 0.7$	2	$[-100,100]$
F2	$f(y,z) = (y + 2z - 7)^2 + (2y + z - 5)^2$	2	$[-10,10]$
F3	$f(y,z) = 2y^2 - 1.05y^4 + \frac{y^6}{6} + yz + z^2$	2	$[-5,5]$
F4	$f(y,z) = 0.26(y^2 + z^2) - 0.48yz$	2	$[-10,10]$
F5	$f(y,z) = 0.5 + \frac{\sin^2(y^2 + z^2)^2 - 0.5}{(1 + 0.001(y^2 + z^2))^2}$	2	$[-100,100]$
F6	$f(y,z) = 0.5 + \frac{\sin^2(y^2 - z^2) - 0.5}{(1 + 0.001(y^2 + z^2))^2}$	2	$[-100,100]$
F7	$f(y,z) = (1.5 - y + yz)^2 + (2.25 - y + yz^2)^2 + (2.625 - y + yz^3)^2$	2	$y\in[-4.5,4.5]$
F8	$f(y,z) = y^2 + 2z^2 - 0.3\cos(3\pi y)\cos(4\pi z) + 0.3$	2	$y\in[-100,100]$
F9	$f(y,z) = \sin^2(3\pi y) + (y - 1)^2(1 + \sin^2(3\pi z)) + (z - 1)^2(1 + \sin^2(2\pi z))$	2	$[-10,10]$
F10	$f(\mathbf{y}) = 0.5 + \frac{\sin(y_1^2 - y_2^2)^2 - 0.5}{[1 + 0.001*(y_1^2 + y_2^2)]^2}$	2	$[-100,100]$
F11	$f(\mathbf{y}) = \sum_{i=1}^{n} y_i^2$	$[2,3,5,10]$	$[-5.12,5.12]$
F12	$f(\mathbf{y}) = \sum_{i=1}^{n} |y_i| + \prod_{i=1}^{n} |y_i|$	$[2,3,5,10]$	$[-10,10]$
F13	$f(\mathbf{y}) = \sum_{i=1}^{n} |y_i|^{i+1}$	$[2,3,5,10]$	$[-1,1]$
F14	$f(\mathbf{y}) = \sum_{i=1}^{n} iy_i^2$	$[2,3,5,10]$	$[-10,10]$
F15	$f(\mathbf{y}) = \sum_{i=1}^{n} y_i^2 + (\sum_{i=1}^{n} 0.5iy_i)^2 + (\sum_{i=1}^{n} 0.5iy_i)^4$	$[2,3,5,10]$	$[-5,10]$
F16	$f(y,z) = 10n + \sum_{i=1}^{n} (y_i^2 - 10\cos(2\pi y_i))$	$[2,3,5,10]$	$[-5.12,5.12]$

F17 $\quad f(\mathbf{y}) = -a.exp\left(-b\sqrt{\frac{1}{n}\sum_{i=1}^{n} y_i^2}\right) - exp\left(\frac{1}{n}\sum_{i=1}^{n}\cos(c y_i)\right) + a + exp(1)$ $\qquad$ [2,3,5,10] $\quad$ [−32,32]

F18 $\quad f(\mathbf{y}) = \frac{\pi}{n}\left\{10\,\sin^2(\pi z_1) + \sum_{i=1}^{n-1}(z_i-1)^2[1+10\sin^2(\pi z_{i+1})] + (z_n-1)^2\right\}$ $\qquad$ [2,3,5,10] $\quad$ [−50,50]

$\qquad + \sum_{i=1}^{n} u(y_i, 10, 100, 4),\ z_i = 1 + \frac{y_i+1}{4},\ u(y_i, a, k, m) = \begin{cases} k(y_i - 1)^m & y_i > a \\ 0 - a < y < a \\ k(-y_i - a)^m & y_i < -a \end{cases}$

F19 $\quad f(\mathbf{y}) = 0.1\left\{\sin^2(3\pi y_1) + \sum_{i=1}^{n}(y_i-1)^2[1+\sin^2(3\pi y_i+1)] + (y_n-1)^2[1+\sin^2(2\pi y_n)]\right\}$ $\qquad$ [2,3,5,10] $\quad$ [−50,50]

$\qquad + \sum_{i=1}^{n} u(y_i, 5, 100, 4)$

F20 $\quad f(\mathbf{y}) = 1 - \cos\left(2\pi\sqrt{\sum_{i=1}^{D} y_i^2}\right) + 0.1\sqrt{\sum_{i=1}^{D} y_i^2}$ $\qquad$ [2,3,5,10] $\quad$ [−100,100]

Table 2 Results of optimization algorithms on F1–F10 benchmark functions.

FID		ALO	IALO	GWO	IGWO	HHO	IHHO	MFO	IMFO	PSO	IPSO	WOA	IWOA
F1	Min	9.63E-12	**3.65E-12**	0	**0**	0	**0**	0	**0**	0	**0**	0	**0**
	Avg	1.63E-10	**1.37E-10**	0	**0**	0	**0**	0	**0**	0	**0**	0	**0**
F2	Min	**3.46E-15**	6.76E-15	3.95E-08	**2.66E-07**	5.24E-07	**5.68E-14**	2.78E-23	**8.38E-26**	1.9E-25	**1.75E-27**	6.32E-06	**2.16E-09**
	Avg	3.24E-13	**2.96E-13**	4.54E-06	**3.46E-06**	0.000307	**5.42E-05**	2.75E-20	**1.27E-20**	4.29E-23	**1.37E-22**	0.001821	**0.000167**
F3	Min	1.83E-16	**4.44E-17**	3.36E-96	**1.1E-99**	1.54E-36	**1.49E-40**	2.54E-27	**1.16E-27**	2.5E-29	**1.15E-32**	4E-40	**3.62E-57**
	Avg	**9.73E-15**	1.33E-14	1.56E-66	**1.57E-71**	1.4E-29	**3.27E-31**	2.75E-23	7.39E-23	5.98E-27	1.42E-26	0.011946	**3.09E-36**
F4	Min	8.42E-17	**5.71E-17**	**6.14E-59**	3.94E-56	**1.27E-50**	3.33E-48	1.48E-23	6.52E-23	9.05E-26	1.57E-25	1.28E-72	**1.85E-75**
	Avg	5.1E-15	**3.62E-15**	1.04E-39	**1.8E-41**	1.1E-34	**3.26E-38**	6.81E-18	**7.7E-19**	1.02E-21	1.41E-21	3.88E-61	**8.25E-65**
F5	Min	**2.22E-16**	6.66E-16	0	**0**	0	**0**	0	**0**	0	**0**	0	**0**
	Avg	2.67E-14	**2.21E-14**	0	**0**	0	**0**	0	**0**	0	**0**	7.07E-05	**0**
F6	Min	8.88E-16	**4.44E-16**	0	**0**	0	**0**	0	**0**	0	**0**	0	**0**
	Avg	3.58E-14	**2.08E-14**	0	**0**	0	**0**	0	**0**	0	**0**	0.000205	**2.68E-06**
F7	Min	6.47E-16	**2.99E-16**	3.87E-08	**1.46E-08**	1.46E-26	**0**	2.5E-20	**4.57E-21**	7.69E-24	7.37E-23	**1.02E-13**	1.37E-12
	Avg	0.091448	**1.99E-14**	1.5E-06	**1.38E-06**	5.38E-09	**1.23E-11**	3.03E-17	7.89E-17	**9.88E-19**	1.39E-17	0.060966	**0.030483**
F8	Min	**1.07E-12**	4.81E-12	0	**0**	0	**0**	0	**0**	0	**0**	0	**0**
	Avg	2.11E-10	**1.65E-10**	0	**0**	0	**0**	0	**0**	0	**0**	0.043663	**0.008733**
F9	Min	**6.79E-15**	4.19E-14	1.47E-07	2.15E-07	6.81E-16	**5.33E-24**	7.06E-25	**1.49E-25**	**5.2E-29**	7.37E-29	**5.77E-11**	9.26E-09
	Avg	2.04E-12	**5.56E-13**	3.98E-06	**3.14E-06**	0.000117	**2.33E-06**	3.27E-22	2.17E-21	**1.95E-25**	1.07E-24	1.1E-05	**6.16E-06**
F10	Min	**2.22E-16**	6.66E-16	0	**0**	0	**0**	0	**0**	0	**0**	0	**0**
	Avg	3.19E-14	**2.19E-14**	0	**0**	0	**0**	0	**0**	0	**0**	0.000246	**2.39E-05**

Table 3 Results of optimization algorithms on benchmark functions in 2D and 3D.

FID		ALO	IALO	GWO	IGWO	HHO	IHHO	MFO	IMFO	PSO	IPSO	WOA	IWOA
2D													
F11	Min	2.67E-16	2.64E-16	1.08E-95	5.84E-98	3.32E-40	1.61E-40	1.09E-27	2.12E-27	4.06E-31	4.6E-31	3.48E-61	1.04E-66
	Avg	1.24E-14	1.67E-14	2.48E-70	4.36E-74	2.65E-31	1.88E-31	1.29E-24	1.27E-24	2.75E-27	5.95E-26	2.22E-42	1.22E-45
F12	Min	2.22E-08	6.57E-08	2.82E-49	5.81E-52	5.27E-21	9.99E-22	5.47E-14	2.91E-14	1.1E-15	2.19E-16	1.11E-31	1.06E-34
	Avg	2.41E-07	2.91E-07	2.72E-37	1.41E-38	4.55E-16	5.28E-17	1.87E-12	2.01E-12	8.07E-14	4.44E-14	1.52E-24	1.89E-28
F13	Min	9.46E-22	4.5E-22	3.9E-104	2.2E-101	5.62E-46	1.87E-50	3.1E-35	2.12E-35	1.65E-37	6.7E-38	1.29E-63	3.23E-69
	Avg	2.18E-16	4.67E-16	4.32E-73	2.82E-73	1.28E-32	9.83E-37	3.22E-30	1.36E-30	9.96E-33	5.2E-33	7.13E-47	1.86E-52
F14	Min	1.46E-15	3.44E-16	3.8E-98	6.2E-102	2.68E-40	5.36E-49	1.65E-26	5.7E-27	7.64E-30	1.2E-30	5.56E-60	1.53E-66
	Avg	5.28E-14	6.56E-14	1.99E-74	9.03E-79	1.15E-31	4.56E-32	9.85E-24	8.2E-24	2.27E-26	8.27E-26	4.52E-41	5.22E-53
F15	Min	2.47E-16	7.49E-17	2.1E-101	6.22E-96	1.44E-38	5.06E-39	1.77E-28	8.83E-28	2.66E-30	2.95E-29	3.19E-47	5.77E-62
	Avg	5.21E-14	6.28E-14	1.9E-66	2.19E-73	1.31E-28	8.14E-30	4.56E-23	4.07E-24	4.27E-26	3.87E-26	1.59E-29	1.01E-42
F16	Min	2.2E-13	3.06E-13	0	0	0	0	0	0	0	0	0	0
	Avg	0.119395	0.079597	0	0	0	0	0	0.039798	0	0	0	0
F17	Min	7.79E-07	7.57E-07	8.88E-16	8.88E-16	8.88E-16	8.88E-16	1.96E-13	3.06E-13	7.99E-15	1.15E-14	8.88E-16	8.88E-16
	Avg	4.38E-06	3.61E-06	8.88E-16	1.03E-15	2.15E-14	3.73E-15	2.03E-11	1.55E-11	1.02E-12	5.69E-13	2.74E-15	2.17E-15
F18	Min	1.78E-14	1.06E-14	9.72E-10	7.29E-10	1.49E-22	3.67E-23	3.57E-25	2.94E-25	1.69E-28	4.33E-29	6.64E-09	6.5E-10
	Avg	3.01E-12	2.59E-12	4.52E-07	2.93E-07	4.6E-06	1.15E-06	5.63E-22	4.27E-22	1.61E-24	1.75E-23	8.18E-07	5.39E-07

Continued

Table 3 Results of optimization algorithms on benchmark functions in 2D and 3D.—cont'd

FID		ALO	IALO	GWO	IGWO	HHO	IHHO	MFO	IMFO	PSO	IPSO	WOA	IWOA
F19	Min	5.86E-14	**6.58E-15**	3.44E-08	**9.06E-09**	1.09E-12	**1.76E-24**	5.45E-25	**1.07E-25**	**2.64E-30**	7.96E-30	1.04E-09	**8.16E-10**
	Avg	3.99E-12	**2.88E-12**	5.13E-07	**3.7E-07**	2.24E-05	**5.07E-06**	**3.9E-22**	1.92E-21	2.63E-24	**1E-24**	9.43E-07	**8.58E-07**
F20	Min	1.1E-07	**7.66E-08**	6.69E-47	**3.18E-49**	8.74E-19	**2.63E-22**	1.72E-12	**6.28E-13**	**5.99E-13**	1.75E-12	5.2E-30	**8.27E-34**
	Avg	0.011985	**2.79E-07**	**0.00799**	0.02397	8.09E-16	**2.35E-16**	0.011985	**0.003995**	0.003995	**4.71E-07**	0.055929	0.039949

3D

F11	Min	3.83E-15	**3.63E-15**	**2.47E-67**	1.62E-65	**4.2E-37**	9.02E-37	**8.52E-18**	1.77E-17	1.99E-24	**6.59E-26**	9.67E-40	**1.35E-42**
	Avg	**1.34E-13**	1.58E-13	7.5E-51	**3.5E-53**	7.57E-27	**4.65E-31**	7.59E-15	**1.68E-15**	6.25E-22	6.39E-22	4.83E-31	**2.69E-34**
F12	Min	2.49E-07	**2.15E-07**	1.89E-31	**3.18E-35**	**4.13E-20**	2.81E-19	1.06E-08	**1.48E-09**	1.54E-12	**3.03E-13**	7.27E-21	**4.12E-22**
	Avg	9.37E-07	**7.52E-07**	4.14E-27	**5.96E-28**	3.73E-15	**1.38E-15**	8.1E-08	**5.46E-08**	2.75E-11	2.82E-11	1.25E-17	**1.83E-19**
F13	Min	7.26E-15	**4.17E-16**	**5.16E-76**	4.28E-75	1.51E-45	**3.25E-54**	2.91E-26	3.07E-26	**6.13E-34**	2.5E-33	9.55E-49	**5.2E-56**
	Avg	8.8E-11	**3.65E-11**	**1.51E-61**	1.57E-61	1.55E-34	**8.66E-35**	**8.52E-22**	1.2E-21	**2.96E-29**	4.03E-29	1.17E-34	**5.35E-40**
F14	Min	1.14E-13	**6.96E-14**	8.16E-64	**2.42E-66**	**9.37E-39**	1.46E-35	**2.68E-16**	4.13E-16	**8.28E-24**	9.3E-24	1.58E-35	**1.21E-45**
	Avg	1.39E-12	**1.02E-12**	2.31E-48	**3.51E-50**	**4.06E-29**	5.32E-29	6.11E-14	**2.72E-14**	9.89E-21	**3.14E-21**	1.71E-28	**2.09E-35**
F15	Min	**4.15E-14**	7.16E-14	1.63E-52	**4.24E-55**	2.99E-34	**3.7E-35**	**8.26E-16**	9.08E-16	**3.53E-24**	3.01E-23	6.26E-22	**2E-32**
	Avg	6.16E-13	**5.53E-13**	1.19E-40	**1.69E-41**	4.91E-28	3.67E-25	2.13E-13	**9.77E-14**	6.39E-20	**4.8E-20**	6.3E-18	**1.62E-24**
F16	Min	**8.89E-12**	1.57E-11	0	0	0	0	1.42E-14	1.14E-13	0	0	0	0
	Avg	2.348102	**1.273548**	**0**	0.039818	0	0	0.318387	0.278589	4.98E-09	3.72E-05	**2.84E-16**	0.119483

F17	Min	2.98E-06	3.52E-06	8.88E-16	8.88E-16	1.74E-08	6.14E-08	5.64E-12	1.47E-11	4.44E-15	8.88E-16	
	Avg	9.83E-06	9.2E-06	3.02E-15	1.88E-15	2.42E-14	6.29E-07	8.21E-07	2E-10	1.98E-10	2.36E-14	7.57E-15
F18	Min	4.63E-12	6.67E-13	1.9E-07	1.03E-07	4.19E-11	7.78E-16	8.8E-22	1.34E-22	3.27E-07	4.29E-08	
	Avg	0.290265	0.331706	1.45E-06	1.34E-06	4.71E-08	3.05E-15	2.32E-20	2.91E-20	5.86E-05	1.26E-05	
F19	Min	4.58E-12	4.97E-13	1.05E-07	5.45E-08	1.78E-07	1.11E-13	1.28E-15	8.56E-24	3.21E-07	2.32E-07	
	Avg	3.1E-10	4.3E-11	2.62E-06	1.84E-06	6.48E-07	2.18E-15	1.11E-22	2.13E-19	5.53E-05	2.22E-05	
F20	Min	4.04E-07	1.03E-06	2.91E-26	4.33E-19	1.16E-07	1.22E-06	1.05E-06	5.19E-06	5.51E-17	6.38E-19	
	Avg	0.087889	0.087889	0.070948	0.084857	0.083899	0.085137	0.087897	0.082027	0.059924	0.075904	

Table 4 Results of optimization algorithms on benchmark functions in 5D and 10D.

FID		ALO	IALO	GWO	IGWO	HHO	IHHO	MFO	IMFO	PSO	IPSO	WOA	IWOA
5D													
F11	Min	2.8E-13	2.13E-13	3.53E-36	2.46E-39	3.22E-35	3.13E-36	1.59E-09	2.13E-09	1.27E-17	2.3E-18	7.94E-29	1.77E-31
	Avg	4.22E-12	3.53E-12	4.18E-32	7.03E-33	6.87E-29	2.71E-29	9.11E-08	6.65E-08	1.13E-15	1.75E-15	1.37E-23	4.37E-26
F12	Min	2.67E-06	2.54E-06	5.32E-20	3.33E-20	4.47E-20	2.37E-19	4.35E-05	9.16E-05	3.48E-09	5.54E-09	3.17E-16	9.33E-17
	Avg	6.43E-06	6.51E-06	5.72E-17	4.89E-18	2.23E-15	5.26E-15	0.000371	0.000371	6.6E-08	5.08E-08	5.95E-14	1.95E-14
F13	Min	6.1E-09	4.73E-10	1.25E-55	1.39E-57	1.28E-46	1.31E-47	1.03E-18	2.49E-18	1.67E-28	8.93E-27	8.61E-47	3.52E-48
	Avg	5.34E-08	4.56E-08	3.55E-48	3.95E-49	3.37E-34	2.21E-34	1.43E-14	1.3E-14	3.6E-24	3.56E-24	2.08E-32	4.44E-38
F14	Min	1.17E-11	1.11E-11	7.65E-36	3.17E-39	2.52E-35	4.36E-35	4.86E-08	4.32E-08	2.74E-17	1.42E-17	1.2E-27	3.13E-29
	Avg	3.28E-10	6.32E-11	1.49E-29	7.78E-32	7.69E-27	7.77E-30	6.36E-07	5.97E-07	1.57E-14	2.87E-14	8.67E-24	2.19E-24
F15	Min	2.39E-12	3.8E-12	1.86E-27	1.29E-28	3.1E-33	4.03E-33	1.94E-07	3.67E-07	4.87E-13	7.4E-13	1.1E-10	3.24E-17
	Avg	1.2E-10	9.86E-11	4.22E-22	3.07E-25	7.74E-26	4.74E-25	4.46E-05	1.86E-05	6.06E-11	1.34E-10	7.9E-05	1.27E-11
F16	Min	5.04E-10	0.994959	0	0	0	0	0.000613	0.000137	5.14E-11	4.36E-08	0	0
	Avg	5.372774	4.974791	0.711767	0.366023	0	0	2.864588	2.138947	0.850729	0.996205	0.720546	1.093191
F17	Min	1.71E-05	7.09E-05	4.44E-15	4.44E-15	8.88E-16	8.88E-16	0.00042	0.000547	1.02E-07	1.84E-08	1.51E-14	1.51E-14
	Avg	4.81E-05	3.9E-05	3.22E-14	1.34E-14	8.19E-14	1.3E-14	0.002421	0.002215	8.59E-07	4.02E-07	3.12E-11	1.45E-12
F18	Min	1.97E-10	1.32E-10	6.16E-07	7.26E-07	2.83E-06	2.79E-07	1.02E-08	3.26E-08	2.69E-16	2.41E-16	3.11E-05	1.49E-06
	Avg	1.0454	0.255534	3.56E-06	3.18E-06	0.000366	8.39E-05	8.78E-05	9.64E-07	6.26E-14	1.51E-13	0.00236	0.000288
F19	Min	2.01E-10	1.56E-10	1.9E-06	3.07E-06	2.41E-06	5.34E-08	3.1E-07	2.9E-08	6.98E-15	4.39E-15	9.58E-06	2.39E-05
	Avg	0.00044	0.000879	8.75E-06	8.52E-06	0.00053	0.000129	2.91E-06	5.95E-06	6.71E-13	3.69E-13	0.003181	0.000275
F20	Min	0.099873	0.099873	0.099873	4.32E-12	1.12E-17	1.64E-18	0.099873	0.099873	0.099873	0.099873	2.12E-11	3.5E-14
	Avg	0.099873	0.099873	0.099873	0.095878	5.34E-14	2.69E-13	0.100125	0.108004	0.099873	0.099873	0.119883	0.103883

10D

F11	Min	1.94E-09	7.46E-11	7.77E-20	1.1E-20	4.79E-36	4.76E-36	0.001307	0.00144	3.95E-09	2.17E-09	1.83E-25	3.55E-26
	Avg	1.63E-08	8.83E-09	1.34E-17	4.65E-19	2.55E-28	5.2E-29	0.010768	0.009506	2E-07	2.01E-07	1.07E-20	3.12E-22
F12	Min	0.000606	0.000449	6.74E-11	4.39E-11	1.08E-17	2.32E-17	0.122223	0.129481	0.000144	0.000224	7.99E-14	8.55E-15
	Avg	3.215276	1.800735	1.88E-09	2.28E-10	3.66E-14	9.86E-15	0.331057	0.283742	0.00154	0.401138	6.83E-12	2.58E-12
F13	Min	4.25E-08	1.4E-08	4.75E-44	2.59E-44	2.2E-52	9.06E-48	1.03E-11	2.24E-11	5.8E-20	1.55E-18	6.06E-42	6.16E-44
	Avg	1.51E-06	6.64E-07	2.71E-37	1.67E-39	1.85E-32	4.54E-35	1.66E-08	9.13E-09	1.79E-15	3.87E-15	1.63E-33	3.37E-37
F14	Min	2.89E-07	1.17E-06	3.06E-19	1.41E-19	2.81E-38	3.03E-32	9.13E-09	0.041459	2.62E-08	1.13E-07	1E-24	1.88E-24
	Avg	0.143754	0.012346	1.4E-16	4.02E-17	7.91E-27	1.77E-27	0.017942	0.136849	4.000003	4.000003	1.66E-20	3.41E-21
F15	Min	0.00039	9.12E-07	5.16E-12	3.07E-12	1.71E-30	7.42E-34	0.570221	0.39989	0.007476	0.007697	4.516456	0.000477
	Avg	2.626344	1.394119	7.34E-10	1.03E-10	3.64E-19	9.72E-23	5.725264	4.046167	1.869961	2.38503	45.37173	0.423929
F16	Min	5.969758	4.974796	1.35E-10	2.59E-11	0	0	7.173798	9.029355	4.831242	3.89318	0	0
	Avg	17.78984	16.83469	3.534622	3.561737	0	0	16.89601	15.35863	9.370488	10.79317	4.097897	8.005167
F17	Min	0.000602	0.000373	3.31E-10	4.07E-10	8.88E-16	8.88E-16	0.752718	0.538108	0.000642	0.000722	2.65E-11	6.12E-13
	Avg	0.795502	0.664168	1.62E-08	4.05E-09	1.29E-13	4.96E-14	1.950155	1.574295	0.003199	0.00375	4.09E-10	1.01E-10
F18	Min	1.32E-05	6.54E-06	3.26E-06	4.06E-06	4.81E-07	4.6E-07	0.009683	0.022533	2.23E-07	2.3E-07	0.00098	0.000768
	Avg	3.133618	2.607253	0.003691	0.000798	0.000217	7.37E-05	0.447726	0.52452	6.76E-05	0.024941	0.08774	0.18555
F19	Min	2.27E-07	5.33E-07	3.15E-05	9.8E-06	1.55E-07	2.73E-07	0.061835	0.106552	1.05E-06	1.51E-06	0.002075	0.002375
	Avg	0.004041	0.002186	0.012079	4.56E-05	0.000742	0.000217	0.558172	0.503186	0.00068	0.000178	0.028187	0.035396
F20	Min	0.199873	0.199873	0.099873	0.099873	3.25E-18	5.04E-17	0.300349	0.403202	0.099892	0.099875	4.65E-11	1.5E-11
	Avg	0.311873	0.235873	0.119873	0.119873	1.21E-12	2.57E-12	0.735177	0.775702	0.214754	0.225	0.115894	0.151878

while IALO and IGWO for 5 and 10 dimensions. Figs. 2 and 3 show the convergence plot of benchmark functions. The curve shows that proposed mutated algorithms obtained superior results in most cases than their base algorithm.

5.4 Experiment 2: Electricity price forecasting

In this experiment, the performance of proposed algorithms: IALO, IGWO, IHHO, IMFO, IPSO, and IWOA are tested on the electricity load price forecasting problem.

Accurate electricity price prediction plays a considerable role in the deregulated energy market as they are responsible for the decision-making process of a power system. Factors such as high frequency and volatility, seasonal dependency, calendar effects, unseen fluctuations, and nonlinear dynamics of electricity prices affect forecasting accuracy. The objective of effective forecasting is the economic benefits and stability of power grids [35]. Therefore, it is much more expected from electricity producers and consumers to develop an accurate price forecasting model. Recently, many price forecasting models have been proposed and applied to electricity price forecasting problems to accurately fit the relationship between historical data and future values. Models such as autoregressive integrated moving average (ARIMA), ANN, support vector machine (SVM), and hybrid models have achieved good forecasting accuracy in the given scenarios. Findings from the literature reveal that hybrid models have proven themselves superior to their base models in terms of accuracy [36].

In the past few years, several hybrid models have been proposed and applied to price forecasting problems. Zhang et al. proposed a hybrid framework by combining COA, SVM, and singular spectrum analysis to mine important information from price signals for electricity price forecasting [35, 37]. Luo and Weng presented a two-stage machine learning model by combining several widely used models such as polynomial regression, SVM, ANN, and deep neural network to forecast the real-time market price [38]. Cheng et al. presented a new electricity price forecasting model for German electricity exchange by combining empirical wavelet transform, SVM, bidirectional long short-term memory, and Bayesian optimization [39]. Bhatia et al. developed an ensemble approach by combining extreme gradient boosted tree, random forest, and Bayesian linear regression for a very short-term electricity price forecasting [40]. Inspired by the literature, we have developed hybrid models (ANN-IALO, ANN-IGWO,

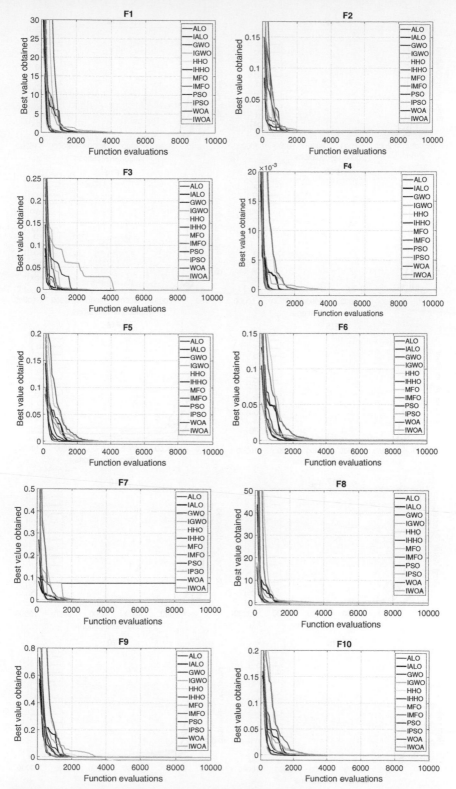

Fig. 2 2-D convergence plot of F1–F10 benchmark functions.

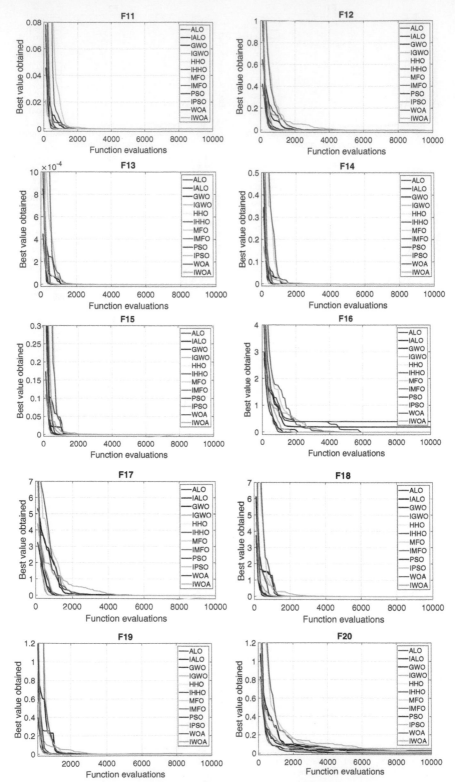

Fig. 3 2-D convergence plot of F11–F20 benchmark functions.

ANN-IHHO, ANN-IMFO, ANN-IPSO, and ANN-IWOA) by combining ANN with improved optimization algorithms to find optimal weight parameters of the network. Also, these hybrid model algorithms are compared with their base models to prove their existence.

Fig. 4 shows the block diagram for the same, which depicts that the weights of ANN updated using the influencer mutation-based algorithm, and after the model is trained for a given number of generation, it is ready to predict the price. The performance of hybrid models ANN-IALO, ANN-IGWO, ANN-IHHO, ANN-IMFO, ANN-IPSO, and ANN-IWOA has been tested on ISO, NEPOOL, and England electricity market data.

5.4.1 Objective function

For the implementation of hybrid models in this work, MSE as an objective function has been considered. The mathematical equation of MSE is given as follows:

$$MSE = \frac{1}{N} \sum_{j=1}^{n} (Y_j - Y_j'')^2 \tag{29}$$

where N is number of elements in training dataset, Y_j is target value, and Y_j'' is obtained value.

5.4.2 Results

For the implementation purpose, Ps is chosen as 100, Max_fe as 100,000, and the mutation rate as 0.2. The convergence plot obtained by ANN-IALO, ANN-IGWO, ANN-IHHO, ANN-IMFO, ANN-IPSO, and ANN-IWOA, along with their base models in the training phase, is shown

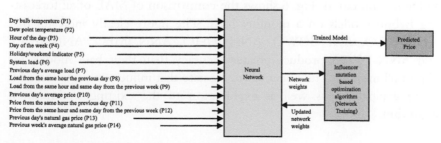

Fig. 4 Framework of the proposed forecasting model.

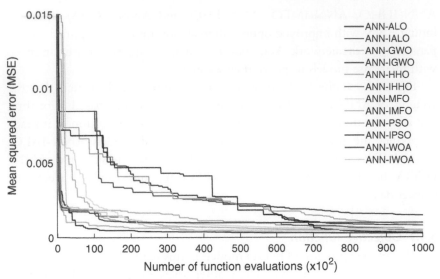

Fig. 5 Convergence curve of the hybrid models.

in Fig. 5. The figure shows a plot between the MSE value of hybrid algo-
rithms with increasing function evaluations. The plot shows that ANN-
IALO, ANN-IGWO, ANN-IHHO, ANN-IMFO, ANN-IPSO, and
ANN-IWOA converge faster and achieve less MSE than their base models.

Table 5 shows the results generated by all hybrid models in terms of
different evaluation metrics. The bold values shown in the table reveal that
hybrid models ANN-IALO, ANN-IGWO, ANN-IHHO, ANN-IMFO,
ANN-IPSO, and ANN-IWOA generate superior results. It can also be
noted that ANN-IGWO obtained a minimum MAPE of 8.068%, while
ANN-IALO had a maximum MAPE value of 13.343% among all forecasting models. Moreover, there has been a 27.4% improvement in MAPE
accuracy of ANN-ALO and 33.41% in ANN-WOA by combining
influencer mutation. Fig. 6 shows the comparison of MAE of all forecasting hybrid models on a monthly basis. The figure clearly indicates that
ANN-IALO, ANN-IGWO, ANN-IHHO, ANN-IMFO, ANN-IPSO,
and ANN-IWOA produce greater accuracy than their base models for
almost all months. The obtained results state that influencer mutation adds
more exploration as well as exploitation to the existing optimization
algorithm.

Table 5 The experimental results (price $/MWh) obtained from various forecasting models.

Algorithm	MAE	MAPE	MSE	RMSE	NMSE	Pearson_cof
ANN-ALO	14.44639	18.39034	413.6058	20.3373	0.552594	0.696234
ANN-IALO	10.73882	13.34321	268.7645	16.39404	0.35908	0.804331
ANN-GWO	6.682112	8.127526	121.0668	11.00304	0.16175	0.917371
ANN-IGWO	6.645133	8.067645	119.4345	10.92861	0.159569	0.919571
ANN-HHO	8.65378	10.22974	198.7611	14.09826	0.265553	0.875154
ANN-IHHO	7.921464	9.58638	163.4406	12.78439	0.218363	0.891215
ANN-MFO	9.564404	12.35544	183.4371	13.5439	0.245079	0.872119
ANN-IMFO	8.709049	11.01771	164.2237	12.81498	0.219409	0.886286
ANN-PSO	12.50582	15.35918	332.7895	18.24252	0.44462	0.758967
ANN-IPSO	6.718147	8.241777	115.7136	10.75702	0.154598	0.919513
ANN-WOA	13.58205	15.52715	404.5352	20.11306	0.540475	0.75573
ANN-IWOA	8.642687	10.33923	191.7351	13.84684	0.256166	0.877314

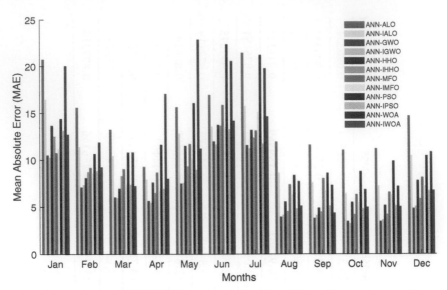

Fig. 6 Comparison of RMSE ($/MWh) value obtained by different forecasting models on a monthly basis.

5.4.3 Limitations

Although the results obtained for these algorithms are reasonable and significant, the accuracy required for real problems has yet to be improved. To achieve this, better prediction models are needed as there is a chance for premature convergence. Also, the proposed mutation operator is parameter dependent, i.e., the randomness within the operator is high.

6. Conclusions and future work

To date, several optimization algorithms have been presented in the literature to solve the local stagnation problem. In this work, we have introduced the concept of an influencer group, as most NIA are updated based on best-performing particles. Here, we have proposed a new mutation strategy inspired by the best-performing particles, the influencer group, which guides the remaining particle to reach optimal value.

The influencer mutation strategy has been combined with ALO, GWO, HHO, MFO, PSO, and WOA algorithms. The modified algorithms are tested on 20 standard benchmark functions. Results show significant improvement in proposed mutation-based algorithms compared to the performance of base algorithms. Further, IALO, IGWO, IHHO, IMFO, IPSO, and IWOA are validated by integrating them with ANN and testing on electricity price forecasting problems. The results show the efficacy of the influencer mutation operator when combined with ALO, GWO, HHO, MFO, PSO, and WOA algorithms. Based on the obtained results, we conclude that influencer mutation is valuable to the literature, which adds both exploration and exploitation to the algorithm. The proposed influencer mutation strategy can be utilized to solve different engineering problems.

References

[1] P. Singh, P. Dwivedi, Integration of new evolutionary approach with artificial neural network for solving short term load forecast problem, Appl. Energy 217 (2018) 537–549.

[2] R. Kottath, P. Singh, Influencer buddy optimization: algorithm and its application to electricity load and price forecasting problem, Energy 263 (2023) 125641.

[3] P. Singh, R. Kottath, Chaos follow the leader algorithm: application to data classification, J. Comput. Sci. 65 (2022) 101886.

[4] D.E. Goldberg, Genetic Algorithms, Pearson Education India, 2006.

[5] P. Singh, R. Kottath, Influencer-defaulter mutation-based optimization algorithms for predicting electricity prices, Util. Policy 79 (2022) 101444.

[6] I. De Falco, A. Della Cioppa, E. Tarantino, Mutation-based genetic algorithm: performance evaluation, Appl. Soft Comput. 1 (4) (2002) 285–299.

[7] A. Biswas, Atom stabilization algorithm and its real life applications, J. Intell. Fuzzy Syst. 30 (4) (2016) 2189–2201.

[8] X.-q. Zhang, Z.-f. Ming, An optimized grey wolf optimizer based on a mutation operator and eliminating-reconstructing mechanism and its application, Front. Inf. Technol. Electron. Eng. 18 (11) (2017) 1705–1719.

[9] S. Zhang, Z. Yongquan, Grey wolf optimizer with ranking-based mutation operator for IIR model identification, Chin. J. Electron. 27 (5) (2018) 1071–1079.

[10] P. Singh, P. Dwivedi, Short-term electricity load forecast using hybrid model based on neural network and evolutionary algorithm, in: Numerical Optimization in Engineering and Sciences, Springer, 2020, pp. 167–176.

[11] S. Gupta, K. Deep, S. Mirjalili, J.H. Kim, A modified sine cosine algorithm with novel transition parameter and mutation operator for global optimization, Expert Syst. Appl. 154 (2020) 113395.

[12] N. Panda, S.K. Majhi, S. Singh, A. Khanna, Oppositional spotted hyena optimizer with mutation operator for global optimization and application in training wavelet neural network, J. Intell. Fuzzy Syst. 38 (5) (2020) 6677–6690.

[13] S. Mirjalili, The ant lion optimizer, Adv. Eng. Softw. 83 (2015) 80–98.

[14] S. Mirjalili, S.M. Mirjalili, A. Lewis, Grey wolf optimizer, Adv. Eng. Softw. 69 (2014) 46–61.

[15] A.A. Heidari, S. Mirjalili, H. Faris, I. Aljarah, M. Mafarja, H. Chen, Harris hawks optimization: algorithm and applications, Futur. Gener. Comput. Syst. 97 (2019) 849–872.

[16] S. Mirjalili, Moth-flame optimization algorithm: a novel nature-inspired heuristic paradigm, Knowl.-Based Syst. 89 (2015) 228–249.

[17] J. Kennedy, R. Eberhart, Particle swarm optimization, in: Proceedings of ICNN'95-international conference on neural networks, vol. 4, IEEE, 1995, pp. 1942–1948.

[18] S. Dhalwar, R. Kottath, V. Kumar, A.N.J. Raj, S. Poddar, Adaptive parameter based particle swarm optimisation for accelerometer calibration, in: 2016 IEEE 1st International Conference on Power Electronics, Intelligent Control and Energy Systems (ICPEICES), IEEE, 2016, pp. 1–5.

[19] S. Mirjalili, A. Lewis, The whale optimization algorithm, Adv. Eng. Softw. 95 (2016) 51–67.

[20] C. Hamzaçebi, Improving artificial neural networks' performance in seasonal time series forecasting, Inform. Sci. 178 (23) (2008) 4550–4559.

[21] M.H. Nadimi-Shahraki, S. Taghian, S. Mirjalili, An improved grey wolf optimizer for solving engineering problems, Expert Syst. Appl. 166 (2021) 113917.

[22] H. Chen, A.A. Heidari, H. Chen, M. Wang, Z. Pan, A.H. Gandomi, Multi-population differential evolution-assisted Harris hawks optimization: framework and case studies, Futur. Gener. Comput. Syst. 111 (2020) 175–198.

[23] I. Aljarah, H. Faris, S. Mirjalili, Optimizing connection weights in neural networks using the whale optimization algorithm, Soft Comput. 22 (1) (2018) 1–15.

[24] N. Panda, S.K. Majhi, Oppositional salp swarm algorithm with mutation operator for global optimization and application in training higher order neural networks, Multimed. Tools Appl. 80 (28) (2021) 35415–35439.

[25] P. Singh, P. Dwivedi, V. Kant, A hybrid method based on neural network and improved environmental adaptation method using controlled Gaussian mutation with real parameter for short-term load forecasting, Energy 174 (2019) 460–477.

[26] S.A.A. Ghaleb, M. Mohamad, E.F.H. Syed Abdullah, W.A.H.M. Ghanem, Integrating mutation operator into grasshopper optimization algorithm for global optimization, Soft Comput. 25 (13) (2021) 8281–8324.

[27] S.K. Majhi, M. Sahoo, R. Pradhan, Oppositional crow search algorithm with mutation operator for global optimization and application in designing FOPID controller, Evol. Syst. 12 (2) (2021) 463–488.

[28] P. Singh, R. Kottath, G.G. Tejani, Ameliorated follow the leader: algorithm and application to truss design problem, in: Structures, 42, Elsevier, 2022, pp. 181–204. vol.

[29] P. Singh, R. Kottath, An ensemble approach to meta-heuristic algorithms: comparative analysis and its applications, Comput. Ind. Eng. 162 (2021) 107739.

[30] P. Singh, R. Kottath, Application of mutation operators on grey wolf optimizer, in: 2021 12th International Conference on Computing Communication and Networking Technologies (ICCCNT), IEEE, 2021, pp. 1–6.

[31] Z. Tan, K. Li, Differential evolution with mixed mutation strategy based on deep reinforcement learning, Appl. Soft Comput. 111 (2021) 107678.

[32] ISO, New England Energy Market Operator, 2008. https://www.iso-ne.com/.

[33] A. Biswas, B. Biswas, Regression line shifting mechanism for analyzing evolutionary optimization algorithms, Soft Comput. 21 (21) (2017) 6237–6252.

[34] A. Biswas, B. Biswas, Analyzing evolutionary optimization and community detection algorithms using regression line dominance, Inform. Sci. 396 (2017) 185–201.

[35] X. Zhang, J. Wang, Y. Gao, A hybrid short-term electricity price forecasting framework: cuckoo search-based feature selection with singular spectrum analysis and SVM, Energy Econ. 81 (2019) 899–913.

[36] F. Wang, K. Li, L. Zhou, H. Ren, J. Contreras, M. Shafie-Khah, J.P.S. Catalão, Daily pattern prediction based classification modeling approach for day-ahead electricity price forecasting, Int. J. Electr. Power Energy Syst. 105 (2019) 529–540.

[37] R. Kottath, P. Singh, A Meta-heuristic learning approach for short-term price forecasting, in: Soft Computing: Theories and Applications, Springer, 2022, pp. 147–156.

[38] S. Luo, Y. Weng, A two-stage supervised learning approach for electricity price forecasting by leveraging different data sources, Appl. Energy 242 (2019) 1497–1512.

[39] H. Cheng, X. Ding, W. Zhou, R. Ding, A hybrid electricity price forecasting model with Bayesian optimization for German energy exchange, Int. J. Electr. Power Energy Syst. 110 (2019) 653–666.

[40] K. Bhatia, R. Mittal, J. Varanasi, M. Tripathi, An ensemble approach for electricity price forecasting in markets with renewable energy resources, Util. Policy 70 (2021) 101185.

About the authors

Priyanka Singh is an assistant professor in Indian Institute of Information Technology, Raichur, Karnataka. She received her B.Tech degree in Computer Science and Engineering from Gautam Buddha Technical University, Uttar Pradesh, India. She completed her M. Tech degree in Information Technology with specialization in Intelligent Systems from Indian Institute of Information Technology, Allahabad, Uttar Pradesh in 2015. In 2020, she received her PhD degree in Computer Science and Engineering from Motilal Nehru National Institute of Technology, Prayagraj, Uttar Pradesh, India. She has published various research papers in reputed SCI journals with good impact factor. Her area of specialization includes Machine Learning, Optimization, Artificial Intelligence, Soft Computing and Time Series Prediction.

Rahul Kottath received B.Tech degree in Electronics and Communication Engineering from the University of Kerala, India, in 2013, M.Tech degree in Advanced Instrumentation Engineering from Academy of Scientific & Innovative Research, India, in 2015, and PhD degree from the Academy of Scientific & Innovative Research, India in 2020. He has publications in renowned international journals and conferences and has authored a book chapter. His research interests include Evolutionary Optimizations, Computer Vision, Motion Estimation, Visual Perception, Artificial Intelligence, Robotics, and Machine Learning.

CHAPTER NINE

Recent trends in human- and bioinspired computing: Use-case study from a retail perspective

Karthikeyan Vaiapury[a], Latha Parameswaran[b], Sridharan Sankaran[a], Srihari Veeraraghavan[c], Meril Sakaria[c], Gomathi Ramasamy[a], and Bagyammal Thirumurthy[b]

[a]TCS Research and Innovation, IITM Research Park, Tata Consultancy Services Limited, Chennai, India
[b]Department of Computer Science and Engineering, Amrita Vishwa Vidyapeetham, Coimbatore, India
[c]Retail Strategic Initiatives, Tata Consultancy Services Limited, Chennai, India

Contents

Abstract

With exponential increases in generation and availability of data, the need for optimal solutions for data-driven use cases is more and more felt. As the definition of problems is becoming fluid, the constraints are becoming dynamic, and information are incomplete, the need for intelligent bioinspired algorithms (BIAs) is being actively explored

Advances in Computers, Volume 135
ISSN 0065-2458
https://doi.org/10.1016/bs.adcom.2023.11.013

by the research community, especially in the areas such as semantic image segmentation, object detection, and video analytics. There are a number of use cases from apparel and fashion industry where extraction or inference of semantic descriptors from images and videos can be put to use for building a solution. In this context, BIAs are found to be effective and efficient and provide a compelling alternative to other types of algorithms. In this chapter, we specifically describe a use-case study using bio- and human-inspired algorithms such as Harris hawk optimization and progressive SpinalNet. The usefulness and potential of human- and bioinspired computing are shown using both real-world own dataset and public dataset and the results are quite promising.

1. Introduction

Human-inspired models ranging from neuromorphic attention [1] till somatosensory models [2] and bio-inspired computing [3] can be used in a wide variety of applications in retail domain such as retrieval [4], recognition [5], advertisement [6], smart shopping [7], drone delivery [8], fashion [9], and surveillance [10]. In this chapter, we review such computing models and their usability in real use cases using multimodal sensors such as RGB and thermal. The human brain and eye have the innate ability to adapt to changes in nature very quickly and easily. They have a very strong and reliable processor for detecting, locating, and identifying objects in any scene.

1.1 Enhancing customer experience: Smart shopping

In the present day, people are much keen to buy all items from household to medicine from their homes through the online shopping website [11]. In particular, in times of this catastrophic epidemic, the number of shoppers and consumers who trade online has increased [12]. Although online shopping comes in a variety of weights, consumer satisfaction can still be enhanced with substantial support from technology.

Consumer shopping experience and buying behavior analysis is an important area in retail, aiming to better understand and explain what, why, when, how often they buy, and so on. With customer being at the center of retail operations, human experience is considered in every aspect of retailing as the basis for running the store planning, merchandizing, and e-commerce functions. The dwell, gaze time of the customer reflects on the indecision of a consumer, or the hand movements reflecting on mal-intentions surveillance aspects provide valuable data to the intelligent algorithms interpret which can provide right signals to the store manager to take the right decisions. Smart AI-based shopping systems are able to not only detect and recognize shopping products but also recommend complementary products wisely based on the customer's previous purchase history, gaze, medical records, etc., which also comes with privacy and ethical concerns.

Biological systems have been modeled by computer scientists as a bioinspired algorithm (BIA), which is much popular to solve problems virtually in any domain. Researchers tend to develop new algorithms based on continuous learning from nature. For example, recently a wing flapping drone with a capability to flap wings has been inspired by the world's fastest bird and can be used in last mile deliveries in e-commerce platforms [8, 13]. In fact, corona virus optimization algorithm [14] has been developed recently for fuzzy clustering.

Inspired by human intelligence, researchers use seamless combination of data from multimodal sensors such as visual cameras (RGB, thermal), audio, text, tactile, and the versatile machine learning algorithms.

Thermal images are useful in retail industry to locate and identify deformed or rotten perishable items, packaged items which have decayed food, fire detection, etc. Physiological activities, such as fever, in human beings and other warm-blooded animals can also be monitored using thermographic imaging. Applying BIAs in thermal images will be a useful study and investigation to perform image analysis.

Deep learning is inspired by human brain architecture, which is also a BIA based on the cerebral cortex which is widely used in research community.

1.2 Biomimicry in fashion industry

Human beings, by their invincible imagination and combined intellect and action, accomplish various essential needs. For example, the costume design incorporates some of the essence of natural creation. The beauty in design of butterfly, rainbow, peacock, and flowers in nature is an amazing creation. Man learns from such design and creates as many new creations as he can. In addition to design and beauty, there are much more that we can learn from nature. For example, to name a few: protective clothing, wearable electronic devices, highly sensitive light sensor, sustainable batteries, etc. are much popular due to its real-time applications. Also, in the military textiles inspired by nature, camouflage protective clothing is used by soldiers to gain protection in a given environment from being overlooked by changing clothing or own products, equipment, etc. Camouflage fabrics is an interesting research area in the community. Biomimicry or biomimetics is one field, which is concerned with emulation of the models, systems, and elements of nature for the purpose of solving complex human problems. Biomimicry is a security feature in the military that reflects the image of animals in human clothing. Creating such clothes is possible through generative adversarial networks (GAN) technology. GAN is used in a variety of ways. Conditional GAN is used to impose a few attributes like seasonal design trends, pattern/print type, etc., specific to demographic conditions.

When it comes to fashion industry, borrowing elements of nature in designing and developing new garments is an old phenomenon. Numerous shapes, abundance of patterns, color combinations, and contrasting naturally occurring materials have become a source of inspiration for designers. The use of such clothing is pervasive among a wide variety of retail end users, whether it is the military personnel who use the camouflaged uniforms to protect themselves, thus mimicking the way animals and plants protect themselves from their predators to weather-friendly garments that protect the human race from the wrath of severe weather. Speedo launched swimwear that reflects the skin of a shark to enhance the performance of swimmers competing in a competition. Bioinfluenced products embrace the values of sustainability.

The fashion industry has been embracing the sustainability movement for a long time, and this leads to large investments by start-ups in this area. A fashion company uses a zero-waste water system by making fabrics from algae, creating a dramatic drop in energy consumption and pollution. Also, algae fabric nurtures and nourishes the skin when users wear it, and the dyes are completely chemical and nonallergenic.

A company from the United Kingdom has developed a new method of dyeing fabrics and protecting the environment. Using bacteria and the fermentation process, they can form dyes that do not fade over time. Most importantly, the process has been proven to use 500 times less water than conventional dyeing methods, making the whole process more sustainable and durable.

A highly durable biodegradable substance has recently been discovered as an alternative to skin, using thread-like cells called mycelium of mushrooms. It can be used as an alternative to synthetic leather. Recently, New York-based company has used Kelp, a variety of seaweeds to produce textiles and fibers. It can be used for Knitting or for 3D printing to reduce waste. The final knitwear is biodegradable and can be dyed with natural pigments in a closed–loop cycle. One company recently developed a bio-degradable and compostable glitter using eucalyptus wood extract, thus avoiding the use of microplastics.

The orange fiber is extracted from the cellulose found in oranges discarded during industrial pressing and processing. The fiber is then enriched with citrus fruit essential oils, creating a unique and sustainable fabric. One major area of reducing wastages and trying to use bio-based alternatives is in the area of packaging. Nautan is made from 100% plant-based materials derived from renewable resources like corn starch, potato starch, and used cooking oil, which is said to be more durable than other types of bioplastics. A Finnish company manufactures bio-based alternative packaging materials made of wood. The resulting material has similar properties of paper and plastic used in retail. Nevertheless, the material has higher tear resistance than the paper and can be recycled alongside a cardboard.

1.3 Organization of the chapter

This chapter is structured as follows. In Section 1, we introduced retail application use cases where human- and bioinspired computing can be investigated and used. In Section 2, we discuss brain-inspired and somatosensory-based models DL RGB imagery, for example, retail product recognition. In particular, we discuss Harris hawks optimization (HHO) algorithm. In Section 3, we provide approaches to recognize retail products. We discuss progressive SpinalNet inspired based on human spinal cord. Finally, conclusions are given in Section 4.

2. Brain-inspired models with user interaction for retail product recognition

2.1 Introduction

Significant progress in the field of thermal imaging technology has brought in more avenues for building applications. This has made it possible to move from just an exploration method in engineering and astronomy into an effective tool in many fields for forming thermograms. Advantages of thermal imaging are noncontact, noninvasive, nondestructive and is a simple technique to capture the temperature distribution of any object. Massive human labors are required for food quality assessment in large-scale retail stores, and it is a time-consuming task. Thermal imaging is much useful in retail industry management to assist the quality of the perishable products and also for product recognition. With improved lifestyle of customers and change in purchasing behavior, a software tool to aid is imperative to help the consumers and the shopkeepers to detect and locate decayed perishable items stacked in a shelf. This tool should aid in product identification and quality inspection of the packaged food including bruise damage detection and deformed or rotten or decayed perishable foods. A case study is presented here using the BIA namely the HHO for product recognition using thermal images.

2.2 Thermal image product recognition in retail stores using convolutional neural network

Product recognition in retail stores is crucial for its management and improves customers' shopping experience. At present, products are recognized using barcode [5], but it has shortcomings that the barcode has to be positioned in the product and human intervention is required to assist the machine in identifying the barcode. Another technology that can be used is radio frequency identification tags, but identification of multiple products poses challenge because radio wave influences or obstructs each other. As retail is evolving, enterprises are focusing on how to use artificial intelligence technology to increase the profit by improving customer's satisfaction and to deliver the best quality. Deep learning in computer vision especially for image classification and object recognition has enhanced drastically making it an ideal solution for retail industry application development. Also deep learning can learn precise features that are useful for product recognition and segmentation of rotten or deformed content from the shelf image. In the meantime, some automated retail stores have emerged, such as Amazon Go and Walmart's Intelligent Retail Lab, which clearly shows that there is interest to fully automate retail with deep learning.

Thermal camera provides a rich source of temperature information, less affected by changing illumination or background clutters. This makes thermal images more suitable to detect objects that are very small and occluded by another object that is common in retail stores [15].

With the radical development in deep learning, the accuracy of object recognition has increased. However, the deep learning model requires extensive processing power and time for training to learn optimum values for parameters, which is vital for a fine-grained object recognition task in real-time scenario. Nature algorithms or BIAs are much significant to solve real-time problems. HHO algorithm can be used for optimizing the CNN model. HHO algorithm can represent the CNN parameters including:

1. Number of CONV/pooling layers (NCPL)
2. Learning rate (α)
3. Kernels number in the CONV layer (KCONV)
4. Kernels size in the CONV layer (FCONV)
5. Stride in the CONV layer (SCONV)
6. Zero padding (P)
7. Type of pooling layer (TPL) including max pooling, average pooling, or L2-norm pooling.
8. Size of kernels in the pooling layer (FPOOL)
9. Stride in the CONV pooling (SPOOL)

The above parameter selection using HHO algorithm can be used to model effective CNN for product recognition in real-time scenario.

2.3 Harris hawks optimization algorithm for learning optimum parameters

HHO [16] is population-based, gradient-free, nature-inspired optimization algorithm. The main inspiration of HHO is the cooperative behavior and chasing style of one of the most intelligent bird Harris hawks, in hunting escaping preys. Several hawks cooperatively track a prey from different directions in an attempt to surprise it (exploration phase). Harris hawks can reveal a variety of chasing patterns based on the dynamic nature of scenarios and escaping patterns of the prey to surprise it and make it tired (exploitation phase). The main advantage of these cooperative tactics is that the Harris hawks can chase the detected prey to exhaustion, which increases its vulnerability. Prey's energy decreases during this iterative process and this can be defined as below:

$$E = 2E_0 \left(1 - \frac{iter}{iter_{max}} \right). \tag{1}$$

where E denotes the escaping energy of the prey, $iter_{max}$ is the maximum number of iterations, and E_0 is starting energy of the prey. In HHO, E_0 value of changes lies in the interval $(-1, 1)$ for each iteration. When the value of E_0 decreases from 0 to -1, it denotes that the prey is exhausted and tired, and when the value of E_0 increases from 0 to 1, it denotes that the prey is becoming active. When $|E| \geq 1$, hawks search for other places to find the location of the prey, i.e., exploration phase is carried out, and when $|E| \leq 1$, exploitation phase is carried out. Four strategies including (1) soft besiege, (2) hard besiege, (3) soft besiege with progressive rapid dives, and (4) hard besiege with progressive rapid dives are presented in the HHO to mathematically model the chasing patterns of the hawk and to learn the escaping nature of the prey. After some iteration, preys' energy is observed to go down step by step. Thus, hawks can intensify the besiege procedure to trap the prey or reach the final global answer.

2.4 Defect detection of fruits and vegetables using thermal images

Defect detection of perishable items in retail store can be done effectively using thermal images. Thermal emission of normal and rotten content is different [17], which makes it effective for its application to segment the rotten or deformed content. Segmentation can be done by using threshold based technique a such as pulse coupled neural network (PCNN) or using deep learning models. The optimum value for threshold can be learned using HHO algorithm [18]. The advantage of PCNN is that it does not require training like CNN and the radial basis function neural network. Thus, it can perform image segmentation more efficiently, but requires initialization of more parameters, which requires human intervention and thorough experiments. HHO algorithm can be used to select optimum values for parameters used in the PCNN algorithm [18] to perform efficient segmentation. Parameters to be optimized using HHO algorithm include grayscale values to perform segmentation. Segmentation of rotten or deformed food using deep learning can be optimized using HHO algorithm. Segmented components of the image can further be classified based on trained features as rotten, deformed, bruised, or good using the classifier.

3. Deep learning and somatosensory model-based retail product recognition

Recently, retail product recognition has become a popular research topic. The benefits of computer vision have transformed many business operations in retail. It is used to analyze the products arranged on the shelf for better placement and to check if there are any items out of stock. It provides retailers with accurate information on the availability of certain products to manage and improve products. This will save manual efforts to verify products and focus on the customer to improve the quality of services. Store-organized products need to be identified and categorized to automate the best customer shopping experience and process. To do all these processes, products must be recognized. But, the challenging part is that many products are identical in terms of shape, color, size, and texture and are placed close to each other and displayed in the same way. Using mere contextual information, it is difficult to identify these types of products. The basic retail product experience will be more or less the same for customers for many years. They go to the store and look for the items they want. But, this is not the same for retailers because they have to analyze customer experience and make decisions based on behavior. The traditional auditing system is complicated, time consuming, and not efficient as it is expensive. Also, this leads to manual error and inadequate analysis. The image recognition approach can be used to overcome the shortcomings of the traditional approach.

There are many state-of-the-art algorithms that include region-based Convolutional Neural Networks (R-CNN) [19], Faster R-CNN [20], single-shot detector (SSD) [21], Region-based Fully Convolutional Network (R-FCN) [22], Mask R-CNN [23], YOLO-v3 (You Only Look Once) [24], EfficientDet [25], YOLO-v4 [26], PP-YOLO [27], PP-YOLOv2 [28], and YOLOR [29]. Most recently, there has been human–inspired algorithm like SpinalNet, progressive SpinalNet that can be used for product recognition.

3.1 Retail image product recognition using EfficientDet

3.1.1 Dataset description

We have created our own shelf product images dataset of pixel resolution 4160×3120. The dataset is available at GitHub link [30]. Sample images and bounding box location are provided in Table 1. Images are manually annotated with item–specific bounding boxes using PASCAL VOC. Our chosen subset consists of 100 images dealing with different products such as Dprotein, Complan, Quaker Oats, Bru, Boost, Horlicks, Surf Excel, Fresh, Himalaya shampoo, Pantene, etc. Each image has 15 instances per image, for a total of 1000 instances of 10 different products. According to the metric used in POSCOL VOC, a detection is correct if the intersection over union between the detected and the ground truth is >0.6.

Table 1 Shelf images' dataset bounding box.

Class	xmin	ymin	xmax	ymax
Dprotein	352	739	796	1616
Complan	758	597	1471	1616
Quaker Oats	1465	800	2016	1652
Bru	2013	1084	2465	1678
Boost	2458	700	2891	1649
Horlicks	3320	758	3684	1620
Surf Excel	504	1781	1065	2778
Fresh	1971	1862	2381	2723
Himalaya shampoo	2384	2078	2610	2771
Pantene	3074	2094	3310	2794

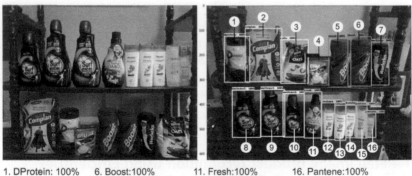

1. DProtein: 100%	6. Boost:100%	11. Fresh:100%	16. Pantene:100%
2. Complan:100%	7. Horlicks:100%	12. Himalaya:100%	
3. Oats:100%	8. SurfExcel:100%	13. Himalaya:100%	
4. Bru: 100%	9. SurfExcel:100%	14. Himalaya:100%	
5. Boost:100%	10. SurfExcel:100%	15. Pantene:100%	

Fig. 1 Products detection results on Shelf (trained and tested on our own dataset).

3.1.2 Experimental results

Initially, we conducted experiments on retail product recognition using SSD EfficientDet (refer Section 3.1.3). Training/test split is 80:20 in the dataset. The results using EfficientDet are provided in Fig. 1, which is visually appealing and quite promising.

Table 2 EfficientNet.

Stage i	Operator F_i	Resolution $H_i \times W_i$	Channels C_i	Layers L_i
1	Conv3x3	224×224	32	1
2	MBConv1, k3x3	112×112	16	1
3	MBConv6, k3x3	112×112	24	2
4	MBConv6, k3x3	56×56	40	2
5	MBConv6, k3x3	28×28	80	3
6	MBConv6, k3x3	14×14	112	3
7	MBConv6, k3x3	14×14	192	4
8	MBConv6, k3x3	7×7	320	1
9	Conv1x1, Pooling, FC	7×7	1280	1

3.1.3 EfficientDet

EfficientDet falls under SSD family has the following two main components.

The backbone network is EfficientNet, while the feature network is weighted Bidirectional Feature Pyramid Network (BiFPN), and the shared class/box prediction network is used. As stated in Ref. [25], the EfficientNet-B0 baseline network is summarized in Table 2.

A BiFPN is a type of FPN network that is used for fast multilevel feature integration. It has both bidirectional cross connection and normalized fusion techniques. It is described at level 6 as given below:

$$P_6^{td} = \text{conv}\left(\frac{w1 \cdot P_6^{in} + w2 \cdot ResizeP_7^{in}}{w1 + w2 + \varepsilon}\right) \qquad (2)$$

$$P_6^{out} = \text{conv}\left(\frac{w1' \cdot P_6^{in} + w2' \cdot P_6^{td} + w3' \cdot ResizeP_5^{out}}{w1' + w2' + w3' + \varepsilon}\right) \qquad (3)$$

In P_6^{td}, the input goes by the top-down approach, and in P_6^{out}, the output goes by the bottom-up approach. All other levels are in the same manner.

BiFPN includes the idea of FPN, PANet, and NAS-FPN. When using efficient and regular connections, information will flow in top-down and bottom-up directions. It enables information to flow in both top-down and bottom-up directions, when using efficient and regular connections. It also has fast normalized fusion techniques. Features with different resolutions are treated equally in a traditional method. So, the contributions of features are unmatched even by those with higher resolution. But in BiFPN, the input features are added with extra weight, helping the network to learn the importance of each feature. Less expensive separable convolutions are used instead of the regular one.

3.2 SpinalNet and progressive SpinalNet

SpinalNet was proposed based on the properties inspired by the backbone of the human somatosensory system.

The SpinalNet tries to reflect the following: (a) gradual input and nerve plexus, (b) voluntary and involuntary actions, and (c) attention to pain intensity.

In the traditional neural networks, the hidden layer receives the input from the input layer and then converts the output from the previous layer to the output layer. In SpinalNet, the hidden layers have three sections: (a) input, (b) intermediate, and (c) output.

The intermediate section contains hidden layers and it receives a portion of the inputs and outputs from the previous layer. Therefore, the weights obtained by the intermediate segment are lower compared to traditional neural networks. SpinalNet is used to deal with issues such as vanishing gradient, computation, number of connections, and number of layers.

The SpinalNet configuration consists of an input section, an intermediate section, and an output section. The input is sent to an intermediate array with multiple hidden layers.

It consists of a four-layer SpinalNet, which is equal to one hidden layer of neural network with four neurons. We will discuss the first two layers [$x1$: $x5$&$x6$: $x10$]. The first layer contains only the linear activation function. Therefore, the first layer of SpinalNet gives only the weighted summation of x1:x5 inputs . The output of the first layer goes to the neurons of the second layer. The second layer receives the weighted summation from x6:x10 and also the weighted summation of x1:x5 from the previous layer. The total weight received by the second layer is x1:x10. Therefore, the two layers of SpinalNet are equal to one layer and the output section of the above model each has two neurons.

The intermediate section's hidden layer has a non linear activation function, whereas the output section's hidden layer has a linear activation function.

3.3 Progressive SpinalNet algorithm

The fully connected (FC) layer plays an important role in deep learning models for input classification based on the features learned from previous hidden layers. FC layers have more parameters, and fine-tuning of these parameters is computationally expensive. This method aims to reduce the number of these parameters with improved performance. It was inspired from SpinalNet and other biological structures.

The size of the architecture is increased in each FC layer. The output of the network is kept the same as the FC layers.

The algorithm has a gradient between the input and output layers, and it solves the diminishing gradient problems in deep neural networks. In this approach, it gets input from the previous layer and CNN layer output and this way all layers contribute to the final layer to make the decision.

3.4 Experimental setup, implementation, and discussion

Intraclass classification is a quite intricate problem, where the task is to identify subclasses of main class or same product brand, but different flavors. To understand the relevancy and usefulness of algorithm, we performed transfer learning progressive SpinalNet SOTA on Pepsi flavor variants of Kaggle dataset. For all experiments, we used HP Omen GPU Laptop GeForce RTX 2060/PCIe/SSE2 with 16 GB RAM for processing. We compared three different algorithms: ResNet18, quantized ResNet18, and progressive SpinalNetWideResNet. The progressive spinal network uses WideresNet, which is top in the benchmark. The results for the Pepsi dataset are provided in Table 3. Although there are potential caveats such as over-fitting and experiments using only meagre data, still results are comparable and promising and could be explored for real-world applications.

Table 3 Results: Pepsi variants (Kaggle dataset).

Model	Train Acc	Train loss	Val Acc	Val loss
ResNet18	0.8981	0.2737	0.9977	0.0181
Quantized ResNet18	0.8928	0.2715	0.9977	0.0127
PSNet (WideResNet)	0.9989	0.0068	0.9955	0.0280

4. Conclusion

In this chapter, we presented numerous promising retail application use cases and trending research directions, where human- and bioinspired computing can be explored. Further, we presented and explained two use-case studies where thermal and RGB images are used. First, we discussed how the user interaction could be used along with brain-inspired models for retail product recognition using HHO algorithm. Next, we explained the deep learning and somatosensory model-based approaches for retail product recognition. Specifically, we discussed EfficientDet and progressive SpinalNet algorithms inspired based on human spinal cord system. We also presented results on real world own shelf dataset product images and Kaggle datasets. We have also shown that there is a huge potential for using human algorithm and BIA in real-world applications with a specific focus in retail industry.

References

[1] G. Indiveri, Neuromorphic selective attention systems, in: Proceedings of the 2003 International Symposium on Circuits and Systems, 2003. ISCAS '03, 3, 2003, https://doi.org/10.1109/ISCAS.2003.1205133. vol.

[2] H.J. Donkelaar, J. Broman, P. Domburg, The Somatosensory system, in: Clinical Neuroanatomy, 2020, pp. 171–255.

[3] A.K. Kar, Bio inspired computing—a review of algorithms and scope of applications, Expert Syst. Appl. 59 (2016) 20–32, https://doi.org/10.1016/j.eswa.2016.04.018.

[4] A. Rahman, E. Winarko, K. Mustofa, Product image retrieval using category-aware Siamese convolutional neural network feature, J. King Saud Univ. Comput. Inf. Sci. 34 (6, Pt. A) (2022) 2680–2687, https://doi.org/10.1016/j.jksuci.2022.03.005.

[5] Y. Wei, S. Tran, S. Xu, B. Kang, M. Springer, M. Panella, Deep learning for retail product recognition: challenges and techniques, Intell. Neurosci. 2020 (2020) 1–23, https://doi.org/10.1155/2020/8875910.

[6] S.S. Barton, B.K. Behe, Retail promotion and advertising in the green industry: an overview and exploration of the use of digital advertising, HortTechnology 27 (1) (2017) 99–107, https://doi.org/10.21273/HORTTECH03578-16.

[7] K.G. Atkins, S.-Y.J. Hyun, Smart shoppers' purchasing experiences: functions of product type, gender, and generation, Int. J. Market. Stud. 8 (2) (2016) 1–12.

[8] S. Perera, M. Dawande, G. Janakiraman, V. Mookerjee, Retail deliveries by drones: how will logistics networks change? Prod. Oper. Manag. 29 (9) (2020) 2019–2034, https://doi.org/10.1111/poms.13217.

[9] H. McCormick, J. Cartwright, P. Perry, L. Barnes, S. Lynch, G. Ball, Fashion retailing—past, present and future, Text. Prog. 46 (3) (2014) 227–321, https://doi.org/10.1080/00405167.2014.973247.

[10] S.R. Rashmi, K. Rangarajan, Rule based visual surveillance system for the retail domain, in: D.S. Guru, T. Vasudev, H.K. Chethan, Y.H.S. Kumar (Eds.), Proceedings of International Conference on Cognition and Recognition, Springer Singapore, Singapore, 2018, pp. 145–156.

[11] F. Huseynov, S. Özkan Yıldırım, Online consumer typologies and their shopping behaviors in B2C E-commerce platforms, SAGE Open 9 (2) (2019), https://doi.org/10.1177/2158244019854639. 2158244019854639.

[12] V. Venkatesh, C. Speier-Pero, S. Schuetz, Why do people shop online? A comprehensive framework of consumers' online shopping intentions and behaviors, Inf. Technol. People 35 (5) (2022) 1590–1620, https://doi.org/10.1108/ITP-12-2020-0867.

[13] H. Chen, Z. Hu, S. Solak, Improved delivery policies for future drone-based delivery systems, Eur. J. Oper. Res. 294 (3) (2021) 1181–1201, https://doi.org/10.1016/j.ejor.2021.02.039.

[14] A. Salehan, A. Deldari, Corona virus optimization (CVO): a novel optimization algorithm inspired from the Corona virus pandemic, J. Supercomput. 78 (2022) 5712–5743.

[15] V. Ghenescu, E. Barnoviciu, S.-V. Carata, M. Ghenescu, R. Mihaescu, M. Chindea, Object recognition on long range thermal image using state of the art DNN, in: 2018 Conference Grid, Cloud & High Performance Computing in Science (ROLCG), 2018, pp. 1–4, https://doi.org/10.1109/ROLCG.2018.8572026.

[16] A.A. Heidari, S. Mirjalili, H. Faris, I. Aljarah, M. Mafarja, H. Chen, Harris hawks optimization: algorithm and applications, Futur. Gener. Comput. Syst. 97 (2019) 849–872, https://doi.org/10.1016/j.future.2019.02.028.

[17] E. Rodríguez-Esparza, L.A. Zanella-Calzada, D. Oliva, A.A. Heidari, D. Zaldivar, M. Pérez-Cisneros, L.K. Foong, An efficient Harris hawks-inspired image segmentation method, Expert Syst. Appl. 155 (2020) 113428, https://doi.org/10.1016/j.eswa.2020.113428.

[18] H. Jia, X. Peng, L. Kang, Y. Li, Z. Jiang, K. Sun, Pulse coupled neural network based on Harris hawks optimization algorithm for image segmentation, Multimed. Tools Appl. 79 (2020) 28369–28392, https://doi.org/10.1007/s11042-020-09228-3.

[19] R. Girshick, J. Donahue, T. Darrell, J. Malik, Rich feature hierarchies for accurate object detection and semantic segmentation, in: Proceedings of the IEEE Computer Society Conference on Computer Vision and Pattern Recognition, November, 2014, pp. 580–587, https://doi.org/10.1109/CVPR.2014.81.

[20] S. Ren, K. He, R. Girshick, J. Sun, Faster R-CNN: towards real-time object detection with region proposal networks, IEEE Trans. Pattern Anal. Mach. Intell. 39 (2017) 1137–1149.

[21] W. Liu, D. Anguelov, D. Erhan, C. Szegedy, S. Reed, C.-Y. Fu, A.C. Berg, SSD: single shot multibox detector, in: B. Leibe, J. Matas, N. Sebe, M. Welling (Eds.), Computer Vision—ECCV 2016, Springer International Publishing, Cham, 2016, pp. 21–37.

[22] J. Dai, Y. Li, K. He, J. Sun, R-FCN: object detection via region-based fully convolutional networks, in: Proceedings of the 30th International Conference on Neural Information Processing Systems, Curran Associates Inc., 2016, pp. 379–387, ISBN: 9781510838819.

[23] K. He, G. Gkioxari, P. Dollár, R. Girshick, Mask R-CNN, in: 2017 IEEE International Conference on Computer Vision (ICCV), 2017, pp. 2980–2988, https://doi.org/10.1109/ICCV.2017.322.

[24] J. Redmon, A. Farhadi, YOLOv3: an incremental improvement (2018), https://arxiv.org/abs/1804.02767.

[25] M. Tan, Q.V. Le, EfficientNet: rethinking model scaling for convolutional neural networks, arXiv:1905.11946 (2020).

[26] A. Bochkovskiy, C.-Y. Wang, H.-y. Liao, YOLOv4: optimal speed and accuracy of object detection, 2020. https://arxiv.org/abs/2004.10934.

[27] X. Long, K. Deng, G. Wang, Y. Zhang, Q. Dang, Y. Gao, H. Shen, J. Ren, S. Han, E. Ding, S. Wen, PP-YOLO: an effective and efficient implementation of object detector, arXiv: 2007.12099 (2020).

[28] X. Huang, X. Wang, W. Lv, X. Bai, X. Long, K. Deng, Q. Dang, S. Han, Q. Liu, X. Hu, D. Yu, Y. Ma, O. Yoshie, PP-YOLOv2: a practical object detector, arXiv:2104.10419 (2021).
[29] C.-Y. Wang, I.-H. Yeh, H.-Y.M. Liao, You only learn one representation: unified network for multiple tasks, arXiv:2105.04206 (2021).
[30] K. Vaiapury, Shelf dataset, 2022. https://github.com/karthikeyanvaiapury/ShelfDataset/.

Further reading

[31] A. Bhargava, A. Bansal, Fruits and vegetables quality evaluation using computer vision: a review, J. King Saud. Univ. Comput. Inf. Sci. 33 (2021) 243–257. https://api.semanticscholar.org/CorpusID:64682463.
[32] M. Tan, R. Pang, Q.V. Le, EfficientDet: scalable and efficient object detection, in: 2020 IEEE/CVF Conference on Computer Vision and Pattern Recognition (CVPR), 2020, pp. 10778–10787.
[33] H.M.D. Kabir, M. Abdar, S.M.J. Jalali, A. Khosravi, A.F. Atiya, S. Nahavandi, D. Srinivasan, SpinalNet: deep neural network with gradual input, arXiv:2007.03347 (2022).
[34] P. Chopra, ProgressiveSpinalNet architecture for FC layers, arXiv:2103.11373 (2021).

About the authors

Karthikeyan Vaiapury is a Research Scientist at TCS Research and Innovation, Tata Consultancy Services Limited. He has done a PhD from University of London QMUL, MS from National University of Singapore NUS, and B.Tech from Bharathidasan University. He has more than 20 years of work experience, which includes both industrial and academics. He holds patents, book chapters, and publications to his credit. He has delivered invited key talks and serves on board of studies, jury in competitions and examiner in doctoral examinations. Research interest includes generative AI, creative AI, computer vision, image processing, multimedia, machine learning, deep learning, and social good.

Latha Parameswaran received the bachelor's degree in mathematics from PSGR Krishnammal College for Women, Coimbatore, India, in 1983, the master's degree in computer applications from the PSG College of Technology, Coimbatore, in 1989, and PhD degree in Computer Science from Bharathiar University, Coimbatore, in 2008. She was in the software industry for 10 years before joining Amrita. She is currently the Honorary Distinguished Professor with the Department of Computer Science and Engineering, Amrita School of Computing, Amrita Vishwa Vidyapeetham, Coimbatore. She serves as a Director of Computational Sciences in PSGR Krishnammal College for Women, Coimbatore. Her research areas include image processing, information retrieval, image mining, information security, and theoretical computer science. She serves in the doctoral committee for many PhD scholars in various universities.

Sridharan Sankaran is currently working as Principal Consultant at TCS Research and Innovation. He did his bachelor's degree in Electronics and Communication from Madras University and master's degree (by research) in Computer Science from IIT-Madras. He has more than 30 years of work experience covering academia, space research, retail, and IT consulting.

Srihari Veeraraghavan is currently working as part of Retail Strategic Initiatives at TCS. He did his bachelor's degree in Mechanical Engineering from Madras University and master's degree in Industrial Engineering from PSG College of Technology, India. He has more than 23 years of work experience covering research and innovation, retail, and IT consulting.

Meril Sakaria served as CTO of Global Retail Unit in TCS. She has been leading Digital Reimagination and Innovation programs for leading retailers worldwide for over 16 years. She brings innovation and deep retail industry experience, together with an ecosystem of technology partners, to help clients reinvent themselves as intelligent enterprises that innovate at scale. She has been instrumental in the incubation of several new services driving radically new growth streams for TCS such as SaaS platform for SMB Market, Products and the Retail Innovation Lab. She holds a master's degree in Management from National Institute of Industrial Engineering (NITIE, Mumbai) and a bachelor's degree in Electronics and Communication Engineering.

Gomathi Ramasamy is currently working as an assistant consultant at TCS Research and Innovation. She did her bachelor's degree in Electrical and Electronics Engineering from Anna University. She has more than 10 years of work experience in the tech industry, mainly focusing on computer vision.

Bagyammal Thirumurthy is an assistant professor at Department of Computer Science and Engineering, Amrita School of Computing, Amrita Vishwa Vidyapeetham. She received a BE degree in Information Technology from Bharathiar University in 2004, and an M.Tech degree in Computer Vision and Image Processing in 2014 from Amrita Vishwa Vidyapeetham. She is currently pursuing a PhD degree with the Amrita Vishwa Vidyapeetham. Her research interests include Image Processing, Computer Vision, and Change Analysis.

PART III

Information and computational systems

> **CHAPTER TEN**

Domain knowledge-enriched summarization of legal judgment documents via grey wolf optimization

Deepali Jain, Malaya Dutta Borah, and Anupam Biswas

Department of Computer Science and Engineering, National Institute of Technology Silchar, Silchar, Assam, India

Contents

Abstract

Extractive summarization of legal documents involves extracting important sentences from documents. In this work, we model the extractive summarization task as an optimization problem in the complete output space, where the goal is to select a subset of important sentences from the document. An effective objective function is proposed

Advances in Computers, Volume 135
ISSN 0065-2458
https://doi.org/10.1016/bs.adcom.2023.11.005

that is infused with domain-specific knowledge along with the exploration of pretrained embeddings for better scoring of candidate summaries. In this work, we have considered a grey wolf optimization-based approach whose objective function formulation contains the legal-specific knowledge along with pretrained embeddings as one of the features of this objective function. The experimental evaluation of the proposed nature-inspired summarization approach is carried out on an annotated Indian Legal Judgment document summarization dataset with the help of ROUGE metrics. From the experimental analysis, it has been observed that the best ROUGE-1 score (0.56034) is achieved by GWO setting with 50 population size and 300 iterations which used the Mini-LM model for finding the pretrained embeddings, whereas the best ROUGE-2 and ROUGE-L scores are 0.30583 and 0.27621, respectively, which have been achieved by GWO setting with 10 population sizes and 300 iterations using the general Legal Bert model. From the experimental results, the improved performances of Legal Bert-based embeddings with a higher number of iterations are observed. Such an approach can have very high practical utility in getting the gist out of lengthy legal judgment documents.

1. Introduction

Nowadays, text documents are freely available on the web and can be accessed by everyone. Such ease of data access can also be seen in the case of legal documents. If enforcement in the court of law is the primary intention behind the creation of a document, then such a document is called a legal document. A legal document can be of various types, such as case judgments, legal bills, patents, and so on. Such documents are also easily available on the internet. The main problem associated with these documents is that they are way too lengthy and have complex structures (especially case judgments from India). Consider a scenario where a lawyer wants to know all the important details of a case, but due to the excessively lengthy nature of the case document, he will have to spend a considerable amount of time reading it. On the other hand, suppose if a judge wants to read the decision of the previous cases, he/she also has to go through the entire document, resulting in waste of time. It is even more difficult for a novice reader to understand a complete legal document by reading it. If such lengthy and complex documents can be effectively summarized, a significant amount of time waste and human labor can be avoided.

This motivates us to perform a summarization of legal documents so that it can help legal practitioners as well as common citizens quickly understand these documents. Summaries can help lawyers stay focused on more important tasks. It will become very transparent if the novice readers can understand the gist of the legal documents by merely reading the summary. In this way, automatic summarization systems are really very helpful in the area of law as well.

There are two ways by which the summarization of legal documents can be achieved: (1) extractive summarization (2) abstractive summarization. The former approach picks the important sentences within the original document to create a summary, whereas the latter method tries to create a summary in such a way that the created summary may not contain the expressions of the original document while retaining the meaning of the original document. In the case of legal documents, most of the approaches that have been proposed by the researchers are extractive in nature. Therefore, in this work, we also propose an extractive approach based on the optimization technique. There are several types of legal documents, such as case judgments, legal bills, patents, contracts, and so on. The most common type of legal document is a case judgment. These are the judgments passed by the courts all around the world. For example, in India, Supreme Court, High Court, and District Courts of India are involved in passing the decision, depending on the case. In 2004, Hachey and Grover [1] proposed a dataset consisting of judgments from the United Kingdom's House of Lords (HOLJ). They have annotated the dataset to perform a rhetorical labeling task in which each sentence of a judgment is labeled with some category with which the sentence is identified the most. Similarly, in 2006, Saravanan et al. [2] have also performed a rhetorical role labeling task on the privately collected case judgments from Kerala High Court. The field was very slow until 2018, but recently researchers have begun to propose publicly available datasets such as plain contracts [3], legal bills from the United States [4], patents [5], and legal–task–specific datasets such as Indian case judgment for rhetorical roles [6] and court judgment prediction dataset [7,8]. Nowadays, researchers are taking more interest in solving legal–specific problems and are also proposing pre-trained models [9,10]. With such a recent surge of legal text analytics-related works, it is high time to propose efficient text summarization approaches that are specific to the legal domain. Keeping the goal of legal text summarization in mind, in this work we have proposed a novel nature–inspired text summarization approach that can effectively summarize case judgments from the Supreme Court of India. We have chosen this problem so that we can create automatic summaries, thereby assisting legal professionals in their quick understanding of case judgments. This problem has been chosen considering the practical problems faced by the legal professionals as well as novice users. More specifically, we consider the grey wolf optimization

(GWO) algorithm [11] for finding the relevant sentences from each of the input documents so that the summary can be formed. The optimization function consists of several features of documents, such as named entity features, sentence similarity scores, part of speech tags, rhetorical labels tags, and so on. These features are used to construct an objective function that gets maximized for finding the relevant sentences in the document. The main idea is that the problem of finding salient sentences from the document is modeled as a discrete optimization problem in the solution space, which can then be solved using GWO.

The primary contributions of this work are listed below:

- A novel GWO-based legal text summarization approach is proposed with a domain knowledge-infused objective function that captures important linguistic features from the document sentences.
- Fine-tuning of GWO hyperparameters is performed using grid search so that better quality summarization is possible.
- Several different neural sentence embedding approaches have been explored during experimentations so that the utility of domain- and country-specific embedding models can be studied.

The rest of the chapter is organized as follows: Section 2 gives a background knowledge on the areas of text summarization, legal document summarization, and grey wolf optimization. Section 3 discusses the related work with respect to legal document summarization along with the optimization approaches that have been used for summarizing legal documents. Section 4 gives a description of the methodology proposed in this work. The experimental results are presented in Section 5 with an analysis of the same presented. Finally, Section 6 concludes the chapter along with future research directions that can be explored.

2. Background
2.1 Text summarization

Text summarization is a research problem that comes under the Natural Language Preprocessing (NLP) field. It is a task of shortening the text from a document by means of automatic techniques such that the shorter version of text contains all the relevant information of the original text. The problem of summarizing text is not new, it has been around for a long time. Text summarization can be performed via two approaches: (1) extractive approaches and (2) abstractive approaches, which are explained in the subsequent sections.

2.1.1 Extractive summarization

Extractive summarization approaches find the important sentences or summary-relevant sentences within the original document. Those important sentences are directly utilized to form a summary. As per the extractive approaches, there are three major steps that are to be followed for getting the important sentences to create a summary. The first step is to represent the sentences of a document in some numerical/vectorized form. For achieving this, hand-engineered features [12–14], linguistic features [15–17], neural network features [18–20], and a combination of hand-engineered and neural network approaches [21–23] can be applied. The second step is to score the sentences based on the sentence representation, followed by picking the sentences as per the desired summary length. There has been a lot of work done in the literature on extractive summarization approaches. Before the application of neural network-based approaches, researchers utilized hand-engineered features to find the important sentences within the original text for summary formation. Classical NLP approaches are leveraged to find the sentences that contain the important features. Based on those features, some sentence scores are assigned to the sentences, and finally, sentences are picked up containing the highest score to get a summary. Nowadays, researchers tend to utilize the deep learning-based approaches whose work revolves around the neural networks to perform the extractive summarization. Researchers consider the task of extractive summarization as whether a particular sentence is important for summary formation or not. In this regard, it is a formulation of a binary classification task [18,19,24].

2.1.2 Abstractive summarization

Abstractive summarization approaches try to create a summary that is an approximation of a human summary, as these summaries may contain the words/phrases that are not present in the original document [25]. This approach is comparatively much harder than extractive approaches as it tries to produce the expressions that may not be present in the original document. There has been a gradual shift from extractive to abstractive approaches in recent years, owing to the advancement in neural network methods. The neural methods originally developed for machine translation [26] have revolutionized the research conducted for natural language generation techniques in the area of summarization [27,28]. The early work of abstractive summarization includes sentence compression [29], sentence fusion [30,31], and sentence revision [32], whereas the full abstractive summarization pipeline includes information extraction, content selection, and surface realization subtasks. While these subtasks are challenging, with the use of neural networks an end-to-end pipeline can be developed which does the summarization task using one network. Taking inspiration from the neural machine translation approach [26], Rush et al. [27] were the first to apply neural networks for the abstractive summarization. After that, several works are performed to achieve abstractive summarization [33–35]. Nowadays, there is a paradigm shift from simple neural network approaches to transformer-based approaches [36]. Researchers have been exploring pretrained models in which they fine-tune/re-train the pretrained models with a minimal change at the end of the network architecture.

2.2 Legal text summarization

When we create the summaries out of the legal documents, it is called legal document summarization. The complex nature of legal documents demands a need for the automatic creation of summaries since these documents are quite large as compared to general documents and possess different kinds of structures as well. Initially, most of the work was done on case judgment documents, but now there are several publicly available legal datasets that consist of different varieties of legal documents and these datasets are task specific [6,7]. Nowadays, researchers' interest in solving legal-specific problems such as rhetorical labeling task [6,37], case judgment prediction task [7,8], legal document summarization [38–42], and so on has increased recently. There have been several works that can be found in the literature that solve legal text summarization problem. For a detailed review, consider the work by Jain et al. [43].

2.3 Grey wolf optimization

It is a swarm-based optimization technique which was developed by Mirjalili et al. [11]. The inspiration comes from grey wolves which strictly follow social dominant hierarchy. Based on the hunting patterns of grey wolves, effective optimization techniques can be developed. Both the social hierarchy-based prey search process and the resulting mathematical model for optimization are discussed below.

2.3.1 Social hierarchy

The social hierarchy is simulated by categorizing the population of search agents into four types of individuals based on their fitness.

1. **Level 1 (Alpha):** Alpha wolves are the dominant wolves in the pack and are mostly responsible for decision-making. They can either be male or female. They might not be the strongest members of the pack, but they are the best in terms of managing the pack. In gatherings, the entire pack acknowledges the alpha wolf by holding their tails down.
2. **Level 2 (Beta):** These subordinate wolves assist in decision-making and other pack activities. The betas are thought to be the pack's advisors and disciplinarians. They are regarded as the best candidates to succeed the previous alpha when he or she dies or becomes very old. Beta ensures that all subordinates follow orders and provide feedback to alpha.
3. **Level 3 (Delta):** These wolves are known as subordinate wolves. They dominate omegas and report to alpha and beta. These wolves are further categorized into the following:
    - **Scouts:** They are responsible for watching the boundaries.
    - **Sentinels:** They are responsible for protecting the pack.
    - **Elders:** These wolves can be alpha or beta, depending on the situation.
    - **Hunters:** These wolves help alpha and beta in hunting.
    - **Caretakers:** Caring for ill, weak, and wounded wolves.
4. **Level 4 (Omega):** They are like the scapegoats in the pack. They are the weakest wolves in the pack and are allowed to eat at last.

Taking inspiration from the social hierarchy that has been followed by the wolves, a mathematical model has been developed by Mirjalili et al. [11] which is explained in the next section.

2.3.2 Mathematical model

In order to represent the social hierarchy of the wolves, the alphas (α) are considered to be the fittest, betas (β) are considered to be the second fittest, and deltas (δ) are considered to be the third fittest solution. The rest of the candidate solutions are considered to be omegas (ω). In the GWO algorithm, the optimization process is controlled by α, β, and δ. The ω wolves follow these three wolves. In order to encircle the prey, we apply the equations as given below:

$$\vec{D} = |\vec{C}.\vec{X}_p(t) - \vec{X}(t)| \tag{1}$$

$$\vec{X}(t+1) = \vec{X}_p(t) - \vec{A}.\vec{D} \tag{2}$$

where t indicates the current iteration, $\vec{A}$ and $\vec{C}$ are coefficient vectors, $\vec{X}_p$ is the prey's position vector, and $\vec{X}$ denotes the grey wolf's position vector.

The calculation of $\vec{C}$ and $\vec{A}$ vectors is done in the following manner:

$$\vec{A} = 2\vec{a}.\vec{r}_1 - \vec{a} \tag{3}$$

$$\vec{C} = 2.\vec{r}_2 \tag{4}$$

where r_1 and r_2 are random vectors in the range of $[0,1]$ and components of $\vec{a}$ are linearly decreased from 2 to 0 throughout the course of iterations.

In order to hunt, grey wolves have the ability to recognize the prey and encircle them. For this purpose, alpha wolves are considered appropriate, followed by beta (β) and delta (δ). Therefore, we consider that the alpha, beta, and delta have better knowledge of the potential prey and end up saving the first three best solutions. Other wolves update their position including the omega wolves on the basis of the best search agents' position. With the help of the following equations, the position updates take place:

$$\vec{D}_\alpha = |C_1.\vec{X}_\alpha - \vec{X}, \vec{D}_\beta = |C_2.\vec{X}_\beta - \vec{X}, \vec{D}_\delta = |C_3.\vec{X}_\delta - \vec{X} \tag{5}$$

$$\vec{X}_1 = \vec{X}_\alpha - \vec{A}_1.(\vec{D}_\alpha), \vec{X}_2 = \vec{X}_\beta - \vec{A}_2.(\vec{D}_\beta), \vec{X}_3 = \vec{X}_\delta - \vec{A}_3.(\vec{D}_\delta) \tag{6}$$

$$\vec{X}(t+1) = \frac{\vec{X}_1 + \vec{X}_2 + \vec{X}_3}{3} \tag{7}$$

Grey wolves stop hunting once the prey stops moving. In order to show it, the value of $\vec{a}$ decreases. With the decrease in the value of $\vec{a}$, there is also a decrement in the value of $\vec{A}$. The individual elements in $\vec{A}$ are random values in the interval $[-2a,2a]$. As the value of a decreases, there is a decrease in the value of $|A|$. When $|A| < 1$, the wolf attacks the prey. These wolves search according to the positions of alpha, beta, and delta. They converge when they see the wolf to attack the prey and diverge from each other to search for fitter prey.

To sum up, a random population of grey wolves is created in the GWO algorithm so that the search process gets started. Alpha, beta, and delta wolves estimate the prey's whereabouts during this search process, and each candidate solution updates its proximity to the prey. There is one parameter a whose value is decreased from 2 to 0 to perform exploitation and exploration. If $|A| > 1$, candidate solutions diverge from the prey, and when $|A| < 1$, the solutions converge toward the prey until the end of iterations. The GWO algorithm terminates as soon as maximum iterations are reached.

3. Related works

As far as legal documents are concerned, there have been several works proposed in the literature whose focus is mainly on extractive summarization. The following sections will provide a method-based literature review on the topic of legal document summarization.

3.1 Rhetorical role-based approaches

Rhetorical role means that each sentence of a legal document has some semantic function which is identified with its rhetorical roles [44]. These roles help in identifying the important sentences from the document and thus also help in summarization. These roles also help in forming structured summary as done by Saravanan et al. [2], where the authors have first identified the rhetorical roles of each sentences and then create a structured summary based on these roles. In another work, Grover et al. [45] explore the use of tenses as linguistic features to find the rhetorical roles for each sentence of a document. Finding rhetorical roles can alternatively be regarded as a thematic segmentation task, where each sentence's themes are determined and more crucial phrases are taken from each theme to produce a table-style summary [46]. Recently, Bhattacharya et al. [37] used deep learning techniques for identifying the rhetorical roles associated with each sentence of a legal document and hence to form a summary.

3.2 Citation-based approaches

This summarization approach is based on a set of citation sentences which help perform citation-based summarization. Galgani and Crompton [47] use both incoming and outgoing citations along with the citances to form a citation-based summary. In another work, Galgani and Hoffmann [48] make use of both the incoming and outgoing citations in order to extract the important catchphrases which help formulate summaries.

3.3 Nature-inspired approaches

These optimization techniques draw their inspiration from how nature responds to difficult situations. For this, several nature-inspired algorithms are proposed in the literature, such as genetic algorithm, particle swarm optimization, and so on. Researchers have been applying these algorithms to solve the problem of legal document summarization. Kanapala et al. [49]

formulated the problem of summarization as a discrete optimization problem in which the goal is to determine the fitness based on the individual statistical features. In order to do so, a mathematical model based on the gravitational search optimization (GSO) algorithm, whose operation is dependent on gravity of law, has been proposed [50].

Apart from these, recently researchers have been proposing approaches based on pretrained models where the task of summarization is considered the binary classification task. With respect to this, Eidelman [4] proposed a supervised pipeline in which first Bert [51]-based embeddings are created, followed by feeding them into a multilayer perceptron (MLP) model to find the important sentences and hence to form a summary. In another work, Jain et al. [52] first utilized the Legal-Bert [9] model to create the embeddings of each sentence of a document, followed by feeding them into an MLP model to predict the summary-worthy sentences for summary formation. There are several comparative analysis works performed in the past to get to know more about the specific behavior of these documents [43,53,54]. There are some general approaches which have been applied to legal documents as well [55,56].

In this work, we have explored the use of the GWO algorithm with the help of a novel objective function where linguistic, neural, and domain knowledge-based sentence features are utilized. Based on the fitness values, we come up with the best solutions and utilize them for summary formation.

4. Methodology

The methodology adopted to carry out this work has been shown in Fig. 1. In this work, we have utilized a GWO optimization algorithm with a domain knowledge-infused objective function that is able to capture linguistic features from the document sentences. We have also utilized several pretrained models to get sentence embeddings so that the utility of domain- and country-specific embeddings can be studied. The main idea behind the proposed approach is to consider the task of extractive summarization as a discrete optimization problem, where we try to make a binary decision of either selecting or discarding each of the sentences present in the document. Once this discrete optimization formulation is available, we can then apply the GWO algorithm for finding the best sentence selection given a document. The primary component in this optimization process is the objective function to be optimized, as GWO will be guided by this function. Hence, an appropriate objective function needs to be designed so that a candidate

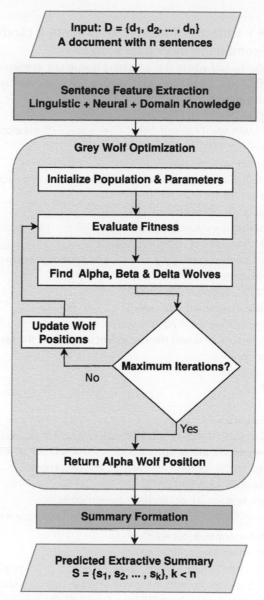

Input: D = {d₁, d₂, ... , dₙ}
A document with n sentences

Sentence Feature Extraction
Linguistic + Neural + Domain Knowledge

Grey Wolf Optimization

Initialize Population & Parameters

Evaluate Fitness

Find Alpha, Beta & Delta Wolves

Update Wolf Positions

Maximum Iterations?

No

Yes

Return Alpha Wolf Position

Summary Formation

Predicted Extractive Summary
S = {s₁, s₂, ... , sₖ}, k < n

Fig. 1 Overall methodology model.

summary with salient sentences gets a higher score. In order to design such an objective function, we have utilized several individual scoring techniques that are finally combined as shown in Algorithm 1. The individual scoring components and Algorithm 1 are discussed in detail in the following sections.

ALGORITHM 1 Objective function to evaluate a candidate solution by combining individual scores.

Input: D={d_1, d_2, d_3, ...d_i} where D is a training document, roleScores, RhLabels, solution, NERScores, NAdjScores, docRoleDist, sentsimScores, sumDocRatio, uniVocab, docUniProb

Output: totalScore

1: allNERS=[ ], allNAdj=[ ], lengthDiff=[ ], selIndices=[ ], solutiontext=[ ], totalScore=[ ]
2: **for** i=1,2,....len(solution) **do**
3:      solution[i]=Round(sigmoid(solution[i]))   ▷ converting each solution in the range of 0 to 1.
4:      **if** solution[i]==1 **then**
5:           totalScore+= roleScores[allRhLabels[i]]   ▷ update the total score with the scores of rhetorical labels
6:           solutiontext.append(d_i)   ▷ adding those sentence which has come as solution 1
7:           allNERS.append(NERScores[i])
8:           allNAdj.append(NAdjScores[i])
9:           lengthDiff.append(Abs(meansentLength-sentLengths[i]))   ▷ it is the difference between the document length and the sentence under consideration
10: selIndices.append(i)
11:      **end if**
12: **end for**
13: cohesionScoreList=[ ]
14: cohesionScoreList=cohesionScore(selIndices,d_i,sentsimScores)
15: posScore=getpositionScore(d_i,solution)   ▷ it is the sentence position score based on the location of sentence in a document
16: lengthDiff= Average(lengthDiff)
17: totalScore= totalScore+ posScore+cohesionScoreList
18: totalScore=$\frac{totalScore}{lengthDiff}$ ▷ totalScore is updated here
19: docRoleDist=normalize(docRoleDist)
20: solRoleDist=normalize(solRoleDist)
21: segRepDiff= getKLD(docRoleDist, soleRoleDist)   ▷ segRepDiff is the difference between the document and solution role distribution
22: totalScore= totalScore+ Avg(allNERs) +Avg(allNAdj)
23: KldVal= getKLD(d_i, solution, uniVocab, docUniProb)   ▷ how much the solution is diverged from the actual document is captured by KldVal
24: totalScore=$\frac{totalScore}{KldVal}$
25: **if** $\frac{sum(solution)}{len(solution)}$ >sumDocRatio **then**

> **ALGORITHM 1 Objective function to evaluate a candidate solution by combining individual scores.—cont'd**
>
> 26: totalScore= $\frac{totalScore}{Abs\left(\frac{sum(solution)}{len(solutio)}\right) - sumDocRatio}$ ▷ totalScore is updated
>
>   concerning the fact that if the length of the solution is greater than sumDoc ratio, it should be penalized by dividing it with the totalScore value
>
> 27: **end if**
>
> 28: totalScore= $\frac{totalScore}{segRepDiff}$ ▷ the totalScore is penalized with segRepDiff
>
> 29: **return** $-\ totalScore$ ▷ since we are minimizing, therefore negative of totalScore is returned

4.1 Individual scoring components of objective function

- **Named Entity Recognition (NER) score:** This feature identifies various named entities, such as name of a person, place, organization, location, product, and so on. In this work, spaCy library [57] is used to find all such NER tokens that are present in each sentence of a document. We count all such tokens in a sentence and divide it by the total length of the sentence to get the final *NERScore* as shown below:

$$NERScore = \frac{Count(NER\ tokens)}{length(sentence)} \tag{8}$$

- **Noun–Adjective score:** Typically, sentences consisting of noun–adjective pairs carry a lot of key information regarding people/places which increases the saliency of the sentences. Due to this, noun–adjective tokens are considered in a sentence for counting, followed by normalization as shown below:

$$NAdjScores = \frac{Count(noun - adj\ tokens)}{length(sentence)} \tag{9}$$

- **Rhetorical score:** Rhetorical roles (*RhLabels*) are the semantic functions that are assigned to each sentence in a document. In this work, we consider seven rhetorical roles as suggested by Bhattacharya et al. [58]. These roles are Ruling by Present Court, Ruling by Lower Court, Facts, Ratio of the Decision, Statute, Precedent, and Argument. We give scores to these rhetorical roles based on their importance as suggested in Ref. [58]. For each sentence of a document, there is some score associated with that particular role. For example, Ruling by Present Court has a score of 10, Ruling by

Lower Court has a score of 9, Facts has score 5, Argument has score 5, Ratio of the Decision has score 2, and Statutes and Precedents are of equal importance with score 2.

- **Kullback–Leibler divergence (KLD) score:**The main idea of KL divergence approach is to add the sentences to a summary set until the KL divergence of the summary distribution with the document distribution increases. Since higher values of KL divergence represent summary sets that are farther away from the document sets, we have considered the inverse of the KL divergence value for making update to our overall summary scores. In this work, we consider the unigram probability distribution for representing the document as well as the candidate summary. Moreover, we also represent the document and summary sets in terms of their rhetorical role distributions. In this way, we get two different types of KL divergence scores with which the overall summary scores are updated.

- **Cohesion score:** The sentences that are selected as part of the summary must have a high degree of cohesion among each other. This can be ensured by considering the pairwise similarity between each of the sentence pairs in the candidate summaries. In order to find this cohesion score, we can precompute the pairwise similarity scores with the help of sentence embedding vectors and the cosine similarity measure. These precomputed scores can be directly accessed during the GWO iterations based on the particular candidate summary under consideration. The overall cohesion score can be calculated as the average of all the pairwise similarity scores for the summary.

4.2 Combining individual components for objective function formulation

The objective function that is to be optimized by GWO is discussed in Algorithm 1. Several of the individual scoring components can be sped up by performing some document-level precomputation and passing the results of the same as input to the algorithm. For example, we can find the *NERScores*, *NAdjScores*, pair-wise sentence similarity scores (*sentsimScores*), sentence rhetorical labels (*RhLabels*), document rhetorical role distribution (*docRoleDist*), document unigram vocabulary (*uniVocab*), and unigram distribution (*docUniProb*) before starting the GWO process. Hence, we provide all of these as inputs to the algorithm. Since we are going to be performing the GWO-based optimization as a minimization problem, we return the negative of the overall summary score found by the objective function. Out of all the different individual scores, there are some for which higher values represent-better quality of summary (henceforth called *positive scores*), whereas some of the other scores represent the opposite (henceforth called *negative scores*). In order to incorporate all such scores, we adopt the strategy of adding the positive scores and dividing by the negative scores. This strategy is reflected in the updates made to the *totalScore* variable in the algorithm. One more important aspect that is to be noted here is that we employ the GWO approach to work in the range of [−10, 10] for each of the elements in the solution vector. However, since the actual objective function evaluation happens in terms of discrete values of 0s and 1s corresponding to each sentence, we perform an initial solution update to convert the solution vector into a binary vector. This is achieved via the application of sigmoid and round-off functions as depicted in the algorithm. Moreover, we also ensure that the algorithm doesn't pick too long or too short summaries by scaling the candidate summary scores with respect to the difference between the desired summary-document ratio (*sumDocRatio*) and the obtained summary-document ratio. Finally, we return (−*totalScore*) as the quality of the candidate summary, which the GWO approach tries to minimize thereby obtaining better solutions iteration by iteration.

4.3 Summary formation

Once we obtain the binary sentence selection vector via the application of GWO, we can then formulate the predicted summary by simply concatenating the selected sentences. Since the optimization process internally takes care of the desired summary length, we do not need to perform any further truncations to the predicted summary.

5. Results and discussion

5.1 Experimental results

The experiments have been performed on Indian case judgments with the help of a GWO-based optimization approach. We consider several settings, such as different population sizes, different number of iterations, and different pretrained models for sentence embeddings.

Table 1 shows the results for different pretrained models with different population size, and different number of iterations. From this table, we can observe that the best ROUGE-1 F1 score is 0.56034 which is achieved with the setting of 50 population sizes, 300 iterations, and Mini-LM Bert.

Table 1 ROUGE metric-based comparison of GWO approach for different population sizes and iterations.

GWO setting		Sentence embedding	FIRE test dataset		
Population size	Iterations		ROUGE-1	ROUGE-2	ROUGE-L
10	100	LM_BERT	0.54404	0.30199	0.27264
20	100	LM_BERT	0.55259	0.29926	0.27384
50	100	LM_BERT	0.56174	0.29979	0.27484
10	200	LM_BERT	0.54840	0.30647	0.27518
20	200	LM_BERT	0.55161	0.29987	0.27373
50	200	LM_BERT	0.55694	0.29624	0.27254
10	300	LM_BERT	0.54746	0.30421	0.27284
20	300	LM_BERT	0.55277	0.30121	0.27347
50	300	LM_BERT	**0.56034**	0.29778	0.27271
10	100	Legal BERT	0.55218	0.3046	0.27479
20	100	Legal BERT	0.55335	0.29832	0.27291
50	100	Legal BERT	0.5597	0.29936	0.27397
10	200	Legal BERT	0.54962	<u>0.30545</u>	0.27446
20	200	Legal BERT	0.55316	0.29921	0.27339
50	200	Legal BERT	0.55872	0.29853	0.27345
10	300	Legal BERT	0.5494	**0.30583**	**0.27621**
20	300	Legal BERT	0.55472	0.30171	<u>0.27561</u>
50	300	Legal BERT	0.55852	0.29823	0.27324
10	100	Indian Legal BERT	0.54465	0.30177	0.27313
20	100	Indian Legal BERT	0.55015	0.29823	0.273
50	100	Indian Legal BERT	<u>0.55946</u>	0.29781	0.27223
10	200	Indian Legal BERT	0.54555	0.30104	0.27213
20	200	Indian Legal BERT	0.55338	0.30057	0.27406
50	200	Indian Legal BERT	0.56138	0.29828	0.27382
10	300	Indian Legal BERT	0.55374	0.30654	0.27461
20	300	Indian Legal BERT	0.55268	0.2985	0.27279
50	300	Indian Legal BERT	0.55936	0.29913	0.27375

The maximum scores are 0.30583 and 0.27621 in terms of ROUGE-2 and ROUGE-L F1 metrics, respectively, which are achieved by the Indian Legal BERT model with 10 population sizes and 300 iterations.

Figs. 2–4 show the trend for different population sizes, considering the different iterations in terms of ROUGE-1, ROUGE-2, and ROUGE-L

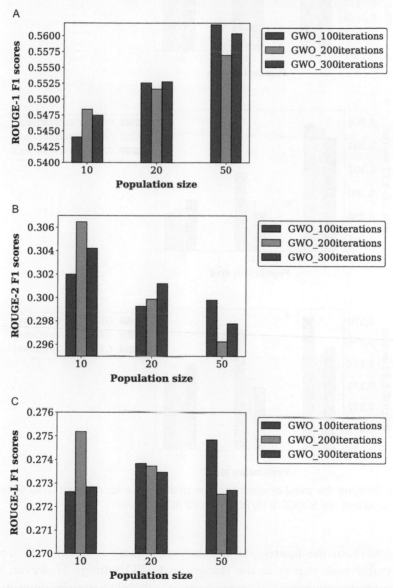

Fig. 2 Studying the trend of summarization performance for different GWO settings with Mini-LM BERT. (A) ROUGE-1; (B) ROUGE-2; (C) ROUGE-L.

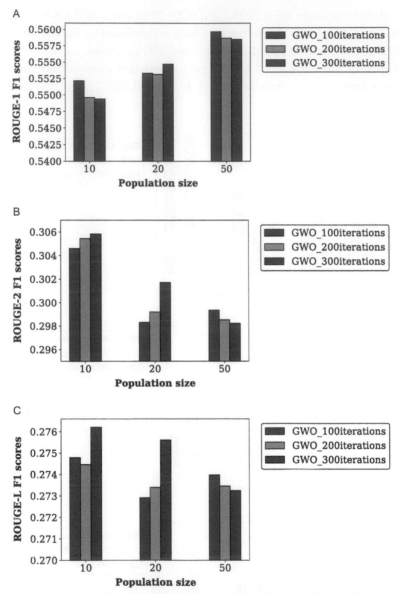

Fig. 3 Studying the trend of summarization performance for different GWO settings with Legal Bert. (A) ROUGE-1; (B) ROUGE-2; (C) ROUGE-L.

metrics. From the figures, we conclude that there is no obvious trend observable with respect to the changes in GWO settings. However, we can say that performance of summarization is slightly better for population size with more number of iterations.

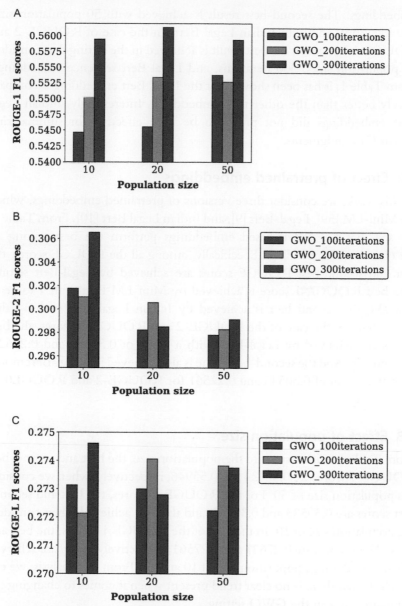

Fig. 4 Studying the trend of summarization performance for different GWO settings with Indian Legal Bert. (A) ROUGE-1; (B) ROUGE-2; (C) ROUGE-L.

After performing the experiments, we found that there is not much difference among these different settings. There is no clear trend visible in the results. However, the best ROUGE-1 F1 score is achieved when we consider 50 population sizes, 300 iterations, and MiniLM as sentence embeddings. The second-best result is achieved with 50 population size, 100 iterations, and with Indian Legal Bert. In the case of ROUGE-2 and ROUGE-L F1 scores, the best result is achieved in the setting that considers 10 population sizes, 300 iterations, and Legal Bert sentence embeddings. From Table 1, it has been shown that the Legal Bert embedding is comparatively better than the other two embeddings. Interestingly, Indian Legal Bert embeddings did not prove to be very effective for summarizing Indian Case judgments.

5.2 Effect of pretrained embeddings

In this work, we consider three versions of pretrained embeddings, which are Mini-LM [59], Legal-Bert [9], and Indian Legal Bert [10]. From Table 1, we see that Legal-Bert-based embeddings perform the best among all the considered embeddings. Specifically, among all the ROUGE scores, the best and second-best ROUGE scores are achieved by Legal-Bert setting. The best ROUGE-1 score is achieved by Mini-LM Bert with a value of 0.56034; the second best is achieved by Indian Legal Bert with a value of 0.55946. In the case of the ROUGE-2 and ROUGE-L scores, the best scores are achieved by Legal-Bert with a value of 0.30583 and 0.27621, respectively. And the second-best value is also achieved by Legal-Bert model with the scores of 0.30545 and 0.27561 for ROUGE-2 and ROUGE0-L, respectively.

5.3 Effect of population size

From Table 1, with respect to the population size, the best and second-best ROUGE-1 scores are 0.56034 and 0.55946, respectively, when we consider the population size of 50. For the ROUGE-2 scores, the best and second-best scores are 0.30583 and 0.30545, and they are achieved when considering population size of 10. In the case of the ROUGE-L scores, the best and second best scores are 0.27621 and 0.27561, respectively. They are achieved when considering a population size of 10 and 20. From this analysis, we can conclude that there is no clear trend present when it comes to changing the population size in the GWO setting.

5.4 Effect of iterations

With respect to the number of iterations, the best ROUGE-1 score (0.56034) is achieved with 300 iterations and the second-best ROUGE score (0.55946) is achieved with 100 iterations. In the case of the

ROUGE-2 score, the best ROUGE-2 score is achieved with 200 iterations and the second-best ROUGE-2 score is achieved with 300 iterations. In the case of the ROUGE-L score, the best and second-best ROUGE-L scores are achieved when considering 300 iterations. Therefore, we can say that with a more number of iterations, better summarization performance can be observed. Further experimental analysis can be performed as part of the future work, considering an even higher number of iterations so that a more robust analysis of the proposed approach can be performed.

6. Conclusion

The availability of a large number of legal documents demands the need for automatic summarization so that quick understanding of these documents can be possible. Keeping this in mind, in this work we have proposed a GWO-based summarization approach in which we have proposed a domain knowledge-infused objective function along with the exploration of several pretrained models to get sentence embeddings. From the results, we found that the GWO setting with 50 population sizes and 300 iterations which uses LM_BERT has achieved the highest ROUGE-1 value of 0.56034, whereas GWO setting with 10 population sizes and 300 iterations has achieved the best ROUGE-2 and ROUGE-L values of 0.30583 and 0.27621, respectively. Interestingly, we could not observe any improvement with the domain-specific Indian Legal-Bert-based embeddings that were expected. Further exploration in this regard is warranted. Overall, the quality of the summarization results obtained is found to be very effective, which is also suggested by the ROUGE-based evaluation. As a part of the future work, further exploration can be carried out with more GWO settings so that a more robust analysis can be performed.

References

[1] B. Hachey, C. Grover, A rhetorical status classifier for legal text summarisation, in: Text Summarization Branches Out, 2004, pp. 35–42.
[2] M. Saravanan, B. Ravindran, S. Raman, Improving legal document summarization using graphical models, Front. Artif. Intell. Appl. 152 (2006) 51.
[3] L. Manor, J.J. Li, Plain English summarization of contracts, arXiv:1906.00424 (2019).
[4] V. Eidelman, Billsum: a corpus for automatic summarization of us legislation, in: Proceedings of the 2nd Workshop on New Frontiers in Summarization, 2019, pp. 48–56.
[5] E. Sharma, C. Li, L. Wang, Bigpatent: a large-scale dataset for abstractive and coherent summarization, arXiv:1906.03741 (2019).
[6] P. Kalamkar, A. Tiwari, A. Agarwal, S. Karn, S. Gupta, V. Raghavan, A. Modi, Corpus for automatic structuring of legal documents, arXiv:2201.13125 (2022).

[7] J. Niklaus, I. Chalkidis, M. Stürmer, Swiss-judgment-prediction: a multilingual legal judgment prediction benchmark, arXiv:2110.00806 (2021).

[8] V. Malik, R. Sanjay, S.K. Nigam, K. Ghosh, S.K. Guha, A. Bhattacharya, A. Modi, ILDC for CJPE: Indian legal documents corpus for court judgment prediction and explanation, arXiv:2105.13562 (2021).

[9] I. Chalkidis, M. Fergadiotis, P. Malakasiotis, N. Aletras, I. Androutsopoulos, LEGAL-BERT: the muppets straight out of law school, arXiv:2010.02559 (2020).

[10] S. Paul, A. Mandal, P. Goyal, S. Ghosh, Pre-training transformers on Indian legal text, arXiv:2209.06049 (2022).

[11] S. Mirjalili, S.M. Mirjalili, A. Lewis, Grey wolf optimizer, Adv. Eng. Softw. 69 (2014) 46–61.

[12] P. Fung, G. Ngai, C.-S. Cheung, Combining optimal clustering and hidden Markov models for extractive summarization, in: Proceedings of the ACL 2003 Workshop on Multilingual Summarization and Question Answering, Vol. 12, Association for Computational Linguistics, 2003, pp. 21–28.

[13] A. Nenkova, L. Vanderwende, The impact of frequency on summarization, Microsoft Research, Redmond, Washington, 2005. Tech. Rep. MSR-TR-2005.

[14] T. Vodolazova, E. Lloret, The impact of rule-based text generation on the quality of abstractive summaries, in: Proceedings of the International Conference on Recent Advances in Natural Language Processing (RANLP 2019), 2019, pp. 1275–1284.

[15] H.P. Luhn, The automatic creation of literature abstracts, IBM J. Res. Dev. 2 (2) (1958) 159–165.

[16] H.P. Edmundson, New methods in automatic extracting, J. ACM 16 (2) (1969) 264–285.

[17] T. Vodolazova, E. Lloret, R. Muñoz, M. Palomar, et al., The role of statistical and semantic features in single-document extractive summarization, Sciedu Press 2 (3) (2013) 35–44.

[18] R. Nallapati, B. Zhou, M. Ma, Classify or select: neural architectures for extractive document summarization, arXiv:1611.04244 (2016).

[19] R. Nallapati, F. Zhai, B. Zhou, Summarunner: a recurrent neural network based sequence model for extractive summarization of documents, in: Thirty-First AAAI Conference on Artificial Intelligence, 2017.

[20] A. See, P.J. Liu, C.D. Manning, Get to the point: summarization with pointer-generator networks, arXiv: 1704.04368 (2017).

[21] Z. Wu, L. Lei, G. Li, H. Huang, C. Zheng, E. Chen, G. Xu, A topic modeling based approach to novel document automatic summarization, Expert Syst. Appl. 84 (2017) 12–23.

[22] J. Xu, Z. Gan, Y. Cheng, J. Liu, Discourse-aware neural extractive text summarization, in: Proceedings of the 58th Annual Meeting of the Association for Computational Linguistics, 2020, pp. 5021–5031.

[23] Y. Huang, Z. Yu, J. Guo, Z. Yu, Y. Xian, Legal public opinion news abstractive summarization by incorporating topic information, Int. J. Mach. Learn. Cybern. 11 (2020) 1–12.

[24] A. Joshi, E. Fidalgo, E. Alegre, L. Fernández-Robles, SummCoder: an unsupervised framework for extractive text summarization based on deep auto-encoders, Expert Syst. Appl. 129 (2019) 200–215.

[25] T. Cohn, M. Lapata, Sentence compression beyond word deletion, in: Proceedings of the 22nd International Conference on Computational Linguistics (Coling 2008), 2008, pp. 137–144.

[26] D. Bahdanau, K. Cho, Y. Bengio, Neural machine translation by jointly learning to align and translate, arXiv:1409.0473 (2014).

[27] A.M. Rush, S. Chopra, J. Weston, A neural attention model for abstractive sentence summarization, arXiv:1509.00685 (2015).

[28] S. Chopra, M. Auli, A.M. Rush, Abstractive sentence summarization with attentive recurrent neural networks, in: Proceedings of the 2016 Conference of the North American Chapter of the Association for Computational Linguistics: Human Language Technologies, 2016, pp. 93–98.

[29] T.A. Cohn, M. Lapata, Sentence compression as tree transduction, J. Artif. Intell. Res. 34 (2009) 637–674.

[30] R. Barzilay, K.R. McKeown, Sentence fusion for multidocument news summarization, Comput. Linguist. 31 (3) (2005) 297–328.

[31] K. Filippova, M. Strube, Sentence fusion via dependency graph compression, in: Proceedings of the 2008 Conference on Empirical Methods in Natural Language Processing, 2008, pp. 177–185.

[32] H. Tanaka, A. Kinoshita, T. Kobayakawa, T. Kumano, N. Kato, Syntax-driven sentence revision for broadcast news summarization, in: Proceedings of the 2009 Workshop on Language Generation and Summarisation (UCNLG+ Sum 2009), 2009, pp. 39–47.

[33] Y. Dong, Y. Shen, E. Crawford, H. van Hoof, J.C.K. Cheung, Banditsum: extractive summarization as a contextual bandit, arXiv:1809.09672 (2018).

[34] M. Zhong, P. Liu, D. Wang, X. Qiu, X. Huang, Searching for effective neural extractive summarization: what works and what's next, arXiv:1907.03491 (2019).

[35] R. Jia, Y. Cao, H. Tang, F. Fang, C. Cao, S. Wang, Neural extractive summarization with hierarchical attentive heterogeneous graph network, in: Proceedings of the 2020 Conference on Empirical Methods in Natural Language Processing (EMNLP), 2020, pp. 3622–3631.

[36] A. Vaswani, N. Shazeer, N. Parmar, J. Uszkoreit, L. Jones, A.N. Gomez, Ł. Kaiser, I. Polosukhin, Attention is all you need, in: Advances in Neural Information Processing Systems, 30, 2017. vol.

[37] P. Bhattacharya, S. Paul, K. Ghosh, S. Ghosh, A. Wyner, Identification of rhetorical roles of sentences in Indian legal judgments, arXiv:1911.05405 (2019).

[38] D. Jain, M.D. Borah, A. Biswas, Bayesian optimization based score fusion of linguistic approaches for improving legal document summarization, Knowl.-Based Syst. 264 (2023) 110336.

[39] D. Jain, M.D. Borah, A. Biswas, A sentence is known by the company it keeps: improving legal document summarization using deep clustering, Artif. Intell. Law (2023) 1–36.

[40] D. Jain, M.D. Borah, A. Biswas, Improving Kullback-Leibler based legal document summarization using enhanced text representation, in: 2022 IEEE Silchar Subsection Conference (SILCON), IEEE, 2022, pp. 1–5.

[41] M.N. Satwick Gupta, N.L.S. Narayana, V.S. Charan, K.B. Reddy, M.D. Borah, D. Jain, Extractive summarization of Indian legal documents, in: Edge Analytics: Select Proceedings of 26th International Conference– ADCOM 2020, Springer, 2022, pp. 629–638.

[42] D. Jain, M.D. Borah, A. Biswas, CAWESumm: a contextual and anonymous walk embedding based extractive summarization of legal bills, in: Proceedings of the 18th International Conference on Natural Language Processing (ICON), 2021, pp. 414–422.

[43] D. Jain, M.D. Borah, A. Biswas, Summarization of legal documents: where are we now and the way forward, Comput. Sci. Rev. 40 (2021) 100388.

[44] S. Teufel, M. Moens, Summarizing scientific articles: experiments with relevance and rhetorical status, Comput. Linguist. 28 (4) (2002) 409–445.

[45] C. Grover, B. Hachey, I. Hughson, C. Korycinski, Automatic summarisation of legal documents, in: Proceedings of the 9th International Conference on Artificial Intelligence and Law, 2003, pp. 243–251.

[46] A. Farzindar, G. Lapalme, Letsum, an automatic legal text summarizing system, in: Legal Knowledge and Information Systems, JURIX, 2004, pp. 11–18.

[47] F. Galgani, P. Compton, A. Hoffmann, Summarization based on bi-directional citation analysis, Inf. Process. Manag. 51 (1) (2015) 1–24.

[48] F. Galgani, A. Hoffmann, Lexa: towards automatic legal citation classification, in: Australasian Joint Conference on Artificial Intelligence, Springer, 2010, pp. 445–454.

[49] A. Kanapala, S. Jannu, R. Pamula, Summarization of legal judgments using gravitational search algorithm, Neural Comput. Applic. 31 (12) (2019) 8631–8639.

[50] E. Rashedi, H. Nezamabadi-Pour, S. Saryazdi, GSA: a gravitational search algorithm, Inform. Sci. 179 (13) (2009) 2232–2248.

[51] J. Devlin, M.-W. Chang, K. Lee, K. Toutanova, Bert: pre-training of deep bidirectional transformers for language understanding, arXiv:1810.04805 (2018).

[52] D. Jain, M.D. Borah, A. Biswas, Summarization of Indian Legal Judgement Documents Via Ensembling of Contextual Embedding Based MLP Models, FIRE, 2021.

[53] S. Gupta, N.L. Narayana, V.S. Charan, K.B. Reddy, M.D. Borah, D. Jain, Extractive summarization of Indian legal documents, in: Edge Analytics, Springer, 2022, pp. 629–638.

[54] D. Jain, M.D. Borah, A. Biswas, Automatic summarization of legal bills: a comparative analysis of classical extractive approaches, in: 2021 International Conference on Computing, Communication, and Intelligent Systems (ICCCIS), IEEE, 2021, pp. 394–400.

[55] D. Anand, R. Wagh, Effective deep learning approaches for summarization of legal texts, J. King Saud Univ. Comput. Inf. Sci. 34 (2019) 2141–2150.

[56] D. Jain, M.D. Borah, A. Biswas, Fine-tuning textrank for legal document summarization: a Bayesian optimization based approach, in: Forum for Information Retrieval Evaluation, 2020, pp. 41–48.

[57] M. Honnibal, I. Montani, S. Van Landeghem, A. Boyd, spaCy: Industrial-Strength Natural Language Processing in Python, Zenodo, 2020.

[58] P. Bhattacharya, S. Poddar, K. Rudra, K. Ghosh, S. Ghosh, Incorporating domain knowledge for extractive summarization of legal case documents, in: Proceedings of the Eighteenth International Conference on Artificial Intelligence and Law, 2021, pp. 22–31.

[59] W. Wang, F. Wei, L. Dong, H. Bao, N. Yang, M. Zhou, Minilm: deep self-attention distillation for task-agnostic compression of pre-trained transformers, in: Advances in Neural Information Processing Systems, vol. 33, 2020, pp. 5776–5788.

About the authors

Ms. Deepali Jain is currently an Assistant Professor at Bennett University, Uttar Pradesh and a PhD candidate (thesis submitted) at the National Institute of Technology Silchar, Assam. Her PhD research work focuses on Legal Document Summarization, leading to the publication of four top-tier SCI indexed journal papers. She has also published numerous international conference papers and achieved top positions in legal text analytics based shared tasks. She also holds an M. Tech degree from NIT Silchar in 2019, where her thesis work focused on Big Graphs. She completed her B. Tech degree from CCS University, Meerut in 2017.

Dr. Malaya Dutta Borah is currently working as an Assistant Professor Grade-I in the Department of Computer Science and Engineering, National Institute of Technology Silchar, Assam, India. Her research areas include Data Mining, BlockChain Technology, Cloud Computing, e-Governance and Machine Learning. She is the author of more than 70 papers including in reputed journals, Book Chapters, various National and International Conferences. She is the Editor in multiauthored books published by CRC Press, Springer and IGI Global. She has three granted patents. She is a member of professional societies like IEEE, CSI and ACM. Her official website: http://cs.nits.ac.in/malaya/

Dr. Anupam Biswas received his PhD degree in computer science and engineering from Indian Institute of Technology (BHU), Varanasi, India in 2017. He has received his M. Tech. and BE degree in computer science and engineering from Motilal Nehru National Institute of Technology Allahabad, Prayagraj, India in 2013 and Jorhat Engineering College, Jorhat, Assam in 2011, respectively. He is currently working as an Assistant Professor in the Department of Computer Science & Engineering, National Institute of Technology Silchar, Assam. He has published several research papers in reputed international journals, conference and book chapters. His research interests include Machine learning, Social Networks, Computational music, Information retrieval, and Evolutionary computation. He has successfully completed Principal Investigator of two DST-SERB sponsored research projects in the domain of machine learning and evolutionary computation. Currently, he is executing three sponsored research projects (one as PI and two as Co-PI) in the domain of AI/DL and Quantum Machine Learning. He has served as Program Chair of International Conference on Big Data, Machine Learning and Applications (BigDML 2019) and currently serving as Publicity Chair of BigDML 2021. He has served as General Chair of 25th International Symposium Frontiers of Research in Speech and Music (FRSM 2020) and co-edited the proceedings of FRSM 2020 published as book volume in Springer AISC Series. He has edited nine books that are published by various book Series of Springer and Elsevier.

CHAPTER ELEVEN

Bio-intelligent computing and optimization techniques for developing computerized solutions

G.S. Mamatha, Haripriya V. Joshi, and R. Amith

Department of Information Science & Engineering, RV College of Engineering, Bengaluru, India

Contents

Abstract

Bio-inspired computing is a field of study that Lois lee knits together subfields related to the connectionism, social behavior and emergence. It is often closely related to the field of artificial intelligence as many of its pursuits can be linked to machine learning. It relies heavily on fields of biology, computer science and mathematics. Briefly it is the use of computers to model the living phenomena and simultaneously the study of life to improve the usage of computer. Biologically inspired computation is a major subset of natural computation areas of research. Some areas of study encompassed under the canon of biologically inspired computing and their biological counterparts are, genetic algorithms, evolution, biodegradability prediction, biodegradation, cellular automata, life emergent system ants, termites, bees, wasps, neural networks, artificial immune systems rendering patterning and animal skins, bird feathers, mollusk shells and bacterial colonies. Linder Mayer systems, plant structures, communication networks

Advances in Computers, Volume 135
ISSN 0065-2458
https://doi.org/10.1016/bs.adcom.2023.11.006

and protocol, epidemiology and the spared of disease, intra membrane molecular processes in living cells, excitable media forest fires the wave heart conditions axons and sensor networks sensory organs.

Optimization techniques takes more bottom-up decentralized approach and often involves the methods of specifying a set of simple rules, a set of simple organisms which adhere to those rules and method of iteratively applying those rules for example, training virtual insect to investigate to an unknown terrain for finding food includes six simple rules which can be adopted. After several generations of rules application, it is usually the case where some forms of complex behavior get built upon complexity until the end results is something markedly complex and quite often completely counterintuitive from what the original rules would be expected to produce. For this reason, most technology-oriented solutions like neural network models, algorithms and other techniques came in to existence for accurate measurements and analysis that can be used to refine statistical inference and extrapolation as system complexity increases. The rules of nature inspired computing are the principle simple rules yet after being used for over millions of years have produced remarkably complex optimization techniques. All these techniques for developing software applications along with optimization techniques are discussed in the chapter.

1. Introduction

Bio-inspired computation techniques are motivating researchers and computing enthusiasts around the globe to come up with new techniques inspired by the methods and techniques used by animals and plants in nature in their everyday life. A whole new frontier of the computational era has begun with the advent of computations based and inspired by natural behaviors, and these have paved the way to become the background for important fields such as artificial intelligence and machine learning and so on. These fields encompass all critical and important aspects employed by living things around the world and thus provide an entire sub collection of studies and use cases that are both fascinating and idealistic in nature. This has led researchers and scientists around the world to come up with techniques that can be implemented in small scale to large scale computational devices (whether remote or cloud connected) and are propelling the way for future machines based on ML (Machine Learning) and AI (Artificial Intelligence). This not only is an important milestone in the way computers operate and behavior is perceived, but also is making an exponential growth owing to its simplicity in approach and effective impact when it comes to its functionality.

To address nature inspired or bio inspired techniques and algorithms, we intend to discuss approaches that match with the natural phenomena of everyday activities. Some existing techniques that are already prevalent in the industry include atomic computing, DNA (Deoxyribonucleic Acid) inspired scientific and molecular level computation, Swarm Intelligence inspired by the behaviors of insects [1], Neural networks employed in training deep learning models that are typically inspired by the way neurons fire and activate essentially mimicking the way a nervous system works in animals. Even the earliest approaches used perceptron and neuron components in the implementation which are essential components in an actual human nervous system neural link.

While one side of bio inspired techniques encompasses moving towards a self-sufficient data science model based on Artificial intelligence and Machine learning. These techniques have drawn their inspiration by nature, the other generic optimization approaches include algorithmic approaches such as ant communication. The functionality of these algorithms are in use even today across communication networks around the world [2,3]. Like Spider Monkey optimization algorithms, Genetic algorithms, cell division, cell multiplication which fall under a branch of computational study called as Cell automata etc. Some techniques such as DNA Based computation is even used in cryptography and network security communication models. Although the implementations are sometimes complex and require a deep dive into the basics, the number of applications of nature inspired engineering techniques are limitless.

The goal of this chapter is to provide users and acquaint them with different approaches used in the world of nature inspired computational activities, and their usefulness in various computerized applications. While traditional computational algorithms and hardcoded logics play an important role in achieving general tasks, it is important to know and understand the need to use nature inspired techniques and algorithms. As these employ an iteratively optimizing approach that yields better results and quicker responses similar to the behavior of various other naturally occurring phenomena around us. Realizing this, in this chapter the objective is to discuss various approaches to nature inspired applications in different domains that are related to computers, and that typically use/or can use bio-inspired techniques to provide solution to problem statements across different use cases of computer networks, image segmentation, character recognition, AI and IoT [4–6]. Some of the use cases include brain-controlled home automation, ant colonization, communication network inspired from epidemiology [7], spider monkey optimization, IoT sensor devices inspired from five senses and events organized based in retina trigger camera designed by occurrence of blinks.

In addition to the above-mentioned techniques, it's also important to understand the importance of various methodologies in existence in the field of Artificial intelligence. Few techniques such as Swarm intelligence, DNA Based computing, Membrane computers that are inspired from intra membrane molecular processes in cells, and Squirrel search algorithm for portfolio optimizations. Optimization is an important step involved while reaching/achieving the end goal of the bio-inspired techniques implementation. This is an important step towards mimicking and replicating the exact functionality as observed in nature. Each technique is drastically different from the other, and the algorithms themselves are optimized in various ways. This diversification provides important exposure to all kinds of domains and provides a wider scope of implementation for use case scenarios.

The chapters in the article are structured as follows: after the detailed introduction to topic about the different techniques inspired by nature, and the importance of such techniques in today's world in the first section, understanding of the current and previous approaches implemented in the field of bio-inspired engineering and computational techniques in literature survey section followed. The scope of existing and previous implementations is helpful to chalk out the plan for future implementations and gives us an insight into what the future potential holds for this accelerating field. The third section illustrates the different implementations proposed earlier for different bio engineering computerized applications. Finally, the conclusion section summarizes the important takeaways from the approaches with applications discussed in this chapter.

2. Background

Researchers across the world are aware of the benefits of bioengineering and nature inspired algorithms. It is widely considered to be the next iteration of computing and advancement in information technology, and computational paradigms. Broadly speaking, the replication of the nature-based phenomenon is regarded as a consequence of evolution. Though there have been various algorithms and approaches in the field of bio inspired and nature inspired engineering techniques, the primary focus of adopting nature inspired algorithms has always been optimization [8]. The reason is that there was a focus towards a convergence point and they are intended to get better and more efficient with each iteration cycle—be it execution or training. The inspiration drawn from nature-based engineering techniques have been based on randomization concepts previously. The main characteristics of such diverse algorithm sets include lower hardcoded values or algorithm centric values, higher optimization rates and non-biased outcomes. This is much helpful when compared to standard algorithms as they offer a benchmark for future comparisons and analysis, provide scalability, diversification and differentiation [9]. The Diverse nature of bio-inspired techniques and implementations stand as benchmarks in addressing the various use cases and problem statements in different domains. One of the most prominent emerging trends include neural networks. Designed to typically imitate the working and firing of normal neurons in human brains and nervous systems, this field relies heavily on deep learning techniques and employs learning models calibrated using multiple epochs in a training process [10]. The network can consist of any number of nodes depending upon complexity required, and activation functions act like firing neurons at different points in the network, working similar to a human brain. This neural network finds its application where convolution is applied and image processing is carried out. Prior to this, traditional image processing such as binarization, contour mapping and typical computer vision algorithms have seen moderate levels of accuracy and success. The need for diversified application and increased precision has demanded the need for nature inspired algorithms to come into play and this is exponentially propelling the work related [11].

Bio-inspired computing is different from that of artificial intelligence and machine learning applications as the traditional artificial intelligence techniques make use of creation-oriented approaches while nature-inspired intelligence and computational paradigms use an evolutionary approach. Another fitting example discussed in the paper published by Ritu Kapur titled, "Review of nature inspired algorithms in cloud computing," gives us insights as to how bio inspired computing can be leveraged to solve NP-hard level problems, and it clearly differentiates the nature inspired computing algorithms from traditional artificial intelligence approaches. The paper also provides a perspective of how these algorithms can be used in combination with existing traditional approaches rather than a standalone approach to solve network and communication related problem statements [12].

One of the most intriguing examples where inspiration is drawn in this paper is from ant's colonization and communication. Ants typically use pheromones to lay down the best possible route, to reach their desired location. In this process, the path where the maximum number of pheromones are present is regarded as the shortest route. This technique is built as an algorithm which improves its optimization in each cycle discussing an important problem statement in communication networks in computer science. The solution thus found is implemented as a logic to solve the problem of best route or missing packets in communication channels in networking devices and routes inspiration [13,14]. Similarly other practices which have vital importance and which derive their inspiration from phenomena occurring naturally in the field of swarm intelligence are discussed in detail here.

A second technique discussed in the paper titled "camera-based eye detection algorithm for assessing driver drowsiness" finds its from one of the most commonly occurring reactions—blink. Previously to address this problem, the car would consist of a live camera feed which monitors the user's face to assess drowsiness. The live feed camera would use traditional computer-based techniques to assess the drowsiness situation, which had an accuracy of up to 65%. This technique however has seen a radical change with the incorporation of bio inspiration algorithms where a blink is associated with the capture of temporal information of the face of the user. The sensing of a blink is a strong signal, and acts as an effective communicative trigger for the drowsiness alert to do its job. If the blink is sustained for a larger period of time, the algorithm inside the system senses this information as an ideal case of driver drowsiness and immediately sounds the alarm. Such triggers are helpful as it saves the driver's life and people commuting in the spatial space around him too [12].

Artificial Intelligence has transformed the way computers behave and the levels at which human computer interaction was established in the yesteryears. Various techniques such as deep learning, neural networks, reinforced learning, machine learning etc. have opened up a new frontier in the computational paradigm. Yet another emerging trend is the advancement of Generative Adversarial Networks (GAN) emerging as another artificial intelligence technique. GAN is primarily being focused on advanced image processing and video processing [15,16] with advanced features. One of the major use cases of GAN that is currently being explored is its ability to replace activation functions in its algorithm with a network of its own which in turn is capable of realizing colors on a black and white image/video. The algorithm is capable of finding its own route to recognizing pixel color intensities, contexts and colorizing the image segments accordingly, similar to a way the human brains can perceive images. This has gained its importance as this is an exact reversal of the image binarization process implemented at multiple levels and depths of images [9].

Reinforced learning can be incorporated with this as the network component in the algorithmic layers of optimization algorithms. Optimization algorithms primarily are inspired by natural execution methodologies, and the latter tends to learn and increase its recognition capability with each execution cycle. This is an effective approach as the end goal of implementing such systems is to achieve optimal results and reduce the amount of time taken during each execution of the algorithm. The problem with these kinds of approaches is the amount of data that is required to train the models, and then there is the problem of deciding the split of test versus training datasets that is needed to get the initial instance of the algorithm working. This work requires a meticulous and careful analysis of the use case, and requires more fine tuning of the algorithm specific to the use case before it can be used [17].

Diving into yet another category, around the world there is a growing need for computational power and processing power. There have been various implementations that are in existence such as increasing cache size, so that the instruction set is always present in the primary memory, thus saving time and increasing efficiency of execution of instruction sets. But as more and more computing power is required, these approaches have to start dealing with the issue of scalability and code portability. The problem cannot be solved by just increasing the capacity of the cache sizes or adding more amounts of memory sticks to the primary memory of the system. Bio inspired approaches such as DNA computing [18]. The idea is to visualize the requirement of computational space in the form of a 100-bit logic gate, and this expanded spatial memory availability contributes to more processing power.

The DNA computing utilizes DNA Strands similar to one existing in nature, and takes the shape of a breadboard circuit with fixed positioning. Such approaches have been proven to be efficient contributors to realization of high-end computing power while at the same time resolving the issues concerned with scalability and portability. This has greatly reduced the computation time by many orders of magnitude. Y. Kon and K. Yabe discuss two primary approaches in the field of DNA Computing namely Hybridization-Ligation and Parallel overlap assembly in their article— "Matrix multiplication with DNA based computing," thus giving us comparisons and contrasts as to how the algorithms on bio-inspired techniques can be applied to solve a basic problem of matrix multiplication. This gives us a fair insight into the scalability factor of how it can be applied in real world application systems and at hardware levels to offer high end computing power [19].

While most of the nature inspired techniques are purely inspiration, some advancements have been also made in the field of integrating computers with actual interactions in the natural environment. For instance, one of the biggest leaps in human-computer interactions has been with respect to interpretation of brain waves by computers. Brainwaves generally are categorized into alpha, beta gamma and delta. With the help of Brain Computer Interface [20], the system is able to interpret and read the brain waves from a human head, and convert them into computer signals which allow a user to control the computer. Adding to this, inspiration has been drawn from the five senses used by mane animals in nature to perceive its surrounding habitat. This is being mimicked in all aspects, one of the biggest being olfactory senses that are being developed artificially [20]. The paper "Embedded e-nose system to identify smoke smell" published by Salahedin Sadeghifard, discusses the implementation of modern electronic noses and their calibration that mimics the working of an actual nose. This is being used to propel automated perfume smelling, and smell sensors are programmed to be sensitive and respond to certain kinds of smell. This typically helps automate and detect odors, volatile and flammable chemical smells that might be dangerous, and can easily escalate if unchecked. From the above literature survey, it can be understood that Bio inspired engineering has come a long way since its inception, and the fact that it can support a diverse variety of domains and implementations assures its continued existence and expansion to provide solutions to many different problem statements at all scales [21].

3. Evolution of bioengineering

3.1 Bio-engineering in computer networks applications

3.1.1 Bio inspired routing algorithms

Looking at randomized algorithms, we consider population-based methods and their stochastic behavior. It takes inspiration from the way that ants operate in this world, essentially. The lesson learnt from ants is that cooperation is a good strategy, which highlights the requirement of communication. This showcases the field of biosemiotics which covers biology and communication, science and symbols. For example, you would have seen Road Science, which says that the U-turn is allowed or school ahead, or various kinds of symbols which we use to communicate essentially. For example, you can see symbols which say animals crossing and so on. Therefore, there is a whole field of semiotics which studies how symbols can be used for communication. If closely observed in the evolution of language, especially the characters, for example, in English, from A to Z, which form words and then sentences and so on. These are just complex semiotic systems where the symbols stand for something and we learn to communicate via those symbols.

The field currently focused in the chapter is called biosemiotics, which is an intersection of biology and semiotics. It encompasses prelinguistic meaning making, which means that we are not looking at it as linguistic symbols or pre linguistic symbols. In biosemiotics the study reveals how real-world chemicals can be used for communication and interpretation of science and codes and so on. One would have seen, for example, animals are frequently marked with a territory.

One might have seen a dog raising a leg and marking authority, saying that this is my iterative. And other animals, especially predators, do quite a bit of this representing a notion of biosemiotics. Similarly, Ant colonies also behave and are complex social structures. Ants have a hierarchy and virtually a car system among them, the Queen being the leader. And then there are drones and workers and all kinds of other hierarchies. They basically do all the work and they go looking for food and they protect their home from predators, all kinds of calamities. The ant colony term [22,23] defines the physical structure in which the ant lives in. It's not just like a housing colony that is often referred to among humans, it also describes the social rules to organize themselves and the work they are indulged in. Let's consider the real-world example of a complex system which emerges from these simple creatures where each ant is a very fairly simple creature. Some workers become foragers and when they mature especially, they are allowed to or then they have to go out of the nest in search for food. Essentially, this cooperation and division of labor, combined with their well-developed communication systems has allowed ants to utilize their environment in different ways approached by few other animals. Of course, human beings are also shining examples of such cooperation because, in fact, we live in much more sophisticated social structures.

All this is possible because of the fact that we lived in evolved societies where, because of cooperation, many different people can specialize into different things to the benefit of all ants in a simpler way. Let's consider this possibility of what happens. How does an Ant colony look for food? Essentially, now, especially living in a tropical country like India, one would have at some time or the other found that one has left a piece of food somewhere.

And after some time, you see there's a whole trail of ants coming and happily taking part of whatever, you have left carelessly around. How does that happen? How do ants invariably find the food that you leave around which is accessible to them? Essentially, the process basically starts with search, the kind of search that we are talking about. So, you can imagine that each Ant is going out looking for food, which the worker ants have the task of going and collecting food for the Ant colony. Each Ant goes and looks for food. But that would not have been too much, except for the fact that when they do find food, it has to be brought back to the nest. And also, they communicate implicitly through biosemiotics to other ants as to where to go and look for food. And this little bit of cooperation results in a massive success for the Ant colony. This kind of signaling is common in other creatures also. For example, bees are known to signal to other bees about where the nectar is and through their dance in the air and that kind of thing and all these creatures, they are not thinking creatures like we consider ourselves to be. They have evolved to communicate through chemicals or biosemiotics and in the process, they have built successful societies. So, let's see what happens when ants are foraging for food and so let us say that we have this. Ant nest and bunch of ants have been sent out to go and look for food and let's assume that for simplicity that there are five ants which we will call as ABCD and E and each of these ants, they go out in search of food so each of them kind of randomly goes off in some direction based on whatever little queues they have and we have named these ants as ABCD and E, they are kind of randomly going off in some direction, essentially and the important factor is that, as they go they leave a trail which is called the pheromone trail. The pheromone is a chemical substance which the Ant drops wherever it walks, essentially. It goes somewhere and it leaves a trail "this is where I am going or this is where I have been" and so on. This trail itself is a very interesting consequence that if another Ant were to encounter a pheromone laid by the first Ant, then it would have a tendency to follow the pheromone. So, this simply means that this is where ants have been through and a chemical that it leaves results in ants' communicating with each other implicitly. Basically, Ant colony optimization technique takes inspiration from bio-semiotic communication between ants. Stochastic greedy method is used by each ant to construct a solution with a combination of heuristic function and pheromone trail. This technique classifies algorithms as swarm intelligence methods.

This is explained using the traveling salesman problem as shown in the code snippet below as per the optimization algorithm explained in Refs. [24,25], where there are M ants and N cities. From the probable distribution each ant selects an edge and updates the best tour. Once that step is done, for each edge in the tour, a pheromone is deposited there by optimizing traveling salesman problems with respect to its time complexity.

Step 0: Initialize:
    a) Set bestroute = nil
    b) Set pheromone level of all edges to r_0.

Step 1: Construct solution for each ant
    While True
        Distribute M ants on N cities
        For each ant construct route do
        repeat
        for n = 1 to N
            ant a selects an edge from distribution probability
        update the bestroute
        enddo
Step 2: for each ant a update pheromone do
        repeat
        deposit pheromone
        stop once pheromone gets over
        enddo
    return bestroute.

3.1.2 Spider monkey routing optimization algorithms

Basically, the algorithm is inspired by the foraging behavior of spider monkeys. Spider monkeys are a particular species of monkeys which live in the deep forest, and they have a very special kind of foraging behavior which is called fission fusion social structure. This particular kind of social structure helps them to find the food source in a very efficient way. This particular kind of foraging behavior is modeled as an optimization algorithm and is used to solve very crucial and complex optimization problems in computer network routing concepts [26,27].

Suppose there is a group of monkeys. Initially they start food foraging and suppose they found food; they have reached the food source. After some time, think of a situation when they started facing the food scarcity in that particular area. In this particular area, the number of monkeys is too many. The availability of the food is less, and therefore they find the food to have scarcity. The common sense the monkeys follow is that they divide themselves into smaller groups, smaller, smaller groups. This process is called the fission.

These small groups now go to different directions and different food sources they found separately. For this smaller group, they have less competition for the food. Food scarcity is now fulfilled here in the evening. Throughout the day, they keep on making this fission process. And in the evening, they return to the same habitat where they used to live. This particular kind of social structure is known as the Fission Fusion social structure. And this algorithm technique is based on this particular behavior.

In the study [28], just imagine that there is an initial group of monkeys or there is the habitat where they are there. Usually, these monkeys live in groups of 50–60 monkeys and they forage together. Whenever they find a scarcity of the food, they divide themselves into smaller parts. Let us start the food foraging. In the morning they start food foraging. Foraging begins in a single group. In the morning they decide among themselves a leader. Usually, this leader is a female leader. In the swarm intelligence algorithm, everybody is equal. All the individuals are homogeneous. This is one of the required conditions in the swarm intelligence that the members of the swarm are all equal. But here this particular member is the leader. At the first site it looks like deviating from the basic assumptions of the swarm intelligence [29].

3.2 Bio-engineering in internet of things applications

3.2.1 Bio inspired sensors

We are going to cover a couple of topics specifically about sensors and processing in biology, how you can create these bioengineered systems, and then how it is that you can use biology in engineered systems for the most part, for the visual, auditory, and somatosensory or motor control systems. The chemo sensory aspect, or being able to touch and smell, if you will, is not as robust as what we've been able to do with the other types of sensors. Within biology, there are many different types of sensors. For instance, oxygen sensing ways of being able to know what it is that your muscles are doing as explained in Ref. [30]. By sensing a change in the length of muscles, hearing, vision, etc., these are all cells that are specialized to work in various ways. The point is that we have a biological way of actually transducing information. That it's sensing to electricity and this information is sent to the brain.

3.2.2 Visual system

The thing that's important to know is that eyes are the first part of your brain as far as the visual system is concerned. You have lens and a pupil in your eye, that allows the light to go through. And the size of that hole can be expanded by changing the muscles that are around it's known as the Iris, the thing that you get your eye color from. By changing that size, you can get different amounts of light coming into the eye. For instance, in the night time, when there's less light available, the hole has to be as large as possible. In animals like humans, we have round pupils. However, some others, like a cat, for instance, don't have round pupils. They have pupils that are such that they can open up that aperture even larger. For instance, cat sizes are slits that can go very fine, very large. Therefore, they can go in a very bright light situation and with a very low light situation also without any problem. Then being to able pick up the light, the light then reaches the back of the eye. That's an area known as the retina. That's about the size of a postage stamp for people. And what the retina does is, it transduces light to electricity. From there, information goes to the center of the brain, nearly known as the lateral geniculate nucleus. The important thing about that is that there are many feedback signals that are sent back to the retina.

So not only is information coming from the retina to the center of your brain, there's also information from the center of your brain going back to the retina to be able to compensate for various scenarios. The visual cortex which lies in the outer layer of the brain, in the back. That's where vision, as we take it for granted, takes place. So how is it that you know what an object looks like? It's because of the visual cortex. There are many pathways where information comes in at the eye and then processes through to the brain. Specifically, within the retina, you have two types of cells, Generally, you have what's kind of in the middle here, rods. And next to that, to the right, cones rods are primarily responsible for a kind of black and white vision, whereas cones are sensitive to different wavelengths of light or different colors. And the way that these cells are distributed in the human retina Is shown there at the top right. There's mostly cones in the center of the retina. And then rods that are distributed more at the periphery. Specifically, you have an area Known as the Fovea that has a very high concentration of these cones. And when you take a look, when you actually use your eyes to be able to see thing as what you're doing is you're concentrating your fovea on something that you'd want to look at, if you will all the cells that go back to the rest of the brain come out in one big bundle and it goes through the blind spot to the rest of your brain now that is a very inefficient design.

The light will hit rods and cones and then you have proteins that will change shape and because these proteins change shape you have ions that flow through these rods and cones and these ions then form electrical responses that are picked up as information. So, every cone is basically a filter and if observed in retinal processing, there is a distribution of the sensitivity of your rods and as you can have your cones and actually cones are fairly responsive to kind of the green area as well. In retinal processing, strictly focuses on cones and if you notice that you have these responses that are same as a filter if you're doing auditory processing or other types of processing where you are sensitive to a particular sub wavelength. For instance, if you have something like a tweeter, it would produce sounds specifically around high frequencies. This is a bit different where here you have cones that are specifically tuned to certain wavelengths of light. At the bottom here we have a similar image except now this is for birds and as you can see for these birds, they have another type of cone and they have one in the ultraviolet regime. We as humans also have the ability to seed light and ultraviolet light.

However, our cornea is that lens at the front of the eye blocks out most UV light. In fact, if you were to have cataract surgery, that lens would be removed and then all of a sudden color would appear to be different from what you're used to. That's something that actually happened to Claude Monet. the great painter all of a sudden, blues were not blue they were an intense purple and didn't know how to handle and to interpret that, so he would paint more paintings and when you look at paintings, that quad when they did after his cataract surgery, you'll notice that they have a definite more purple hue.

With little more background information about what's going on within the retina, you have an assembly of cells and the processing is done in the way that the cells are connected to each other. We have some photoreceptors or rods and cones. The way that you transduce light and then because the way because of the way that you connect these cells together, you get processing in the retina of an assembly of rods and you can see points out what cells are doing? what operations and when light is received in the center of the field what happens is that particular cell will respond in a particular way. The way the cells are connected you could do the same thing instead of simply looking at light versus dark you can do the same thing with cones so you have cones that are receptive for red in the center.

For instance and cones that are receptive for green in the surrounds and you get that same sort of effect so then what you can do is you can use this information where you put all of this together if you look at the far right there where you put multiple arrangements of these center surround areas together what you end up with is a patch of information that now not only gives you specifics about what color is where or where's their light versus where is their dark you can also get information about the location of edges and lines so again because of the way things are connected you can get higher-order pieces of information specifically.

Here now what you can tell is light and also the direction in which the light is moving so for instance getting information from the way that this receptive field is created if you look at the top here you have a visual stimulus that's going to be presented when the light is here the output is not particularly strong however when the light is moving from left to right you get a very strong output from this particular receptive field whereas on the other side where the light is moving in the other direction the response is still there but it's far more muted so you know that there's some motion however the direction is given to you by the strength of the response so you can get Direction as well strictly from how it is that you connect your components together so there was some work done in an area known as neuromorphic engineering.

There is a work created in which a circuit known as the silken retina, where a usage of a photo detector is arranged in this particular way to be able to emulate the human retina. This circuit was able to show that indeed with a very low power applied it detects light and processes information in a way similar to what we see in biology. Furthermore, there's work that was done by Andre von Schack and others in the development of Logitech trackball. In this invention movement of light, comparison of 2 snapshots images are shown and the optical flow through image changes using Pixar aid are considered.

With some other circuitry it was shown what a static image looks like and the change in the image looks so normally. For instance, in movies, images are numbered in sequence, the operation of a mouse is concerned is based on this concept changing from the point A to B. This is how the receptive fields can be connected together for numbering and direction of that movement. We would be concentrating more on kind of showing what goes on in biology and how is it that we can implement that with engineered systems and hopefully with what's presented here again to emphasize that you can take some of this as inspiration although this particular circuitry has information of custom-made Asics which is something that you all may not have access to. However, those concepts of being able to have a sensor and using the way the sensor is connected with other components to build in your processing as a design method.

3.2.3 Auditory system

Let us discuss information about the auditory system so what you already used to is what you see on the outside in the Oracles also known as a pin up so in humans you just have this fleshy part here it kind of just sits the side of your head and doesn't do very much what you could look at these is kind of like little satellite dishes where it's shaped so that you could pick up sound from multiple directions and concentrate it on the area that's doing the actual sensing work that area the first portion is the eardrum. If something bad happens, for instance if it's perforated or if it's broken, that is simply a membrane that will move when there's a pressure change between the inside and the outside of the membrane. That is the first step in transduction of sound is simply pressure waves that are moving around the in the air you go from a pressure wave to actual physical motion with these three bones these are the smallest bones in the body known as the ossicles and there are some muscles connected to those bones and those muscles allow you to have some automatic gain control so when you have a very loud sound those muscles will protect those bones and prevent them from moving very far and if you are listening to very soft sound those muscles will permit those bones to move as much as possible there for transducing as much of those very small pressure waves as a can from there those bones connects to a portion known as the cochlea and that is where the actual transduction from information from the outside world to the brain itself takes place, so the cochlea is the first part of your brain as far as your auditory system is concerned and with the cochlea. You have changes in pressure and an internal fluid inside of the cochlea that will then produce changes in cells that then allow ions to flow in and out therefore causing an electrical response in the brain. How is it that we know what different sounds are and how is it that we process that information so within the cochlea if you were to roll that out so cochlea just going to point to it again there is that kind of snail shaped organ in fact it comes from the Greek word meaning snail if we to roll that out you would have an area that is made for processing low frequencies and you have an area that's made for processing high frequencies so with this physical orientation of your cells in this organ you are able to get information about different specific frequencies that you hear now generally speaking how does this matter it's the difference between knowing that you can hear the sound of your mother's voice versus a truck honking versus crickets chirping etc. as shown in Fig. 1.

Within the natural world is very interesting but we as humans can now also interpret something a bit deeper so specific information is transmitted

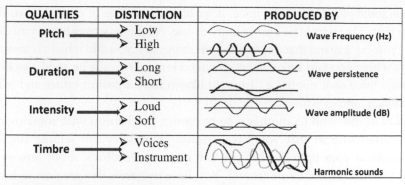

QUALITIES	DISTINCTION	PRODUCED BY
Pitch	➢ Low ➢ High	Wave Frequency (Hz)
Duration	➢ Long ➢ Short	Wave persistence
Intensity	➢ Loud ➢ Soft	Wave amplitude (dB)
Timbre	➢ Voices ➢ Instrument	Harmonic sounds

Fig. 1 Distinguishing sounds samples.

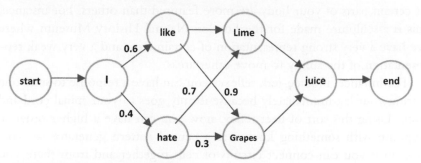

Fig. 2 Hidden Markov model sample.

with sounds in the form of speech for instance one form of music etc. and with the way that the cochlea is set up what you can do is you can process information again strictly because of where things are physically within this organ so for instance the sort of arrangement of words for instance, I and father it's a particular sound on this phony the frequency distribution is what you see there and then in a word like he is in he'd you have a frequency distribution in the fashion that is displayed there as you can see just looking at this analysis.

As possible in your filter Bank, this Fig. 2 represents just one filter and what you want here are many filters. One can separate all of the components and a frequency domain. One thing that you could do is you can use as hype of analysis known as a hidden Markov model as shown in Fig. 2, which is explained with apple and tomato juices. For explanatory purpose the juices are replaced with lime and grapes.

Systems like Siri can interpret different words that can be expanded in two different sentences. For instance, some people in this world hate grapes juice, some people like lime juice. If we had Clamato on it, everyone would hate that. That would be a probability of one. All the way over but as you can see, you can expand that out to different things so if you said it had a sentence now where you said I like grapes, the next sentence might be and I also like orange juice or it might be "I would like to have a peanut butter and jelly sandwich" etc.

Let us move on to another type of sensing mechanism with somatosensory processing so that you have a series of muscles and sensing organs throughout your body so that you can get some feedback about how it is that you can kind of move around. You have two areas that are next to each other. Motor cortex and somatosensory cortex in the outer layer of your brain. if you look at a map of this, you would see that the representation of certain parts of your body are more featured than others. For instance, this is a sculpture made for the London Natural History Museum where we have a very strong representation of certain areas and a very weak representation of the ability to move other areas.

For instance, in knee-jerk reflexes you can have a response to someone hitting your leg immediately because it only goes to your spinal cord and back. Using this sort of fleet reflex, now you can have a higher order of response with something known as a central pattern generator so what you do is you can connect circuits of cells together and from there you can get patterns. so instead of simply considering a knee-jerk reaction, you can have other reactions. For instance, what you can do is connect some of these components together based upon the way that they're connected. You can end up with it just by slightly changing one input, you can have different gates and this is some work that was done by researchers who are exploring the neurological basis of movement.

3.3 Bio-engineering in brain controlled system applications

The notion of interface between the brains and the robots caught human imagination for a long time. The technology knows how of the Brain Computer Interface (BCI) is meant to build an interface between the brain and any electronic or electrical gadget. Like smart home appliances, wheel chair, robotic equipment etc.; using electrical encephalogram (EEG), non-invasive method to measure electrode electrical potential on the scalp produced by the intelligence. This EEG technique is utilized to set up portable BCI controllers which are synchronous and asynchronous in nature. Due to non-invasive, low price, practicality, portable and easy to use features, non-invasive EEG BCI's are the most promising user interface in the field of human beings with extreme motor disability.

For some handicapped patients with body incapability or paralysis when the characteristic of the brain is still normal, have massive brain focus and thinking, the severely broken muscle and fearful gadget can communicate with the external environment and work separately every day. Their lives are highly unpleasant and will have a certain influence on their repair process, which has led to significant body and cerebral damage. The aim for many years in scientific rehabilitation is to restore or to embellish the abilities of handicap patients to operate and communicate with outside environment. BCIs can therefore be utilized to enable patients with severe cognitive problems or muscular injuries, recover their ability to speak to the outside environment instantaneously through a brain electrophysiological response. BCI might be useful for the elderly as superior assistance, rehabilitation and younger-strong technologies to operate video games or robotic arm for different purposes. The Majority of the typical brain computer interface equipment is expensive, large and laborious, making it tough to popularize the brain computer research in real life.

One of the highlights of the intelligence computer interface is that the portable brain–computer interface is easier to carry, use, safe and dependable. BCI techniques know-how the brain is usually split into two sorts of brain and their measuring effort. They are invasive BCI and not invasive BCI, depending on the electrical talent measurement method. Invasive BCI, for example, may cause an immune reaction which seriously harms the user and, due to the fact that the approach requires engaged surgery, it is hardly generally applicable to persons with disabilities, and the equipment price is huge and not yet reached through many governmental means.

It's been found that the non–invasive brain computer interface is less accurate than the invasive interface, it is nonetheless far less expensive and can easily be delivered by everything, as opposed to all various ways. The use of sign features and the most common are motor imaging, P300 wave, continuous kingdom visual evoked potential (SSVEP) for building realistic brain/computer interface systems. The excessive signal-to-noise ratio and resilience have been the reason why several methods are currently extensively used.

Human brain basically works on electric powered signals conveying all over the physique to direct the statistics in command to function the physique parts. While rotating the eyeball body upsurges or diminishes the confrontation in the nearby eye area. This version in electric powered alerts can

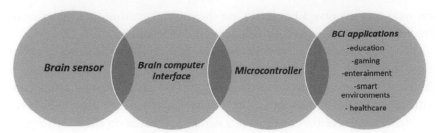

Fig. 3 Block diagram of overall system for BCI based smart home control [28].

be measured by the use of electrodes or the myoelectric sensors. By imposing these alerts processors, one can interface specific gadgets to manage on demand. Hence proposed device in Fig. 3 is designed to manipulate laptop and hardware device the usage of Genius waves electric powered signals. Projected schemes will detect the editions in electric sign energy through voltage degrees close to the eye region and generate a wireless radio frequency alert in order to manage the domestic automation prototype model. By enforcing this system one can lengthen it to bio permitted human body components to regulate via intelligence waves.

Electroencephalography (EEG) is the maximum considered attainable non-invasive interface, generally due to its first-rate temporal resolution, ease of use, movability and low cost set up cost. But as well as the technology vulnerability, some other giant blocks the usage of EEG as a BCI. In trials skilled brutally incapacitated human beings to self-regulate the gradual cortical abilities in the EEG to such degree that signals ought to be used as a binary signal to manipulate a PC cursor. The test noticed 10 sufferers were educated to cross a pc cursor by governing their brainwaves.

The technique was sluggish, necessitating greater than time required for patients to write one hundred characters with the cursor, while coaching frequently grabbed many months.

3.4 Bio-engineering in artificial intelligence applications

Optimal architecture for neural networks using bio inspired algorithms is growing rapidly. Today, neural networks, it's such a buzzword and everyone's talking about it and everyone wants to learn about it. Deep learning has become such an important part of our daily lives. So many lives depend on it. For weather. Farmers across the world who use it for their crops, people who run their own businesses and how much they depend on it to increase their sales and that it can offer a better service [31,32].

In so many ways, neural networks are important, and in a lot of ways, we're dependent on it. The main use of neural networks is because they're in our own brain, and our brain is basically complicated. This massive neural network that is performing so many computations every second so that one can make the right perceptions and make the right decisions. We just thought it would be better if we could appreciate how important neural networks are even just to our own lives right now. As and when one reads this, your brain tries to understand that and to come up with the perception and to understand it, we think, is something that happens so quickly without one even realizing that it's happening. And so that is something that's just fascinating. Very fortunate, and we should be really excited about the times we live in. So, all this is possible in the brain because the brain has been wired to have the right architecture.

So that's the beauty about it, that your brain has the exact perfect architecture that it requires to make all these complex decisions and all these complex computations so that you can make the right decision in the quickest time possible. So that's kind of like the intuition behind why these matters and why you need the best architecture for a neural network. So that today when we can go and apply neural networks to problems that are new and that come up every day, we need to know what is the best architecture so we can optimize and so that we can make sure we can prevent any sort of underfitting or overfitting. Let's point out a couple of things about a biological neuron before we can go further. This is how a neuron looks in Fig. 4 which is similar to the neuron as shown in Ref. [30]. This is how the neurons in our body look. So, the dendrites are basically like the point of contact for the inputs. And then the body of the neuron sort of does the computations, and the terminals are sort of like the outputs, and then the

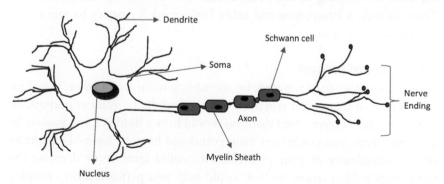

Fig. 4 A sample structure of a neuron.

outputs. The terminal is then connected to other dendrites of other neurons that are around this neuron, and it passes on information in a chain-like manner. The outputs produced by this neuron are then passed to the next neuron, and then that neuron passes on information to the next neuron. And so that's how the information sort of flows through our network. The dendrites are used for the inputs and you would have some computation happening as it travels through the neuron and then you would have the outputs coming out of the terminals.

In a mathematical sense, it's been sort of converted into a very simple way as (Inputs + Weights) fed to a transfer function through a activation function to get activation output. It's been converted into a sort of mathematical model. Just like how we had the dendrites for the inputs here we just have input zero, input one, input two. These would just be numbers. These are basically these could be anything, maybe it could be pixels of an image, or it could be features that describe a certain feature of an object, or it could be just about anything, but it would be basically, it would be a number. And so basically you divide your inputs into different numbers and you pass in each number through the inputs and then you have weights.

Weights are basically they tell us how important the corresponding input is. If you think maybe input zero maybe is more important than input two, then you would assign weight zero to be greater than much greater than weight two. They tell us the relative importance of the inputs. Then a simple computation happens on the inputs. Let us not go deep into it, but it's just a simple sum of the input and the weights and then that's passed on to the inactivation function. That's how a neuron passes on information. It takes in the inputs, multiplies it by its own weights, takes the sum of it, passes it to an activation function and that function returns as an output. That's the basic functioning of one neuron box. There are many variants of this. There is even a Perceptron and other basic models that try to mimic the working of a neuron.

3.4.1 A neural network

One can build a neural network by assembling neurons together. Based on the number of inputs that you need to have and the number of outputs, an input and output layers. And then you would have a hidden layer, you could have one layer, you could have two, you could have 10 layers. Depending on the complexity of your problem, you could increase or decrease the number of hidden layers. So that would help you perform more complex predictions or classifications that you are trying to solve. Each node in the network represents one neuron. As the input comes in, it also has outputs coming out of it and each output would have weight associated with it. The outputs of one layer are fed as inputs for the next layer.

3.4.2 DNA computing

Short history of DNA: Leonard Alderman in 1994 realized that DNA nucleotide bases could be used to solve logical problems and created a proof of concept to use DNA to solve many computational problems. Later in 1995, the idea of DNA-based memory appeared when Eric Ball proposed an immense amount of data that could be stored in DNA strands. How did DNA-based computers even work well normally with the classical computers? The ones that are being used right now have logic gates. For example, we have diodes or transistors that consist of simple switches with DNA. The way the molecules are used is very similar. They can be triggered to bind and release from each other to create a circuit of sort of logic of their own and one method called DNA strand displacement.

It has an input strain of DNA that binds to a DNA logic gate and this displaces the third strand of DNA which serves as an output. DNA strand in another method input DNA binds to a DNA logic gate and releases enzymes to cut other strings of DNA which go and bind and release and this creates sort of a chain reaction. Why should we use DNA based computers? After all, don't we have perfectly capable laptops and computers sitting at our desks right now, well yes as of now DNA based computers are still relatively new and have not seen use as substitutes for classical computers. This does not even come close to our current models however they show promise of capabilities way beyond our normal classical computers. If DNA can store an entire genome in the nucleus of a cell, think about how much we could store if we use it to store computer data the density of info would be unimaginable compared to our current electronics. Also, our current computers from silicon chips can only perform one operation at a time. It's been refined so small and so fast that you don't notice it normally but as problems with higher complexity arise more and more computational power is required to get these done and the greatest promise of DNA computing is that it can have parallel computing capability independent of each other.

DNA strands can simultaneously represent different possible solutions and this will overcome the problem of needing computational power and some may be thinking, isn't quantum computing supposed to be the same exact thing with this parallel computing? yes. That's true but quantum computers require extremely precise conditions they need to be at almost absolute zero temperature to name one and they're tremendously expensive whereas DNA molecules can stay stable at room temperature and synthesizing DNA molecules has been getting cheaper and cheaper as time goes on and we've been researching DNA for so long that we have expert techniques like Crispr to manipulate all this DNA. Of course, it will be a long time if ever if DNA computers actually replace our current classical computers.

3.4.3 Recent successes of DNA computing

With DNA computing being a relatively new field, many practical uses have not been found for this new technology. However, proof of concept exists given that a few major breakthroughs have been made in the field; the first of these recent successes occurred in 1994 when the Hamiltonian path problem was solved by Leonard Edelman. The fact that this NP-complete complex problem was able to be solved using DNA served as the basis of the rest of the field and as a proof of concept of its ability to be used in many currently explored parallel computing applications. A few of the most important breakthroughs given in Fig. 5, similar to the explained in Ref. [33] that have occurred since the field's inception are as follows:

- DNA based cryptography and steganography
- DNA dominating domino operators and
- Self-assembling repro reprogrammable DNA

DNA based cryptography has seen much work done because of the possible speed ups that can be seen given the efficient parallel programming utility which DNA based computing provides novel algorithms for public key encryption as well as conventional steganography have been explored and utilized in small cases, this application of DNA computing can provide significant benefits to us as more efficient secure cryptographic methods are always greatly appreciated perhaps the most important technology enabling DNA computing to possibly proliferate domino operators in their discovery in 2017 by researchers with a monumental success.

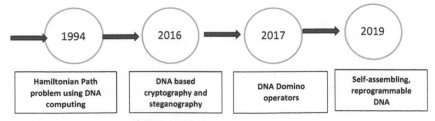

Fig. 5 Recent successes of DNA computing.

For DNA based competing methods, without getting into the nitty-gritty details of the processes that allow dominant operators to function in essence, they provide us with methods to efficiently monitor memory states and DNA without sequencing it furthermore, these domino operators enable the recording of and monitoring of cellular events and can be used to build forms of logic through layering this advancement represented one of the largest steps.

In this field, since its birth in the 1990s the last major breakthrough occurred when researchers at uc Davis formulated DNA which self-assembled allowing it to be essentially programmed the inputs for any of the algorithms ran using this method was initially limited to six bits. For this research trial however the very fact that this was achieved shows the promise of DNA based computing and one can always hope for many future advancements to be made to allow for more practical uses of DNA computing. Before, DNA computing was a very new development and different applications are being explored constantly. Molecular nanotechnology researchers have looked into DNA computing and have discovered DNA sequences with functional properties for preventing genes that could be harmful showing how the use of DNA computing creates potential for curative applications other research has shown that DNA computing can be used to advance understanding in biology by using the computations to decipher the sequence of large proteins and by using the DNA computing, it will take much less time because DNA computing is able to work so fast it can hold lots of information it could even be used to compute connecting flight paths around the world or to crack long passwords.

Some are even looking into using DNA computers for video games which will provide a video and audio quality higher than the real world. DNA computing can be applied to multiple different uses and is very helpful. Using DNA you can perform millions of operations at once allowing complicated problems to be solved in seconds. Another advantage using DNA computing is its ability to operate quickly but also have less operations. For example, a mixture of 1018 strands of DNA can operate at 10,000 times the speed of advanced super computers and by doing less processes DNA molecules can also store much more information the data density of DNA molecules is way more than the latest computer hard drives can store in the same space. However even though DNA computing can be very helpful it also has some lacking. In order for DNA computing to be useful one needs a lot of resources meaning that if a very complex problem is going to be solved a lot of DNA'S is needed. Also, DNA is liable to errors including mismatches, insertions and deletions thus it is not always accurate and of course DNA computing is new and still being explored so not everything is known yet.

4. Conclusion

NIA or Nature inspired algorithms and techniques are paving way for the emergence of a new set of computational paradigms. The inspiration drawn from naturally occurring phenomena not only lets researchers understand the physical, chemical and biologically interacting world better, but also provides deep insights into how the same can be replicated and artificially implemented across various engineering domains. Simultaneously this presents new challenges for the implementations as they cannot rely on standard system architectures and traditional ecosystems anymore and have to adopt/implement novel ideas and approaches to the way computation devices are designed and developed.

The second most important phase shift to be noted is the way the systems are decommissioned and maintained. Novel algorithms inspired by nature and its working may be relatable but methods, architectural frameworks employed to realize the exact functionality and replicate it, may always not be easy to understand or might not even be similar. As a result, it is important to understand the basic methodologies that are currently in practice providing solutions to problems, and also the differentiating factor setting the nature inspired or bio inspired algorithms apart from traditional computational practices. A whole new frontier for nature inspired algorithms, and its related work in the category of optimization, needs to be opened up in this segment, similar to the work being carried out in the domain of artificial intelligence and machine learning, and even possibly include them under a single roof so that all components are relatable at a convergence point. This is not only useful for existing researchers but provides a comfortable onboarding for new interested researchers who are getting into the field of Bio-inspired engineering.

Another important challenge for an engineer aspiring to work in this field is that he needs to be familiar with computer science related topics as well as biology related concepts. Only then will a researcher be able to correlate and understand the incorporation of both the concepts and understand the intention of adapting such an approach while solving problems. While there is a lot of background work to be carried out before beginning any implementations in this category, the results are very promising and favorable. The nature inspired algorithms are highly efficient algorithms and they are purpose driven and specific to each domain. The end goal of each algorithm is to optimize and reduce the effort of computation, by providing a well-defined standard of architecture, design, approach and efficiency especially in the areas of software and product development.

Parallelly the advancements in the field of nature inspired and bio intelligent software and algorithms have also paved the way for scientists to come up with novel ways in which a living component can develop ways to communicate with non-living entities such as computers. This provides opportunity for establishment of strong correlation between the entities and fosters strong interaction between living and on living components. Thus, it can be concluded that the interaction and integration of bio-inspired algorithms with biological components enables computers and living objects to co-exist and leverage the benefits of each while moving towards a sustainable computing future.

References

[1] K. Kaur, Y. Kumar, Swarm intelligence and its applications towards various computing: a systematic review, in: 2020 International Conference on Intelligent Engineering and Management (ICIEM), 2020, pp. 57–62.

[2] V. Conti, C. Militello, L. Rundo, S. Vitabile, A novel bio-inspired approach for high-performance management in service-oriented networks, IEEE Trans. Emerg. Top. Comput. 9 (2021) 1709–1722.

[3] A. Gopalakrishnan, P. Manju Bala, An advanced bio inspired shortest path routing algorithm for SDN controller over VANET, in: International Conference on System, Computation, Automation and Networking, 2021.

[4] I. Fischer, Near real time performance of population-based nature inspired algorithms on cheaper and older smartphones, in: 5th International Conference on Soft Computing and Machine Intelligence, 2018.

[5] K. Aggarwal, S.K. Yadav, Nature inspired approach integration in cloud environment for performance elevation with internet of things, in: 2020 International Conference on Computing and Information Technology (ICCIT-1441), 2020, pp. 1–5.

[6] S. Sharma, B. Kaushik, A comprehensive review of nature-inspired algorithms for internet of vehicles, in: 2020 International Conference on Emerging Smart Computing and Informatics (ESCI), 2020, pp. 336–340.

[7] A. Ali, Y. Hafeez, S. Muzammil Hussainn, M.U. Nazir, BIO-INSPIRED COMMUNICATION: a review on solution of complex problems for highly configurable systems, in: 2020 3rd International Conference on Computing, Mathematics and Engineering Technologies (iCoMET), 2020, pp. 1–6.

[8] Z. Liu, Y. Wang, S. Yang, K. Tang, An adaptive framework to tune the coordinate systems in nature-inspired optimization algorithms, IEEE Trans. Cybernetics 49 (2019) 1403–1416.

[9] H. Li, X. Liu, Z. Huang, Z.P. Zeng, Z. Chu, J. Yi, Newly emerging nature-inspired optimization—algorithm review, unified framework, evaluation, and behavioral parameter optimization, IEEE Access 8 (2020) 72620–72649.

[10] M.Z. Khan, S.K. Pani, Artificial neural synchronization using nature inspired whale optimization, IEEE Access 9 (2019) 16435–16447.

[11] V. Bharti, B. Biswas, K.K. Shukla, Recent trends in nature inspired computation with applications to deep learning, in: 2020 10th International Conference on Cloud Computing, Data Science & Engineering (Confluence), 2020, pp. 294–299.

[12] R. Kapur, Review of nature inspired algorithms in cloud computing, in: International Conference on Computing, Communication & Automation, 2015, pp. 589–594.

[13] I.U. Khan, I.M. Qureshi, M.A. Aziz, T.A. Cheema, S.B. Shah, Smart IoT control-based nature inspired energy efficient routing protocol for flying ad hoc network (FANET), IEEE Access 8 (2020) 56371–56378.

[14] C. Miao, G. Chen, C. Yan, Y. Wu, Path planning optimization of indoor mobile robot based on adaptive ant colony algorithm, Comput. Ind. Eng. 156 (2021) 107230.

[15] N. Prabhat, D. Kumar Vishwakarma, Comparative analysis of deep convolutional generative adversarial network and conditional generative adversarial network using hand written digits, in: 2020 4th International Conference on Intelligent Computing and Control Systems (ICICCS), 2020, pp. 1072–1075.

[16] H. Jeong, J. Yu, W. Lee, Poster abstract: a semi-supervised approach for network intrusion detection using generative adversarial networks, in: IEEE INFOCOM 2021—IEEE Conference on Computer Communications Workshops (INFOCOM WKSHPS), 2021, pp. 1–2.

[17] A. Luntovskyy, O. Nedashkivskiy, Intelligent networking and bio-inspired engineering, in: 2017 International Conference on Information and Telecommunication Technologies and Radio Electronics (UkrMiCo), 2017, pp. 1–4.

[18] H.H. Hussien, DNA computing for RGB image encryption with genetic algorithm, in: 2019 14th International Conference on Computer Engineering and Systems, 2019, pp. 169–173.

[19] N. Rajee, Y. Kon, K. Yabe, O. Ono, Matrix multiplication with DNA based computing: a comparison study between hybridization-ligation and parallel overlap assembly, in: 2208 4th International Conference on Natural Computation, 4, 2008, pp. 521–525.

[20] S. Jin, Y.L. Byun, S. Byun, Analysis of brain waves for detecting behaviors, in: 2018 International Conference on Intelligent Informatics and Biomedical Sciences (ICIIBMS), 3, 2018, pp. 114–116.

[21] S. Sadeghifard, L. Esmaeilani, A new embedded e-nose system to identify smell of smoke, in: 2012 7th International Conference on System of Systems Engineering (SOSE), 2012, pp. 253–257.

[22] Y. Zhai, L. Xu, Y. Yanxia, Ant colony algorithm research based on pheromone update strategy, in: 2015 7th International Conference on Intelligent Human-Machine Systems and Cybernetics, 1, 2015, pp. 38–41.

[23] X. Liu, Sensor deployment of wireless sensor networks based on ant colony optimization with three classes of ant transitions, IEEE Commun. Lett. 16 (2012) 1604–1607.

[24] F. Ji, M. Jiang, Tabu annealing lion swarm optimization algorithm, in: 2021 International Conference on Computer Engineering and Artificial Intelligence (ICCEAI), 2021, pp. 422–426.

[25] S.K. Yevale, P.K. Bharne, Data clustering algorithms based on swarm intelligence, in: 3rd International Conference on Electronics Computer Technology, 2019.

[26] W. Firgiawan, S. Cokrowibowo, A. Irianti, A. Gunawan, Performance comparison of spider monkey optimization and genetic algorithm for traveling salesman problem, in: 2021 3rd International Conference on Electronics Representation and Algorithm (ICERA), 2021, pp. 191–195.

[27] D. Bhawarthi, G.V. Chowdhary, Performance evaluation of spider monkey optimization for congestion control in optical burst switched network, in: 2020 International Conference for Emerging Technology (INCET), 2020, pp. 1–7.

[28] V. Sharma, A. Sharma, Review on: smart home for disabled using brain–computer interfaces, J. Inform. Sci. Comput. Technol. 2 (2) (2015) 142–146.

[29] R.C. Eberhart, D.W. Palmer, M. Kirschenbaum, Beyond computational intelligence: blended intelligence, in: 2015 Swarm/Human Blended Intelligence Workshop (SHBI), 2015, pp. 1–5.

[30] A. Moreno-Domínguez, O. Colinas, T. Smani, J. Ureña, J. López-Barneo, Acute oxygen sensing by vascular smooth muscle cells, Front. Physiol. 14 (2023) 1142354, https://doi.org/10.3389/fphys.2023.1142354. PMID: 36935756; PMCID: PMC10020353.

[31] G.K. Parimala, R. Kayalvizhi, An effective intrusion detection system for securing IoT using feature selection and deep learning, in: 2021 International Conference on Computer Communication and Informatics (ICCCI), 2021, pp. 1–4.

[32] P. Bhadane, P. Ravikesh Bhaladhare, Optimized deep neuro fuzzy network based automatic approach for segmentation and food recognition, in: 2021 5th International Conference on Information Systems and Computer Networks (ISCON), 2021, pp. 1–4.

[33] D. Fan, J. Wang, E. Wang, S. Dong, Propelling DNA computing with materials' power: recent advancements in innovative DNA logic computing systems and smart bio-applications, Adv. Sci. (Weinh) 7 (24) (2020) 2001766, https://doi.org/10.1002/advs.202001766. PMID: 33344121, PMCID: PMC7740092.

Further reading

[34] J.C. Bansal, H. Sharma, S.S. Jadon, M. Clerc, Spider monkey optimization algorithm for numerical optimization, Memet. Comput. 6 (2014) 31–47.

About the authors

Dr G.S. Mamatha, currently working as a Professor & Associate Dean (PG studies) in ISE Department, RV College of Engineering, Bangalore. She is working in the areas of Cloud computing, AI, IoT, Software Engineering and Networks. She has around 54 publications to her credit in International Journals and conferences. She is currently involved in consultancy and research work with companies and agencies for the submission of proposals. She is also responsible for establishing "Women in Cloud Center of Excellence in India" in association with WiC, USA for training in cloud technologies. She is also a part of research advisory committee for many internal and external research scholars for guidance.

Haripriya V. Joshi was a student of MTech in Information Technology ISE Department, RV College of Engineering, Bangalore, currently working in Intel as Validation Engineer. She presented three papers on deep learning in various Scopus indexed journals and her interest areas are AI, programming and cloud.

R. Amith was a student of MTech in Information Technology ISE Department, RV College of Engineering, Bangalore. He is currently working as Senior Software developer working in Daimler Truck Innovation Center India. He has over 7 years of experience in IT Services sector for companies like Cognizant, Unisys.

CHAPTER TWELVE

Optimizing the feature selection methods using a novel approach inspired by the TLBO algorithm for student performance prediction

Suja Jayachandran[a,b] and Bharti Joshi[a]

[a]Department of Computer Engineering, Ramrao Adik Institute of Technology, D Y Patil Deemed to be University, Navi Mumbai, Maharashtra, India
[b]Vidyalankar Institute of Technology, Mumbai, Maharashtra, India

Contents

Abstract

Over recent years, there has been a lot of advancement in the education domain. Data captured through Learning Management System, Management Information system, and any online platforms, contributes to research in the domain of learning analytics and educational data mining. Analyzing the educational data generated through these resources has become a challenging task. But all the educational data in its various formats are helpful for educational institutes to improve teaching–learning methodology, employability, student retention, and to make optimal decisions for the success of the institute. We provide a new technique in this study inspired by the teaching–learning-based optimization

Advances in Computers, Volume 135
ISSN 0065-2458
https://doi.org/10.1016/bs.adcom.2023.11.007

algorithm to optimize feature selection methods. We have successfully found the optimal set of features and evaluated them using different machine learning algorithms. It is observed that, based on the dataset, Gradient Boosting Classifier, Decision Tree Classifier, Support Vector Classifier, Adaptive Boosting, and XGB Classifier have outperformed other classification methods like Logistic Regression, Random Forest, and K Nearest Neighbor Classifier. By testing the proposed methodology with Gradient Boosting Classifier, the AUC score improved by 10% and in Decision Tree Classifier by 27% compared to the model tested without the proposed feature selection method. Based on these results, we have concluded that student performance prediction depends on current academic record, past results, attendance, and socio–demographic factors.

1. Introduction

- In the past few years, educational institutions have been using Management Information systems or Learning Management Systems to store student data like enrollment records, attendance records, and internal and external marks of each student which helps them track the student's progress effectively. Educational data mining has acquired considerable recognition over the last few years as it analyzes the data to help educational organization formulate better educational strategies, in turn enhancing the quality of education. It uses data mining techniques to facilitate instructors and learners in analyzing the learning process. Data mining algorithm helps in extracting hidden information from the educational data, and the knowledge extracted from this helps the institute improve its teaching method and learning process. Along with it, learning analytics (LA) uses statistical and visualization tools which help in retrieving, exploring, and summarizing the data about the learner and optimizing academic affairs. The customized teaching–learning process can be devised for slow and fast learners to foster better performance. Early detection of such types of students helps the instructors maintain student retention. With advances in machine learning algorithms, we can provide customized classroom teaching by providing real-time feedback to students based on their behavior in class [1].
- Student academic performance prediction helps the institute classify students into low, average, and high performers which can later be considered for customized teaching–learning methodology. There is a major impact on features selected from the dataset on the student performance prediction model. Preprocessing steps comprising the FS algorithm are chosen as suitable features for ML model. It helps improve the machine

learning model. Different types of FS algorithms are filter, wrapper, and hybrid models. Filter methods are carried out during the preprocessing step and are independent of any ML algorithm; however, they rely on all features. Their score is calculated based on a statistical correlation with the target variable. The wrapper method uses learning algorithms to evaluate the features. Hybrid feature selection combines the properties of both filter and wrapper methods.

- The TLBO is a teaching–learning process-inspired algorithm proposed by Rao and Patel [2] based on the impact of the teacher on the performance of students in the classroom. It works in two phases: 1. Teacher's Phase and 2. Learner's Phase. In this, learners act as a population, the design variables of the optimization problem are different subjects offered to the learners, and the fitness value is a learner's result. The best solution for the population will be the teacher. It is summarized in Fig. 1.

- Teacher's Phase: During this phase, the teacher provides knowledge to the learners and tries to get better results for the class. Here, we need to

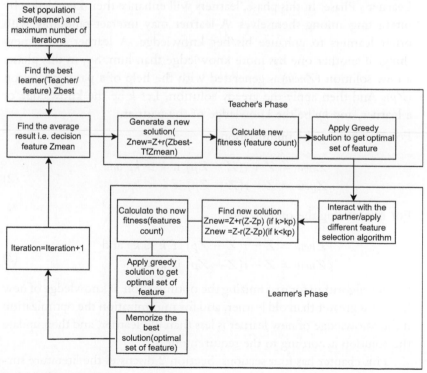

Fig. 1 TLBO algorithm.

calculate the best solution, i.e., the best learner from the population. The first step is to generate a new solution ($Znew$ for each Z) based on the best solution obtained and the mean result of the population. Second, we will update the fitness function and then we will apply greedy selection to accept the new solution, provided it is better as the goal is to increase the mean result of the population. It will update the knowledge of all learners. Let's say Z is the current solution, $Znew$ is the new solution that has the potential to be a teacher, $Zbest$ is the best teacher who has put his/her best effort into increasing the mean solution, Tf is the teaching parameter (same for all variables either 1 or 2), $Zmean$ is the mean result of the learner, and r is the random number (between 0 and 1).

$$Znew = Z + r(Zbest - Tf\,Zmean)$$
$$\text{where } Tf = round(1 + random) \tag{1}$$

These accepted values which are retrieved from the teacher's phase will act as an input to the learner's phase.

- Learner's Phase: In this phase, learners will enhance their knowledge by interacting among themselves. A learner may interact randomly with other learners to enhance his/her knowledge. A learner learns new things if another one has more knowledge than him. So, in this phase, a new solution ($Znew$) is generated with the help of a partner solution (Zp). And then apply the greedy solution. Let k be the knowledge of a learner, and kp be the knowledge of the partner.

- For maximizing the optimization

$$Znew = Z + r(Z - Zp) \text{ if } k > kp \text{ and}$$
$$Znew = Z - r(Z - Zp) \text{ if } k < kp \tag{2}$$

- For minimizing the optimization

$$Znew = Z + r(Z - Zp) \text{ if } k < kp \text{ and}$$
$$Znew = Z - r(Z - Zp) \text{ if } k > kp \tag{3}$$

Greedy algorithm for maximizing the optimization, if knowledge of new learner is greater than old learner, and for minimization the optimization if the knowledge of new learner is less than old learner, and then update the solution according to the requirement.

This chapter has four sections: Section 2 discusses the literature survey related to educational data mining, the feature selection algorithm used in educational data mining, and the TLBO algorithm. Section 3 emphasizes the proposed method. Section 4 shows the experimental setup and results. Section 5 provides the conclusion.

2. Related work

Optimization is the way to find the best or most beneficial value from a pool of values based on preset criteria. There are many optimization algorithms, most of which are nature-inspired algorithms. These algorithms include evolutionary algorithm (EA), genetic algorithm (GA), particle swarm optimization (PSO), and ant colony optimization (ACO). Biswas et al. [3,4] have proposed a new optimization algorithm called the atom stabilization algorithm which is designed based on the atomic model. Kennedy and Eberhart [5] developed the population-based optimization-based optimization method known as PSO. The algorithm mimics social animal behaviors like flocking of birds and schooling of fish. In Ref. [6], authors have discussed a strategy to enhance PSO performance by preventing particle's ineffective motion. The impact of the cognitive avoidance component gives the movement of the particle an extra boost as it moves in the direction of the optical solution. In Ref. [7], authors have discussed an extensive application domain where swarm-based technique is applied to find the optimal solution. Sarkar et al. [8] have discussed a genetic algorithm-based approach to enhance the performance of a deep learning model by selecting appropriate features from visual data. An evolutionary algorithm is a heuristic stochastic process used in evolutionary optimization. The algorithm generates the next state based on the current state, and the process keeps on until a set of predetermined requirements are satisfied. In Ref. [9], visual analysis method is proposed to compare the performance of EA. This performance is determined with the help of visualizing the angle and shifting of the regression line with respect to the neutral line. There are many applications where these optimization algorithms can be applied to improve performance. In this section, we have discussed work done to improve student performance.

- To optimize the quality of teaching–learning methodologies, many studies are going on in the educational domain. With the aid of machine learning, deep learning, and data mining algorithm, we can predict student performance and employability and devise a customized teaching–learning environment that will help the institute's growth. Asthana and Hazela [1] discussed the application of machine learning in the tailor-made educational setting. They considered socio-demographic factors, feedback systems, and behavioral aspects. Identification of fast and slow learners in the initial stage will help instructors change teaching methodology. Similarly, Ko et al. [10] have applied ML techniques to find important attributes that a good learner exhibits. From this finding, teachers can provide a learning environment like an online assessment quiz/test for students to self-evaluate themselves at any time. As per their study, Naive Bayes is the most appropriate method to predict student performance. Xu et al. [11] used gradient boosting decision tree (GBDT), K-nearest neighbors (KNNs), and decision tree (DT) methods to identify student performance in the flipped classroom and an online course. The prediction of student performance in online mode will help the teacher identify poor and good performers, which can later be used to customize the teaching methodology during face-to-face classroom teaching. Identification of such students should be done at an initial stage. Li et al. [12] have proposed the SPDN model, which uses online learning activities and merges the offline information to forecast the result of the students based on bi-long short-term memory. They were able to predict the result in the beginning, of course, itself which helps instructors find students who are not performing well.

- Recently learning analytics has become one of the key fields of educational technology, attracting researchers to explore new ways to improve academic affairs. Lemay et al. [13] have reviewed papers and compared learning analytics and educational data mining to predict student performance and model student behavior. Learning analytics papers investigate how to improve the engagement of students in a class, different open-source tools for teaching, and student engagement in social network, whereas educational data mining (EDM) papers focus more on techniques and methods of data analysis. Aldowah et al. [14] reviewed papers on educational data mining and learning analytics in higher education. They were able to identify papers suggesting different models predict student performance based on computer-supported learning analytics (CSLA), computer-supported predictive analytics (CSPA), computer-supported behavioral analytics (CSBA), and computer-supported visualization analytics (CSVA).
- In recent times, there has been tremendous growth in the online teaching–learning process. However, despite its advantages and popularity, the problem of student retention cannot be ignored. Prediction of the students who were low performers at the initial stage help reduce the dropout rate. Waheed et al. [15] have discussed how demographic

characteristics and student's clickstream activity affect student performance. The authors have used the deep learning model to predict at an early stage to help instructors take corrective measures. Long et al. [16] have discussed how lecture video data analytics can be used as an instructional resource in classroom teaching. The paper discusses the difference between the usage of lecture videos in classroom teaching and flipped classroom format. Czibula et al. [17] have launched a classification model SPRAR based on relational association rules to binary classification into pass (+) class and fail (−) in a particular course at the end of the exam. Khan and Ghosh [18] have performed a systematic survey of papers based on educational data mining. It was observed that during the course, performance was estimated, but prediction improves if we work on the start of the course and compare it with current performance. Adekitan and Salau have used information miner data mining algorithms to analyze the progress of undergraduate students during the first 3 years of study to predict the final year result [19].

- Feature selection is a crucial step in data preprocessing. Zaffar et al. have analyzed the performance of the FS algorithm and classifier on two different student datasets to predict student performance [20]. It shows how the feature count affects the performance of the model. Punlumjeak and Rachburee [21] have compared four FS methods: genetic algorithm, support vector machine, information gain with four supervised classifier-naive bays, decision tree, K-nearest neighbor, and neural network. As discussed in Ref. [22] feature selection is a dimensionality reduction technique to choose relevant features from the original feature set to improve learning accuracy and lower computational cost. Miao and Niu have surveyed different FS method-filter method, wrapper method, and embedded method. In Ref. [23], Li et al. have discussed different FS methods based on information: theoretical-based, sparse-learning-based, and statistical-based methods. Factor analysis is discussed by Hassan et al. [24] and the result shows that socio-demographic factors, student's attitudes before and after attending a class, and their approach to understanding the topics discussed in the class affect the student performance. Exploratory and confirmatory factor analyses are two powerful statistical techniques for factor analysis. Ozturk [25] performed a confirmatory analysis of the educators' attitude toward educational research. Alraddadi et al. [26] and Allam and Nandhini [27] have used wrapper feature selection method called binary TLBO and machine learning algorithm to predict, and it shows better accuracy compared to a model designed with a basic FS method.

3. The proposed method

3.1 Motivation

Being in the education domain for a long time, we feel the success of an institute depends majorly on the teaching–learning process. Predicting students' performance is a difficult task as it depends on a lot of aspects, such as performance throughout the academic year, socio-demographic factors, behavioral factors, unbiased grading approach, and so on. Prediction of low and high performers at the initial level will help in improving the overall result of the class and motivate students to perform better.

3.2 Design

In this research article, we will estimate the performance of FS methods and different classifiers. We will enhance the performance of the model based on inspiration from the TLBO algorithm. TLBO is a population-based meta-heuristic algorithm that finds a global solution from solutions obtained from the population [28]. In TLBO, the population is a group of students present in the class. As discussed earlier, TLBO has two phases: the teaching phase and the learner phase. In the teaching phase, learners get knowledge from the teacher, and learning phase involves the learning process between learners.

In this research article, the comparison between different FS algorithms and classifiers is applied to student datasets. Here, we will try to answer the following questions:

(a) RQ1: Which are the key features to predict student performance? Which category do they belong to—low or a high performer?

(b) RQ2: Which is the best combination of FS algorithm and classifier to forecast the performance of students?

To answer these questions, we have applied a different FS algorithm to a dataset downloaded from the UCI ML repository. The effect of these algorithms on the ML model is compared based on accuracy as the performance metric. The proposed method is explained in Fig. 2.

As shown in Fig. 2, the proposed method is as follows:

- Step 1: Import the student dataset containing academic records and socio-demographic details.
- Step 2: As part of data preprocessing, we will first check for null/missing values and them transform the categorical values into numerical values.

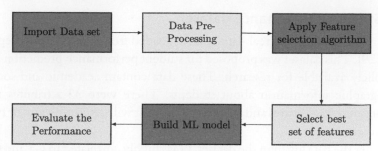

Fig. 2 Proposed method.

Then we need to perform exploratory factor analysis to check whether the data are under sample or over the sample. The last step under data preprocessing will be normalization to standardize the numerical value.

- Step 3: Next step is to apply different FS algorithms to find the rank of each feature based on their correlation with the target variable. Here, we will set a threshold of 0.4, and all features that correlate more than the set threshold will be selected. Now based on the count, we will give rank to each feature. Features with more than or equal to 3 are considered for further processing, and all other features will not be considered.
- Step 4: Here, we will apply the TLBO algorithm. As discussed earlier, TLBO is a nature-inspired algorithm. Based on Eq. (1) which is used to improve the overall result of the classroom by identifying the best teacher in the teaching phase and learning phase. So here we will consider features as populations, and the best set of features will be selected randomly through 50 iterations.
- Step 5: We will model the system using an ML algorithm and evaluate the performance based on metrics like accuracy, recall, precision, F1 score, and AUC score. In each iteration, all 8 algorithms will be executed based on the randomly selected set of 15 features and one with the highest AUC score will be considered the best classifier, and the collection of features selected by that model will be considered the set of the best features based on Eq. (2).

4. Experimental setup and results

In this research, all experiments have been implemented in the python programming language, and various libraries such as NumPy, Pandas, Scikit-Learn, and Seaborn are used.

4.1 Student performance dataset

Here, we have used the real dataset downloaded from the UCI ML repository [29]. This dataset was proposed for student performance prediction and is publicly available for research. These data contain academic and socio-demographic information about students. There were 33 attributes that we considered as features and a target feature that will determine low or high performers.

As per the data given in the dataset, G3 is highly correlated to G1 and G2. So ideally we can predict results based on G3 alone or an average of G1 and G2. In this chapter, we have dropped G1 and G2 as we have done binary classification of students according to G3 scores. Students are categorized as high performers if they have a G3 score greater than 10 or else as low performers.

4.2 Data preprocessing

In the ML process, data preprocessing includes data cleaning, data normalization, and data transformation. The selected dataset is complete as there are no null or missing values present. For further processing, we implemented a few steps:

1. Transformation: As some of the features had categorical values, it was difficult for the classification algorithm to operate on label data. As a result of using one-hot encoding, all categorical values were transformed into a suitable numerical value and new binary variables were added.

2. Factor analysis: Here, we have applied exploratory factor analysis to bring down the number of variables and examine the linear correlation among the variables. As per the guidelines for sample size discussed by Comrey and Lee [30], for factor analysis let the minimum sample size be N. Total N greater than 200 is recommended and if $N = 50$ then sample size is very poor, if $N = 100$ then sample size is poor, if $N = 200$ then fair if $N = 300$ then good and if N greater than 1000 then excellent sample size. The article also suggests that the minimum relation between variable/feature size and sample size should be 1:5, whereas ideally, it should be 1:20. Now, in our dataset, we have 38 features, so ideally, we should have $38 \times 20 = 760$ samples. After performing one-hot encoding and then aggregating new variables, we have 26 features, which would shrink to $26 \times 20 = 520$ samples. But we have 395 samples, so we have 75% of the ideal sample size. But if we consider the

minimum sample size required, i.e., 1:5, then we have more than the required sample size. So, we will not perform any oversampling or under-sampling methods as the data fulfill the minimum requirements of factor analysis.

3. Normalization: To standardize the value of numeric features, a Min-max scaler is used. It scales and translates each feature individually such that values are in the range of 0 and 1.

This completes our data preprocessing step. Now our data are ready for modeling different classification algorithms based on selected features.

4.3 Feature ranking

In this experiment, we have applied different feature selection algorithms such as Pearson, Chi-square, KMO, linear regression, Ridge, Lasso, and Random Forest and found the correlation between each feature with the target variable. As discussed, different FS methods such as the filter method, wrapper method, and hybrid method. Here under the filter method, we have used Pearson correlation and Chi-square. Under the wrapper method, we have used Random Forest, linear regression, recursive feature elimination (RFE), gradient boosting method (GBM), and KMO method, and lastly, under the hybrid method, we have used Lasso and Ridge method. Table 1 shows the correlation of each feature with target feature based on different methods.

As seen in the table, the correlation of each feature with the target feature is drawn through different feature selection methods. In this experiment, to find out the best features we have counted the features based on the correlation value, i.e., threshold above 0.4. Table 2 shows the count of features selected based on threshold.

For better feature reduction, features with more than or equal to 3 are considered for further processing, and all other features are discarded. Fig. 3 shows the rank of each feature

Now based on the rank of each feature, our proposed algorithm will select randomly 15 features, and this selection will be called for 50 iterations. As it is inspired by the TLBO algorithm, this proposed methodology will help us find the best set of features and their performance will be evaluated with respect to performance metrics like accuracy, recall, precision, F1 score, and AUC score. This selection of the best features will help us improve student performance prediction.

Table 1 Feature selection algorithm.

Feature	Chi-square	KMO	Lasso	LinReg	Pearson	Random forest	RFE	Ridge	GBM
Failures	0	0.88	1	0.9	1	0.72	1	1	0.14
Gender	0.88	1	0.48	0.64	0.26	0.11	0.89	0.65	0.29
Higher	1	0.89	0.14	0.77	0.49	0.08	0.92	0.6	0.03
Schoolsup	0.04	0.66	0.49	0.71	0.2	0.13	0.84	0.68	0.1
Famsup	1	0.88	0.35	0.46	0.08	0.08	0.7	0.49	0.19
Romantic	0.27	0.48	0.45	0.55	0.34	0.08	0.78	0.57	0.16
Medu	0.67	0.78	0.22	0.22	0.59	0.17	0.49	0.25	0.29
Fjob–teacher	0.14	0.51	0.28	1	0.24	0.07	0.86	0.77	0.05
Mjob–teacher	0.64	0.39	0.07	0.45	0.13	0.04	0.51	0.4	0.07
Mjob–health	0.17	0.11	0.6	0.69	0.3	0.03	0.97	0.67	0.04
Mjob–services	0.2	0.03	0.52	0.5	0.19	0.1	0.95	0.54	0.12
Age	1	0.6	0.15	0.14	0.43	0.24	0.19	0.16	0.36
Address	0.99	0.73	0.15	0.21	0.27	0.04	0.62	0.24	0.06
Famsize	1	0.41	0.27	0.36	0.2	0.06	0.59	0.39	0.11
Pstatus	1	0.71	0	0.14	0.13	0.05	0.76	0.16	0.06
Fedu	0.89	0.83	0	0.02	0.4	0.15	0.03	0	0.31
Studytime	0.99	0.87	0.24	0.26	0.25	0.19	0.54	0.29	0.3
Paid	0.51	0.89	0.07	0.16	0.26	0.05	0.43	0.19	0.33
Activities	0.97	0.69	0.07	0.17	0.01	0.1	0.41	0.19	0.28

Internet	1	0.85	0	0.23	0.25	0.04	0.73	0.21	0.04
Goout	0.88	0.74	0.29	0.29	0.35	0.27	0.57	0.34	0.28
Health	0.9	0.54	0.09	0.07	0.14	0.2	0.05	0.09	0.48
Absences	0	0.65	0.03	0	0.07	1	0	0.01	1
Fjob–health	0.94	0.29	0	0.5	0.13	0	0.81	0.34	0
Reason–course	0.87	0.15	0.16	0.43	0.25	0.05	0.68	0.37	0.21
Traveltime	0.99	0.88	0.11	0.07	0.3	0.15	0.14	0.1	0.14
Nursery	1	0.84	0	0.09	0.12	0.05	0.16	0.1	0.05
Famrel	1	0.41	0.09	0.09	0.11	0.14	0.11	0.11	0.13
Freetime	1	0.8	0.13	0.14	0	0.2	0.38	0.17	0.27
Dalc	0.45	0.79	0	0.11	0.12	0.07	0.24	0.11	0.15
Mjob-at-home	0.16	0.91	0	0.15	0.3	0.08	0.46	0.11	0.06
Reason–home	0.98	0	0	0.38	0.03	0.05	0.65	0.3	0.05
Guardian–mother	1	0.42	0	0.3	0.03	0.03	0.32	0.2	0.08
Guardian–father	0.98	0.43	0	0.33	0.06	0.05	0.35	0.18	0.08
Walc	0.04	0.78	0.04	0.11	0.12	0.15	0.22	0.11	0.38
Fjob-services	0.53	0.14	0	0.09	0.01	0.05	0.27	0.04	0.02
Fjob-at-home	0.92	0.3	0	0.33	0.01	0.04	0.3	0.21	0
Reason-reputation	0.44	0.12	0.05	0.12	0.24	0.05	0.08	0.02	0.09

Table 2 Feature selection count.

Feature	FSCount
Failures	7
Gender	6
Higher	6
Schoolsup	5
Famsup	5
Romantic	5
Medu	4
Fjob–teacher	4
Mjob–teacher	4
Mjob–health	4
Mjob–services	4
Age	3
Address	3
Famsize	3
Pstatus	3
Fedu	3
Studytime	3
Paid	3
Activities	3
Internet	3
Goout	3
Health	3
Absences	3
Fjob–health	3
Reason–course	3
Traveltime	2
Nursery	2
Famrel	2
Freetime	2

Table 2 Feature selection count—cont'd

Feature	FSCount
Dalc	2
Mjob-at-home	2
Reason-home	2
Guardian-mother	2
Guardian-father	2
Walc	1
Fjob-services	1
Fjob-at-home	1
Reason-reputation	1

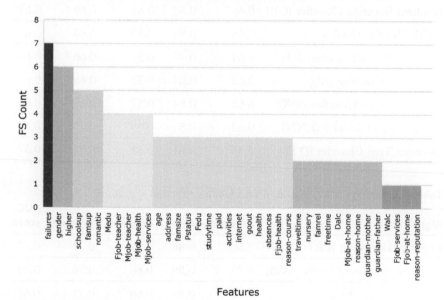

Fig. 3 Feature ranking.

4.4 Evaluation of machine learning algorithm

We have assessed the performance of eight ML algorithms like Logistic Regression, K-Nearest Neighbours Classifier, Support Vector Classifier (SVC), Decision Tree Classifier (DT), Random Forest Classifier, Gradient Boosting Classifier, Adaptive Boosting Classifier, and XGB Classifier.

As discussed in the design phase, we have implemented these eight ML algorithms two times: one on the dataset except for the proposed FS algorithm and the other on the same dataset with the proposed FS algorithm. It is seen that we have an improved model with better accuracy, recall, precision, F1 score, and AUC score. Table 3 shows the performance metric of ML algorithm without feature selection algorithm, i.e., all 25 features are considered.

Table 4 shows performance metrics of ML algorithm with feature selection algorithm. It has randomly selected 15 features. Table 4 shows the highest performance when features like Medu, Fedu, activities, Fjob–health,

Table 3 Evaluation of ML algorithm without feature selection algorithm.

Algorithm	Accuracy	Recall	Precision	F1 score	AUC score
Logistic Regression (LR)	0.68	0.88	0.61	0.72	0.69
Gradient Boosting Classifier (GB)	0.66	0.82	0.61	0.69	0.67
XGB Classifier (XGB)	0.66	0.84	0.59	0.69	0.67
Random Forest Classifier (RF)	0.64	0.75	0.59	0.66	0.64
Ada Boost Classifier (AB)	0.63	0.84	0.57	0.68	0.64
Support Vector Classifier (SVC)	0.62	0.84	0.57	0.68	0.63
K Neighbors Classifier (KNN)	0.64	0.5	0.65	0.57	0.63
Decision Tree Classifier (DT)	0.56	0.61	0.53	0.57	0.57

Table 4 Evaluation of ML algorithm with feature selection algorithm.

Algorithm	Accuracy	Recall	Precision	F1 score	AUC score
Logistic Regression (LR)	0.66	0.88	0.59	0.71	0.67
Gradient Boosting Classifier (GB)	0.73	0.89	0.66	0.76	0.74
XGB Classifier (XGB)	0.66	0.88	0.60	0.71	0.68
Random Forest Classifier (RF)	0.64	0.80	0.58	0.68	0.65
Ada Boost Classifier (AB)	0.65	0.89	0.58	0.70	0.66
Support Vector Classifier (SVC)	0.63	0.89	0.57	0.69	0.65
K Neighbors Classifier (KNN)	0.56	0.41	0.55	0.47	0.56
Decision Tree Classifier (DT)	0.71	0.80	0.65	0.72	0.71

Mjob-services, Fjob-teacher, failures, absences, study time, gender, schoolsup, famsize, famsup, paid, and Mjob-teacher have the highest correlation with the target variable.

4.5 Result and discussion

From the outcome of the experiment, clearly, the Gradient Boosting Classifier, Decision Tree Classifier, Support Vector Classifier, Adaptive Boosting, and XGB Classifier have performed well with few features in comparison to models designed with all features. Fig. 4 shows the comparison of the accuracy score graphically and indicates that Gradient Boosting Classifier and Decision Tree Classifier show 15.8% and 25% improvement, respectively. Similarly, Fig. 5 shows the comparison of the AUC score and indicates that the AUC score is improved by 10% in Gradient Boosting Classifier, in Decision Tree Classifier by 27%, in Support Vector Classifier by 2%, in Ada Boost by 3% and in XGB Classifier, the AUC score has improved by 2% compared to the model tested without the proposed feature selection method. The proposed feature selection algorithm shows an improvement in the performance of the model. This answers our two research questions initiated at the beginning, i.e., which are the important

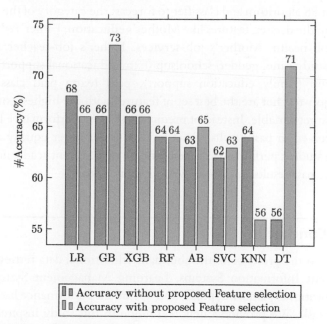

Fig. 4 Comparison of accuracy score.

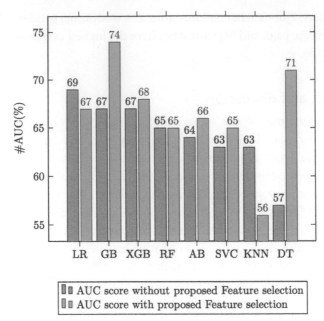

Fig. 5 Comparison of AUC score.

features to predict student performance and which is the most effective combination of FS algorithm and classifier to forecast the category of the student. So based on the dataset, features like Mother's education, Father's education, Father's job-health, Mother's job-services, Father's job-teacher, failures, absences, study time, gender, schoolsup (extra educational support), family size, famsup (family education support), paid (extra paid classes), and Mother's job-teacher are the best set of features with the highest correlation with the target variable. Instead of predicting student performance based on grades received in past results, we should consider other equally affecting factors. So student performance depends on his/her current academic record, attendance, past results, and socio-demographic factors.

5. Conclusion

Due to the presence of the abundant educational data retrieved from Management Information Systems, Learning Management Systems, and online educational platforms, prediction of student performance has become a difficult task. We provide a new technique in this study inspired by the TLBO algorithm. Its major objective is to boost the average result of the class. As the average result of the class has a direct correlation with the performance of an individual student, prediction of their performance at an initial stage will help to improve it. In this study, we have found a minimum number of features required for optimal prediction. It is observed that based on the dataset, Gradient Boosting Classifier, Decision Tree Classifier, Support Vector Classifier, Adaptive Boosting Classifier, and XGB Classifier have outperformed other classifiers like Logistic Regression, Random Forest, and K Nearest Neighbor Classifier. Moreover, the result shows the proposed methodology enhanced the performance of the Gradient Boosting Classifier, Decision Tree Classifier, SVC, Ada Boost, and XGB Classifier. We can extend this study in the future for a bigger dataset and compare the results of the ML model for the same.

References

[1] P. Asthana, B. Hazela, Applications of machine learning in improving learning environment, in: Multimedia Big Data Computing for IoT Applications, Springer, 2020, pp. 417–433.

[2] R. Rao, V. Patel, An elitist teaching-learning-based optimization algorithm for solving complex constrained optimization problems, Int. J. Ind. Eng. Comput. 3 (4) (2012) 535–560.

[3] A. Biswas, Atom stabilization algorithm and its real life applications, J. Intell. Fuzzy Syst. 30 (4) (2016) 2189–2201.

[4] A. Biswas, B. Biswas, K.K. Mishra, An atomic model based optimization algorithm, in: 2016 2nd International Conference on Computational Intelligence and Networks (CINE), IEEE, 2016, pp. 63–68.

[5] J. Kennedy, R. Eberhart, Particle swarm optimization, in: Proceedings of ICNN'95-International Conference on Neural Networks, vol. 4, IEEE, 1995, pp. 1942–1948.

[6] A. Biswas, A. Kumar, K.K. Mishra, Particle swarm optimization with cognitive avoidance component, in: 2013 International Conference on Advances in Computing, Communications and Informatics (ICACCI), IEEE, 2013, pp. 149–154.

[7] A. Biswas, B. Biswas, Swarm intelligence techniques and their adaptive nature with applications, in: Complex System Modelling and Control Through Intelligent Soft Computations, Springer, 2015, pp. 253–273.

[8] D. Sarkar, N. Mishra, A. Biswas, Genetic algorithm-based deep learning models: a design perspective, in: Proceedings of the Seventh International Conference on Mathematics and Computing, Springer, 2022, pp. 361–372.

[9] A. Biswas, B. Biswas, Visual analysis of evolutionary optimization algorithms, in: 2014 2nd International Symposium on Computational and Business Intelligence, IEEE, 2014, pp. 81–84.

[10] C.-Y. Ko, F.-Y. Leu, Examining successful attributes for undergraduate students by applying machine learning techniques, IEEE Trans. Educ. 64 (1) (2020) 50–57.

[11] Z. Xu, H. Yuan, Q. Liu, Student performance prediction based on blended learning, IEEE Trans. Educ. 64 (1) (2020) 66–73.

[12] X. Li, X. Zhu, X. Zhu, Y. Ji, X. Tang, Student academic performance prediction using deep multi-source behavior sequential network, in: Advances in Knowledge Discovery and Data Mining, vol. 12084, Nature Publishing Group, 2020, p. 567.

[13] D.J. Lemay, C. Baek, T. Doleck, Comparison of learning analytics and educational data mining: a topic modeling approach, Comput. Educ. Artif. Intell. 2 (2021) 100016.

[14] H. Aldowah, H. Al-Samarraie, W.M. Fauzy, Educational data mining and learning analytics for 21st century higher education: a review and synthesis, Telematics Inform. 37 (2019) 13–49.

[15] H. Waheed, S.-U. Hassan, N.R. Aljohani, J. Hardman, S. Alelyani, R. Nawaz, Predicting academic performance of students from VLE big data using deep learning models, Comput. Hum. Behav. 104 (2020) 106189.

[16] R. Long, T. Tuna, J. Subhlok, Lecture video analytics as an instructional resource, in: 2018 IEEE Frontiers in Education Conference (FIE), IEEE, 2018, pp. 1–7.

[17] G. Czibula, A. Mihai, L.M. Crivei, S PRAR: a novel relational association rule mining classification model applied for academic performance prediction, Procedia Comput. Sci. 159 (2019) 20–29.

[18] A. Khan, S.K. Ghosh, Student performance analysis and prediction in classroom learning: a review of educational data mining studies, Educ. Inf. Technol. 26 (1) (2021) 205–240.

[19] A.I. Adekitan, O. Salau, The impact of engineering students' performance in the first three years on their graduation result using educational data mining, Heliyon 5 (2) (2019) e01250.

[20] M. Zaffar, K.S. Savita, M.A. Hashmani, S.S.H. Rizvi, A study of feature selection algorithms for predicting students academic performance, Int. J. Adv. Comput. Sci. Appl. 9 (5) (2018) 541–549.

[21] W. Punlumjeak, N. Rachburee, A comparative study of feature selection techniques for classify student performance, in: 2015 7th International Conference on Information Technology and Electrical Engineering (ICITEE), IEEE, 2015, pp. 425–429.

[22] J. Miao, L. Niu, A survey on feature selection, Procedia Comput. Sci. 91 (2016) 919–926.

[23] J. Li, K. Cheng, S. Wang, F. Morstatter, R.P. Trevino, J. Tang, H. Liu, Feature selection: a data perspective, ACM Comput. Surv. (CSUR) 50 (6) (2017) 1–45.

[24] S. Hassan, N. Ismail, W.Y. Jaafar, K. Ghazali, K. Budin, D. Gabda, A.S. Samad, Using factor analysis on survey study of factors affecting students' learning styles, Int. J. Appl. Math. Inform. 1 (6) (2012) 33–40.

[25] M.A. Ozturk, Confirmatory factor analysis of the educators' attitudes toward educational research scale, Educ. Sci. Theory Pract. 11 (2) (2011) 737–748.

[26] S. Alraddadi, S. Alseady, S. Almotiri, Prediction of students academic performance utilizing hybrid teaching-learning based feature selection and machine learning models, in: 2021 International Conference of Women in Data Science at Taif University (WiDSTaif), IEEE, 2021, pp. 1–6.

[27] M. Allam, M. Nandhini, Optimal feature selection using binary teaching learning based optimization algorithm, J. King Saud Univ. Comput. Inf. Sci. 34 (2018) 329–341.

[28] P. Sarzaeim, O. Bozorg-Haddad, X. Chu, Teaching-learning-based optimization (TLBO) algorithm, in: Advanced optimization by nature-inspired algorithms, Springer, 2018, pp. 51–58.

[29] P. Cortez, A.M.G. Silva, Using data mining to predict secondary school student performance, 2008.

[30] A.L. Comrey, H.B. Lee, A First Course in Factor Analysis, Psychology Press, 2013.

About the authors

Suja Jayachandran has done her bachelor's and master's degree in Computer Engineering from University of Mumbai, India. She has been teaching in Vidyalankar Institute of Technology since year 2011. During her 12 years of teaching experience, she has guided many undergraduate students. Under her guidance, the students have done explicitly well and presented their work at various national level hackathon bagging prizes and accolades. Her areas of interest are Learning Analytics, Machine Learning and Data Science. At present, she is guiding three postgraduate students in the field of Machine Learning and Data Science. She is life member of ISTE and EdTech Society, India.

Bharti Joshi has done her bachelor's degree in Electronics and Instrumentation and ME in Computer Engineering from S.G.S.I.T.S Indore, India. She has been teaching in various engineering colleges in India and had been department head in three institutes. She completed her PhD (2014) from Department of Computer Science and Information wTechnology, VNSG University Surat Gujarat, India. She has 30 years of teaching experience in reputed engineering colleges in Indore, Bangalore and Navi Mumbai. She is associated with research projects and consultancy work with BARC Mumbai and IIPh Gandhinagar. She has registered one patent and has one copyright in her name. Her area of interest is Artificial Intelligence, Data Science and Machine Learning. At present, she is guiding five research scholars in the field of Data Science, Quantum Computing and Automated Guided Vehicle. She is life member of ISTE and fellow member of Institute of Engineers.

Communication and networking systems

PART IV

Communication and networking systems

CHAPTER THIRTEEN

Applying evolutionary methods for the optimization of an intrusion detection system to detect anomalies in network traffic flows

A.M. Mora[a], P. Merino[a], Diego Hernández[a], P. García-Sánchez[b], and A.J. Fernández-Ares[c]

[a]Department of Signal Theory, Telematics and Communications, ETSIIT-CITIC, University of Granada, Granada, Spain
[b]Department of Computer Engineering, Automatics, and Robotics, ETSIIT-CITIC, University of Granada, Granada, Spain
[c]Department of Languages and Computer Systems, ETSIIT-CITIC, University of Granada, Granada, Spain

Contents

Advances in Computers, Volume 135
ISSN 0065-2458
https://doi.org/10.1016/bs.adcom.2023.11.008

Abstract

Cybersecurity is a major concern nowadays, involving big amounts of resources in security companies, as well as in the academia. One of the main research lines in this scope are Intrusion Detection Systems (IDSs), which are programs or methods designed to supervise (or analyze) network traffic in order to identify suspicious patterns or clear attacks to any node in the monitored network. MSNM (Multivariate Statistical Network Monitoring) is one of the state-of-the-art algorithms, able to detect different security threats inside real network traffic data with a very high performance in most types of attacks. However, semi-supervised MSNM strongly depends on a set of weights whose values are normally defined using a rather simple optimization algorithm.

This chapter proposes the application of different Evolutionary Algorithm approaches in order to optimize these set of variables, aiming to increase the performance of MSNM against several types of attacks, including port scanning and botnets. To this end, we have considered a dataset, UGR'16, containing real network traffic flows, specially designed to test IDSs. In addition, we have analyzed the performance of a Particle Swarm Optimization approach.

The obtained results are very promising and lead us to conclude that EAs are a great tool to improve the performance of this IDS.

1. Introduction

Cybersecurity has become a critical issue at present, as the network traffic has grown exponentially and most of threats are focused on computer networks resources and services. According to Anderson [1], an intrusion is an unauthorized attempt to access, manipulate, or provoke a system to get useless. A computer attack or *cyberattack* is any circumstance or event that can negatively impact an organization's operations, assets, or users, through unauthorized access, destruction, disclosure or modification of information, and/or a denial of services [2]. These network attacks normally gain privileged access to a host by taking advantage of known vulnerabilities. Thus, cyberattacks must be prevented or detected as soon as possible in order to avoid a big loss in the affected companies. To this end, there is a big effort and investments put on research on this domain, in which the design of effective Intrusion Detection Systems (IDSs) is one of the main topics.

An Intrusion Detection System (IDS) [3] is a defense method that detects hostile activity on a network. Their aim is to detect and prevent activities that could compromise the security of the system, or an ongoing hacking attempt, including the unauthorized recognition or data collection phases performing, for instance, port scans. A key feature of IDSs is their ability to provide insight into unusual activity and produce warnings to notify administrators, as well as block suspicious connections. These systems can be either HIDSs, that monitor critical *hosts in the network* (accesses to their resources, modification of internal data or running programs); or can be NIDSs, which are focused on monitoring *data transported by the network* (streams, datagrams, packets). NIDSs are, essentially, algorithms able to identify malicious or suspicious patterns inside network traffic analyzed in real time. They are normally pre-trained and tuned using datasets of gathered network traffic flows and belongs to one of two main types: *sign-based detectors* (consider a database of already known attacks) or *anomaly-based detectors* (consider a model of normal traffic and try to detect unusual variations on the monitored traffic).

One of the state-of-the-art anomaly-based IDS is called MSNM (Multivariate Statistical Network Monitoring) [4]. It applies a variation of Principal Component Analysis [5] (e.g., Multivariate Statistical Process Control) to detect anomalies inside a huge dataset of network traffic flows, namely UGR'16 [6]. Thus, this approach analyzes a set of data related to normal traffic (non-attacks) and defines a set of thresholds so, when new network traffic data arise, they could be categorized as anomalous if the derived variables go beyond these thresholds. This method works fine even in the case of having a large number of input variables. However, it has a limitation, since it values all the input variables equally -it performs an auto-scaling of them-, making it difficult the detection of attacks that alter only some of these variables. To deal with this problem, a semi-supervised version was developed [7] by the authors of MSNM algorithm. It assigns different weights to each variable in order to set its importance in the detection. This, in turn, generated the problem of optimizing these weights for each type of attack. In this line, José Camacho developed a specific optimization algorithm, named *run to run PLS* [8], which obtained very good results in three of the four types of attacks included in the dataset (DENIAL OF SERVICE and two kinds of PORT SCAN) on which it was tested. It had also an acceptable detection performance in the case of the fourth attack (BOTNET), but considerably worse.

The present study tries to improve the results obtained by MSNM IDS enhancing the optimization process of the aforementioned variable weights. To this end, different Evolutionary Algorithms (EAs) [9] variations will be designed and applied, as well as another bio-inspired approach such as Particle Swarm Optimization (PSO) [10]. EAs are a family of metaheuristics inspired in the Darwinian natural evolution of the species. They have been widely used in the literature as very effective optimization tools in a big number of different domains, as well as by the authors of this work [11–14]. PSO is inspired in the behavior of particles in nature, as well as some social conducts followed by some species, such as birds or fishes when they move in groups (flocking strategies). This metaheuristic has been also applied successfully as an optimization method in many different problems [15–18]. Thus, the work describes the adapted approaches regarding their codification, genetic operators, and fitness functions. Then, all the methods are tested in different experiments also using the UGR'16 dataset, in order to compare the results with those obtained by the standard MSNM method.

The rest of this chapter is structured as follows: next section introduces background and preliminary concepts to the reader (EAs, PSO). Section 3 explains MSNM algorithm and its semi-supervised variation, while Section 4 briefly presents the UGR'16 dataset. The proposed EA and PSO approaches are described in Section 5, while the obtained results applying them and analyzed in Section 6. Finally, reached conclusions and future lines of research are commented in Section 7.

2. Preliminary concepts

This section presents the two main metaheuristics applied in this work, namely Evolutionary Algorithms (Genetic Algorithms) and Particle Swarm Optimization.

2.1 Evolutionary algorithms

Evolutionary Algorithms (EAs) [9] are a type of metaheuristic studied within the scientific field of Evolutionary Computation. These algorithms are stochastic optimization techniques based on the process of natural selection. They work with a population of *individuals* (or *chromosomes*), which are possible solutions for the problem to solve. Each solution is evaluated using an objective function to obtain a score, called *fitness*. The fittest individuals are more likely to reproduce and generate new solutions (*offspring*) that will inherit part of their structure. After several iterations (called *generations*) the selective pressure will produce better solutions. In each generation different operators are applied to the parents chosen to recombine (*crossover*) producing offspring. The *selection* of parents is performed following a specific criterium. There is also the possibility of modifying individuals (*mutation*). At the end of each generation the worst individuals are eliminated. This process continues up to a certain stopping criterion, such as a fixed number of generations. Fig. 1 shows the general EA process.

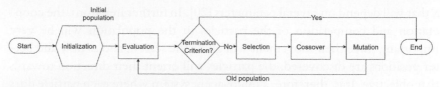

Fig. 1 Evolutionary Algorithm process diagram.

One of the advantages of this type of algorithms is that they can find optimal or near-optimal solutions to problems that are difficult to solve by human experts. On the other hand, they can provide these solutions in a reasonable time, even in high dimensional problems.

There are several types of EAs, the most famous being Genetic Algorithms (GAs) [19]. In this case individuals are encoded as a vector of genes to which crossover and mutation operators are applied. Other EAs evolve different structures, such as Genetic Programming, in which trees (representing operations or programming code) are evolved; or Evolution Strategy, which evolves genes together with the probability of mutating each of these genes.

In this work we will apply GAs, which can be found mainly in two different models:

- Steady-state: in this approach just two individuals are selected as parents in every generation to which the genetic operators are applied. This is an elitist process, since the worst individuals are replaced by the generated offspring and thus, the diversity in population is reduced step by step.
- Generational: in it, a big part of the population, normally half of it, is selected as parents. So, the other half is generated as offspring. All together compose the new population for the following generation having higher diversity.

2.2 Particle swarm optimization

Particle Swarm Optimization (PSO) [10] is one of the most well-known and applied population-based optimization metaheuristics. It is inspired by the social behavior of animal movement in nature, such as bird flight or fish schools. This metaheuristic has been widely used in different scientific fields, such as: humanities, engineering, medicine, or advanced physics. Thus, there have been proposed countless different approaches of PSO.

The PSO algorithm is initialized with a random population of candidate solutions, formed by several particles (particle swarm) that move through the search environment with a velocity v and with a given direction

x that will depend on several parameters [20]. In further iterations, the cooperation and competition of the particles in the population will be very important, since their movement will guide the search, and as new and better positions are discovered, the particles will orient their direction towards that objective. It is, therefore, a multi-agent system where each particle does not act alone individually, but it needs social interaction with the rest of the individuals, sharing, for instance, the best solution found to the moment.

The basis of the algorithm will be to calculate the speed. The population responds to the quality parameters p_{best} (local optimum) and g_{best} (global optimum), both individually and as a group. The changes made to p_{best} and g_{best} ensure the diversity in the population as the algorithm remains dynamic. Thus, the population changes its state (behavioral mode) only when g_{best} changes, fulfilling the stability principle. In addition, this also means that the population is adaptive. The overall objective will be the optimization of a problem from the initial population, where each particle will base the search having as reference both: its best-found position and the best global positions found by the rest of particles, as they move through the search environment.

It will always try to converge to good solutions avoiding local optima. Despite this, it is not guaranteed that the algorithm will not fall into such local optima, as it will depend on several factor. Besides, since PSO is a metaheuristic, it cannot be guaranteed that the global optimum will always be found.

3. MSNM as IDS

MSNM is the result of the evolution of some previous anomaly detection methods. Therefore, to understand how it works, we will give an introduction of these. The first method is Statistical Process Control (SPC), developed by Walter Andrew Shewhart [21], whose objective is to discriminate between common and anomalous causes of variation in a system. SPC establishes two phases for the construction of an anomaly detector; the first phase consists of detecting all the abnormal causes of variation of a system and correcting them, once this is done, we will say to have the system under statistical control. The second phase consists of establishing control limits or Upper Control Limits (UCL), which are the maximums of the variation of a system under statistical control. The main problem with SPC is that it only works on one variable. Multivariate Statistical Process Control (MSPC) is an extension of SPC for the consideration of multiple variables, however, it fails to interpret the relationship between variables. So, for cases where there are a large number of variables it is necessary to apply a method like Principal Component Analysis (PCA) [5].

PCA works on two-dimensional data, usually mean-centred and sometimes also auto-scaled, where there are M variables for N observations. This method aims to find the subspace of maximum variance in the M dimension, transforming the original variables (possibly correlated) into what are called principal components (PCs), which are variables without linear correlation. This makes it possible to reduce the dimensionality of the data without losing information. The advantage of this methodology is that it can handle a large amount of data, which is especially interesting in the case of network anomaly detection where a large amount of data can be relevant in a single observation.

Multivariate Statistical Network Monitoring (MSNM) is the proposal of José Camacho et al. [4] to adapt PCA-based MSPC to the computer network environment. It is relevant to note that PLS was conceived for industrial control systems, where most of the data are easily quantifiable measurements (temperature, pressure, etc.). However, in network monitoring, we are faced with a large number of logs and other heterogeneous data. Thus, MSNM, based on previous works [22,23], overcomes this difficulty by considering two classes of variables:

- Variables as event counters that take place in a period of time (packets sent and received, TCP connections on a specific port, etc.).
- Variables that quantify the dispersion or deviation of the previous ones, allowing a reduction in the number of variables and avoiding problems in the interpretation of results.

This is an unsupervised system, in which only a dataset with no labels is provided. The great advantage of this type of system is that it does not require prior knowledge about the attacks to be detected. So, it is possible to detect any kind of anomaly, being it a known attack, an unknown attack or a system failure. In summary, the MSNM methodology consists of the following steps:

- It converts logs and other observed data into counters, with the possibility of synthesizing part or all of these counters into aggregate variables by means of histograms. This dataset is referred to as the calibration or training dataset.
- Once the variables to be used are available, they are centered and optionally normalized (auto-scaled).
- Anomalies are eliminated to obtain a system under statistical control.

- The number of principal components is selected and PCA is applied, using the scores and residuals to extract the Q-statistics and D-statistics, on which Upper control limits (UCL) are defined.
- Once it is defined the model and control limits, they can be applied on a test dataset to check their effectiveness. It should be noted that the test set should be preprocessed in the same way as the calibration set.

Even if the "classic" MSNM algorithm works reasonably well in many cases, there are anomalies that require a more specific (or supervised) method to be detected. However, *semi-supervised MSNM* [7] aims to establish a hybrid method, which takes advantage of the benefits of both types of systems. It was observed that the detection of certain attacks is influenced to a greater extent by certain variables, while others are not relevant. Therefore, the idea of this methodology is to replace the auto-scaling described in the second step of the previous list by assigning a weight to each variable. The problem it causes is the decision of the values of those weights in order to maximize the detection capacity of a specific attack or anomaly.

Camacho et al. [7], implemented a method to set the weights optimally, named *Run-to-Run PLS* (*R2R-PLS*). Partial Least Squares (PLS) optimization is a multivariate regression technique, which optimizes the fit of MSNM features, providing a semi-supervised learning approach.

In the present work, different metaheuristics are proposed and tested to optimize this set of weights, and thus, enhance the performance of MSNM method in the detection of anomalies.

4. Dataset used: UGR'16

This dataset contains a big amount of network flow traces collected by researchers of the University of Granada (UGR), monitoring during 5 months the traffic in a real network belonging to an Internet Service Provider (ISP). The data collection was conducted in two phases using Netflow sensors. First capture was carried out between March and June of 2016 under normal conditions, i.e., the network was used normally by the ISP clients. The aim was to model and study the normal behavior of the network users, and to detect certain anomalies such as SPAM campaigns. After this, the flows of the dataset were "manually" labeled indicating if they correspond to "background" (legitimate flows), or "anomalies" (non-legitimate flows). This is the *CALIBRATION* part of the dataset, used to train the models. Second capture was conducted between July and August of 2016. There were launched some "controlled" (or synthetic) attacks aimed to obtain a test dataset for validation of anomaly detection algorithms. To do this, 25 virtual machines were deployed within one of the ISP sub-networks. Five of these machines attacked the other 20. This is the *TEST* part of the dataset, used to validate the trained models.

Four types of attacks were done: Denial of Service (*DOS*), port scanning from one attacking machine to one victim machine (*SCAN11*), port scanning from four attacking machines to four victim machines (*SCAN44*), botnet traffic (*NERISBOTNET*). These attacks were launched during 12 days in different periods of time, following either planned or random scheduling, and within real background traffic.

Fig. 2 presents a graphic summary of the UGR'16 capture process, including the generated attacks. There are two border routers monitoring/gathering all the traffic going to/from Internet. There are one network with attacking machines and two networks with victim machines (one without the protection of a firewall and another one that is protected). There were 5 attacking machines, 15 victim machines in the core network, and 5 victim machines in the inner network. The general characteristics of

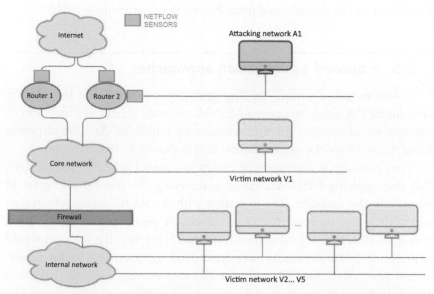

Fig. 2 Representation of Network Attacks in the dataset UGR'16, and different affected machines. It can be seen the different networks (core not protected and inner protected by the firewall) and the different machines involved in the generated attacks (Ax and Vx).

Table 1 Features of UGR'16 dataset Calibration and Test patterns.

Feature	Calibration	Test
Capture start	10:47h 03/18/2016	13:38h 07/27/2016
Capture end	18:27h 06/26/2016	09:27h 08/29/2016
Attacks start	N/A	00:00h 07/28/2016
Attacks end	N/A	12:00h 08/09/2016
Number of files	17	6
Size (compressed)	181GB	55GB
# Connections	$\approx 13,000M$	$\approx 3,900M$

the dataset are provided in Table 1. As it can be seen, there is a huge amount (thousands of millions) of captures.

This dataset has been used in many different works, so it is widely referenced in the cybersecurity literature. Its main advantage is that it contains data collected from a real network, so it is very interesting to test and validate IDS approaches (or other methods). It also contains periodic or cyclostationary data, as the background traffic follows day/night and weekday/weekend patterns. The dataset can be downloaded from: https://nesg.ugr.es/nesg-ugr16/.

5. Proposed optimization approaches

This section describes the four approaches presented in this chapter to optimize the Semi-supervised MSNM method, specifically they aim to improve set of weights for the variables on which its decision depends. All of them follow the same scheme, as it is shown in Fig. 3.

They consider as a measure of quality the *Area Under the Curve (AUC)* [24] after applying MSNM method considering the corresponding set of weights for the variables. This is related with the ROC curve, which represents the proportion of true positives vs false positives in a classification system, so, the AUC is a value between 0 and 1. If it is 1, the solution would be perfect and MSNM would have detected all the anomalies in the dataset. In case the value is 0, it indicates that no related anomaly has been detected. A value equal to 0.5 indicates a random classification. The AUC can also be defined as specificity vs. sensitivity: sensitivity is the ability of the detector to identify as anomalies traffic that is actually normal; specificity is the ability of the detector to identify as normal traffic, anomalies.

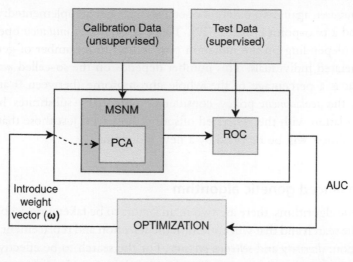

Fig. 3 Optimization scheme of a weight vector for Semi-supervised MSNM.

Three datasets are used in this process: a first one for the calibration of MSNM model, containing no labeled data and none anomaly; a second one including background traffic and anomalies, which will be used to optimize the weights; and finally, a third file which will be used to test the performance of the whole Semi-supervised MSNM method (and to compute the AUC).

The four proposed algorithms are a basic Genetic Algorithm, an enhanced version of it using advanced Selection and Replacement mechanisms, a Memetic Algorithm, and a PSO implementation.

5.1 Simple genetic algorithm

Every *individual* in the GA is an array of 134 values, each of them in [0,1]. These correspond to the associated weight to each variable, being "0" a non-considered variable and "1" the most relevant. The *fitness* is a function of the quality of every possible solution, namely the aforementioned AUC after applying MSNM method considering the corresponding set of weights for the variables that the individual has. Thus, the aim is to maximize this objective function. The population is initialized randomly. The first approach implemented is a *generational* one, so half of the population is selected as parents for the following generation. Two different *selection* mechanisms are implemented: a random selection of parents and a selection based on their fitness, where the best individuals are chosen. With regard to the *crossover*, again two different operators have been implemented: a uniform and a two-point crossover [9]. There is a random *mutation* operator, applied (depending on the mutation probability) to a number of genes of the generated individuals. This number depends on the so-called *mutation rate*, that is a percentage of the whole chromosome (between 0 and 1). Finally, the *replacement* policy considered, potentially substitutes half of the population with the generated offspring, however, just those that have a worse fitness will be replaced by a new individual.

5.2 Enhanced genetic algorithm

In genetic algorithms, there are two main factors to be taken into account to guide the search and that will condition the selection and replacement of the population: *diversity* and *selective pressure*. For the search to be effective, we rely on the search criterion, or directly related to the selective pressure of each individual, which gives them a higher probability of being selected for reproduction, mutation and therefore, survival. This is very important because if there is no selective pressure, the algorithm will be random to certain point and will not offer any possibility of improvement in the search. Moreover, the diversity of the population is extremely important, since the lack of diversity will result in a stagnation of the search at a non–global optimum, a phenomenon called premature convergence, which is one of the most important issues in GAs.

Diversity and selective pressure are inversely proportional. The more diversity exists, the more disparate the individuals are, so the selective pressure is lower, and when the pressure increases because for example the best individuals are always chosen, the diversity decreases, since the similarity index of the individuals will be high. When there is an average balance between the two, we will say that we are in *useful diversity*.

To this aim first Enhancement will be focused on the **Selection policy** of the parents, so three different models have been proposed:

- *Random Selection (RS)*: A random element of the population is selected as a parent. All individuals have the same probability of being selected, therefore it does not generate any selective pressure.
- *Linear Ranking (LR)*: The individuals are ordered depending on their fitness value and a probability of selection is associated to them depending on their order, which will be the higher the better positioned they are in the list obtained. That is to say, the worst element of the population will have a probability of being chosen (not null) and the best element of the

population will have the highest probability of being chosen. This allows the population to keep the fittest individuals but gives some margin for individuals with lower fitness to be selected as well [25].

- *Negative Assortative Mating (NAM)*: One parent is randomly selected. The other parent is selected from a group of x elements, so that the individual belonging to the group with fitness farthest from the first parent will be chosen. This is oriented to generate diversity, since the parents will be remarkably different. According to the study by Fernandes et al. [26], it is necessary to indicate a method of similarity between two individuals, so we have considered as similar parents with close fitness values.

The second enhancement will be focused on the **Replacement policy** of the individuals with respect to the generated offspring. There are different replacement strategies that, among other objectives, aim to improve diversity with a consequent increase in the quality of the solutions. In many of them, offspring are more likely to replace their most similar parent, based on genotypic similarity. This aims to minimize the fact of having several very similar individuals in the population by replacing one by another. On the other hand, clustering methods help to create populations close to the optimal solutions, maintaining several local optima, which in multimodal problems (as is the case) is a good way to converge to good solutions. These are the implemented approaches:

- *Replace Worst Strategy (RW)*: The new individual replaces the worst individual in the population. It generates a high selective pressure, even when the parents are chosen randomly, which may result in a high similarity index and limiting the diversity of individuals.
- *Restricted Tournament Selection (RTS)*: Having a child h, w individuals are randomly selected from the population, and the element that most resembles h is chosen. The child h will replace that individual in the current population.
- *Deterministic Crowding (DC)*: The offspring replaces the most similar parent, so diversity is maintained since they are not classified by any criteria and the new genes of the new offspring, although being similar to the parent, may bring to the population some difference and diversity.
- *Worst Among Most Similar (WAMS)*: Groups of variable eligible size $w = 3, 5 \dots$ are created where w individuals are chosen randomly. Then, the individual from each group that most resembles the generated child is chosen, obtaining w candidates to replace the parent. From this group, the element with the lowest fitness is chosen and replaced by the new child. There is some similarity with the RTS replacement; with the

difference that RTS tends to maintain the highest diversity while WAMS chooses to keep the fittest individuals in the population. The value of w controls the selective pressure of the algorithm. If it is increased, the probability of eliminating the worst individuals will also be high, and as a consequence the pressure will increase as well. Conversely, the lower the value of w, the higher the competition between individuals from different groups will be.

5.3 Memetic algorithm

Memetic algorithms [27] are population-based metaheuristics composed of an evolutionary framework and a set of local search algorithms that are activated within the generation. It is a combination of several concepts borrowed from other metaheuristics such as population-based search (genetic algorithms, for example), and the application of an improvement to one or several solutions in a local way (such as local searches) [28]. These algorithms try to address the shortcomings of other algorithms by hybridizing two different metaheuristics. For instance, evolutionary algorithms are good explorers, while local search algorithms are bad explorers and vice versa, so the evolutionary part of the algorithm will vary the population and create new solutions while the local search will help the algorithm to converge to global optima. Therefore, they are considered an advanced version of genetic algorithms, since they maintain the ability of population-based metaheuristics and all the genetic operators, so that a set of candidate solutions is stored for the problem under consideration, and around certain candidate solutions a local search will be applied to improve it.

Thus, local search will be applied to the offspring before replacement, in order to "refine" the generated solutions to be even better. The local search to be applied here will be *Local Search First Best*, where the first neighbor that improves the current solution will be selected. This is an effective and very fast method widely used in the literature. It can be applied in different ways, so there is no general rule to search for a nearest neighbor. In this case, we have considered that individual A is a neighbor of B if they have most of the same genes, but some of them are in reverse order. Thus, to generate a neighbor, we just swap two genes randomly each time. According to the work of Vrajitoru [29] it is worthwhile to apply local search on every individual if the derived computational complexity is relatively low. Diversity in the population is crucial for a memetic algorithm, because if we lose diversity, i.e., all individuals in the population have a high similarity index, we will fall into a region of local optima.

5.4 Particle swarm optimization

This metaheuristic has been adapted to the resolution of this problem, namely, finding the set of weights that will yield the best fitness (AUC) in the detection of different attacks inside network data flows.

Taking the analogy of PSO natural movement to its algorithmic representation, a particle is composed of the following elements:

- Vector x: Stores the current position of the particle in the search space (weights). In our case the position will be called *fitness* from now on, since it is the environment through which we will move and the best position to find will be the best fitness.
- Vector p_{best}: Stores the position (fitness) of the best solution that the particle has found so far.
- Vector v: Stores the velocity that the particle will follow. The velocity will be either a positive or a negative value to be able to move through the whole search space [10].

At first, a random solution will be generated, both for the fitness vector x and for the velocity vector v. In this case, the velocity will be bounded in a domain between $[-1.1]$. While generating the particles, their fitness and velocity, the global optimum will be calculated by calculating the fitness of each particle. In this way the algorithm from the beginning will be focused on the global optimum needed to apply the directionality formula for the particle.

An invariable number of iterations is set as a stopping criterion and the algorithm will run that number regardless of the result. Another frequently used option is to detect if in a specific number of iterations, the result has not been improved (we have reached a local optimum), then algorithm stops, in order to save computational time. As output, we will obtain a fitness vector corresponding to vector x which will be the best global solution found. For each i-th element of the position vector (fitness) its value will be varied according to the velocity, so that a value (positive or negative) corresponding to the current velocity for that element will be added, obtaining a new position vector.

At this point, once the position of the particle has been updated, we evaluate if the previous fitness has been improved, then the vector p_{best} will also be updated. The flight speed of the particles will be dynamically adjusted at each iteration for the particular particle, being conditioned by both the individual and the community at large. This adjustment depends on several factors, namely:

- xi (i-th element of the position vector): The velocity will always depend on the location (fitness belonging to the domain) where the particle is located.

- *random()*: The velocity has a strong random component since at each iteration the velocity is affected by a random number in the range [0,1].
- *gi*: Position of the best-known particle, that can be locally (particle environment) or globally (the whole cloud).
- $p_{best}i$: Best solution found by the particle individually.
- *ω*: Inertia factor: Positive constant that usually decreases linearly at each iteration. It describes how the previous velocity influences the current velocity. By changing its value, the global and local search capacity can be adjusted.
- ϕ_1, ϕ_2: Ratios or learning rates. These are non-negative weights or constants that control the cognitive and social components. They are usually similar or equal values.
- *lr*—Learning rate. Positive constant that affects the position (fitness) to control the speed that is applied. A learning rate too high will cause the particle to move through the environment too fast, making it difficult to reach an optimum. However, a learning rate too small will cause the particle to move too slowly through the environment, so that the exploration ability will be overshadowed and the algorithm is likely to stall at a local optimum. It affects the algorithm in the same way as the inertia factor.

These factors are combined in a formula with two parts: *cognitive part* (how the particle learns from its environment) and *social part* (how the particles interact). Both parts must cooperate in order to achieve good results, since the effectiveness of the algorithm depends on its cooperative nature. If the cognitive component is much higher than the social component, each particle is attracted to its own position, which will result in excessive wandering in the environment. On the other hand, if the social component is much larger than the cognitive one, the particles are attracted by the best global position and will converge very fast.

6. Experiments and results

Several experiments have been conducted, analyzing the performance of all the proposed optimization methods combined with MSNM. They are also compared with the standard approach and with its semi-supervised variation (optimized by R2R-PLS). It is important to note that semi-supervised MSNM was one of the state-of-the-art approaches to detect anomalies in network traffic flows, so improving its performance is not an easy task.

We have not considered the whole UGR'16 dataset in these experiments, because, as it can be seen in Section 4, it has millions of patterns. Thus, we have used a representative subset of 18,000 captures in which there are background traffic and attacks. It is composed by 12,000 patterns of Calibration (6000 for initial calibration and 6000 for optimization of weights) and 6000 patterns of Test datasets.

The considered metric in all the results will be AUC, as explained in Section 5, but a different run must be done for each type of attack. This means that the algorithms will be run to get specialized weights to detect one of the attacks each time. Moreover, given that EAs are stochastic methods, in all the experiments (every approach for every type of attack), 10 runs are performed and the average result is shown in the graphs.

The used configurations in each algorithm has been set after a process of exhaustive experimentation, i.e., several runs were conducted with different configurations (mutation and crossover probabilities, population size, number of generations, mutation rate) and those with offer the best results in average have been selected. This has implied hundreds of runs for each approach.

6.1 Simple GA study

Some different experiments have been conducted considering the simplest EA approach. First of all, we will test its evolutionary behavior, this means that the algorithm has a proper convergence tendency, so solutions in the population should be better every generation. The configuration of the GA is: generational model, 40 individuals, 100 generations, crossover probability 1, mutation probability 0.5, mutation rate 0.04 (genes to be mutated in the individuals).

Fig. 4 shows the evolution of the best (or maximum) and average fitness (of the whole population) in different runs of the algorithm. There is an example run for each type of attack. As it can be seen, there is a clear maximization tendency in both, the maximum and the average fitness of the population, with some expected oscillations in the average, as in the case of SCAN44 run. Since GAs are stochastic methods, it is normal that some runs start from a worse population, which takes several generations (or maybe never) to get to a "stability" in the quality of the individuals of the population. Anyways, these results lead us to think that the algorithm is working properly.

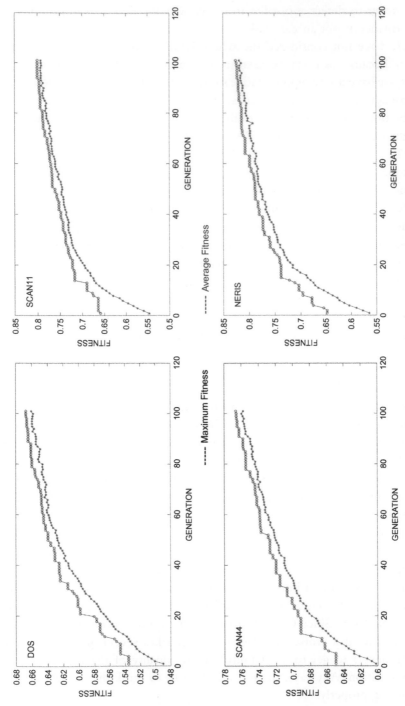

Fig. 4 Maximum and Average Fitness evolution in one example run of the Simple GA per type of attack.

The second experiment tries to analyze the diversity in the population, and how it evolves along the generations. GAs (and all the EAs) need to maintain always a certain degree of diversity, which would allow the algorithm to move to new zones of the search space and thus, escape from local optima. On the contrary, if getting to a certain generation, almost all the individuals are very similar, the algorithm will be probably stagnated and the following generations will not improve the current solution (or not do it too much).

So, in order to study this effect, we have defined a measure of the similarity between individuals and also of the whole population. We will consider this computation:

0. Distance between individuals is the number of genes with a different value.
1. Sum the distance from one individual to the rest of the population (St).
2. Sum all the St from all the individuals (TSt).
3. Divide TSt by the number of individuals (Mt).
4. Divide Mt by the number of genes (Mg).
5. Sum all the Mgs of all the individuals (Dm).
6. Divide the Dm by the number of individuals $\rightarrow$ SIMILARITY.

This measure will be a value between 0 and 1, giving us an indicator on how similar are the individuals in average. A value close to 0 indicates that the individuals are pretty different, while a value close to 1 means all of them are almost equal.

We have conducted the experiment with the following configuration for the GA: generational model, 40 individuals, 100 generations, crossover probability 1, mutation probability 0.5, mutation rate 0.8. We have set a big mutation rate to ensure there are considerable differences in the last generations.

Fig. 5 shows the obtained results for a run focused in each one of the attacks. As it can be seen, the similarity is reduced in every generation as it would be desired, i.e., the individuals are improving and thus, they tend to be closer to the optimal solution regarding their values (weights for the variables). However, the high value of the mutation rate makes it possible to have still big differences (in average) between individuals in the latter generations.

After these positive preliminary studies, we finally tested the Simple GA approach for optimizing the weights for the Semi-supervised MSNM method. We tested first a Steady–State approach with the following configuration: 40 individuals, 500 generations, crossover probability 1, mutation probability 0.5, mutation rate 0.02. Fig. 6 shows the obtained results

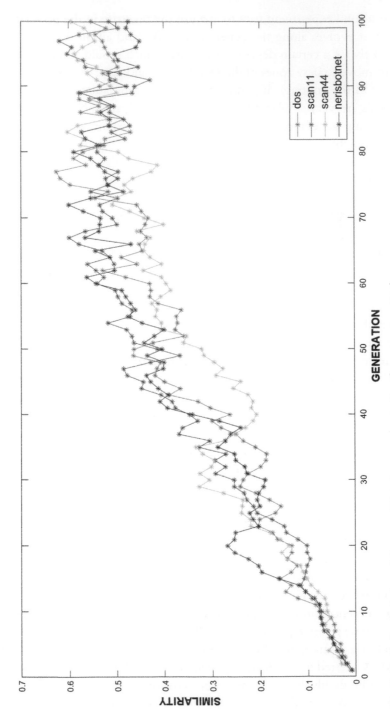

Fig. 5 Similarity evolution of the population in every generation for a run per attack.

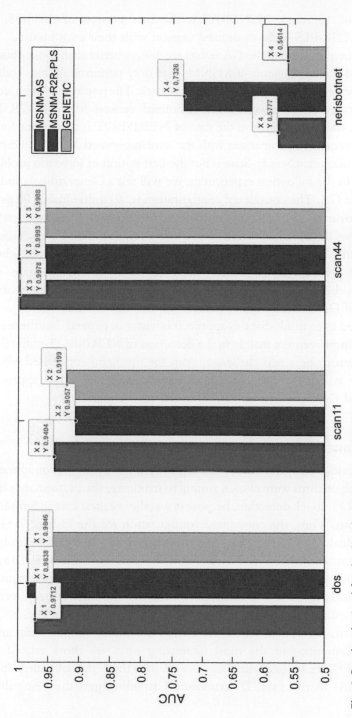

Fig. 6 Results obtained for the Simple GA approach (GENETIC) considering a Steady-State model, 10 runs.

(average of 10 runs). MSNM-AS is the first version of the IDS, and MSNM-R2R-PLS is the optimized version with their own method.

It can be noticed that the GA results are always better than those obtained by the first implementation of MSNM, that only performs an auto-scale on the variables, except in the case of SCAN11 attack. The results are also improved in some cases with respect to the optimized version MSNM-R2R-PLS. However, the performance in the case of NERISBOT is quite poor for this genetic approach in comparison with the semi-supervised MSNM. This fact, lead us to think that Steady-State is not the best option to solve this problem.

Thus, in the following experiment, we will test a Generational model of the Simple GA. The considered configuration is: 40 individuals, 100 generations, crossover probability 1, mutation probability 0.5, mutation rate 5. Fig. 7 shows the obtained results (average of 10 runs) focused in the different types of attacks. As it can be seen, the Generational model gets 5% better results in NERISBOT, which was the flaw in Steady-State case. However, the results with respect to the other methods are a bit worse in the case of DOS and SCAN11 attacks.

This led us to think that this approach is better in general, but there is still room for improvement mainly in the detection of NERISBOT traffic flows, which seem to be a real challenge, even for the Semi-supervised MSNM IDS. We will try to improve these results using the other approaches (Enhanced GA, Memetic and PSO) in the next experiments.

6.2 Enhanced GA study

Given previous results, for the following experiments the best configurations for these algorithms were chosen aiming to maximize the performance in the NERISBOT attack detection, because it was the weakest case in Simple GA experiments. Thus, the considered configuration for the Enhanced GA is: 50 individuals, 100 generations, crossover probability 1, mutation probability 0.05, mutation rate 0.5. All the results refer to the average of 10 runs.

The different enhancements will be tested in the following experiments. First, *Random Selection* is evaluated, considering the different replacement policies explained in Section 5.2. The results are plotted in Fig. 8. As it can be seen, the results for all the attacks are pretty similar using any of the replacements, but the most interesting ones are those related with NERISBOTNET, because these are better than the obtained in the Simple GA. In this case, Deterministic Crowding gets the best value in average.

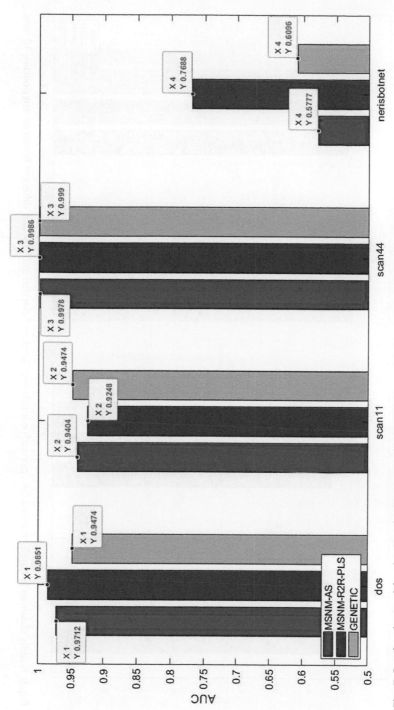

Fig. 7 Results obtained for the Simple GA approach (GENETIC) considering a Generational model, 10 runs.

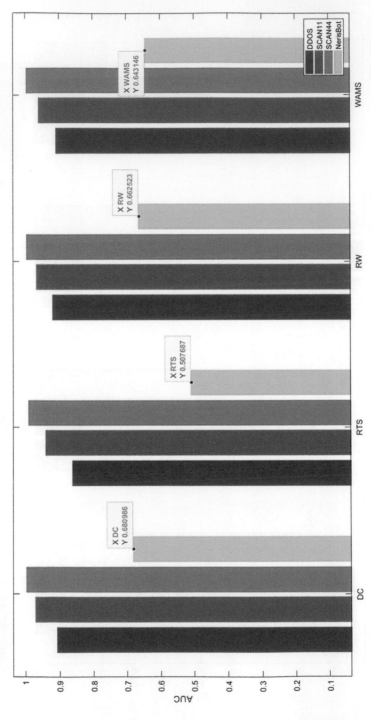

Fig. 8 Results obtained for the Enhanced GA approach considering Random Selection and the four different Replacement policies, 10 runs.

Second experiment, testes the *Negative Assortative Mating* Selection approach. Fig. 9 shows the obtained results using all the replacement policies. This time, the results are quite similar to those obtained in previous experiment, but this time the best combination is the use of the Replace Worst Strategy.

Third experiment, testes the *Linear Ranking* Selection approach. Fig. 10 shows the obtained results using all the replacement policies. As it can be seen, results with this approach are a bit worse than in the first experiment (RS), but this selection policy together with Deterministic Crowding gets a higher AUC value than the two previous selection methods.

6.3 Memetic algorithm study

Premature convergence is tremendously important in memetic algorithms, since the local search leads us to local optima, so we must maintain a certain diversity in the population to avoid this phenomenon. Having this as reference, in this section we will carry out a study in which local search will be applied to a certain number of individuals chosen in different ways and we will check which of them is the best for the problem that is being addressed. We will consider the same configuration as in the experiments of Section 6.2, but we will check these Local Search approaches:

- LS1: Every 2 generations, LS Will be applied to all the individuals in the population.
- LS2: Every 2 generations, LS Will be applied to a subset of individuals randomly selected with probability 0.1.
- LS3: Every 2 generations, LS Will be applied to the 10% of the best individuals.

Results for each of these approaches are presented in Fig. 11. As it can be seen, results for the first three attacks are pretty similar in all the cases, but NERISBOTNET results vary a lot. The application of LS to all the individuals do not yield the best results, as it could be expected, but the worse. This is because this method reduces drastically the diversity in the population, which seems to be very important to get good solutions in this problem. So, the best approach is the application of Local Search to a set of random individuals, which obtains results close to those of MSNM-R2R-PLS in NERISBOTNET.

6.4 PSO study

As described in section 5.4, PSO depends on several parameters, that were studied in order to set the best combination. After an exhaustive

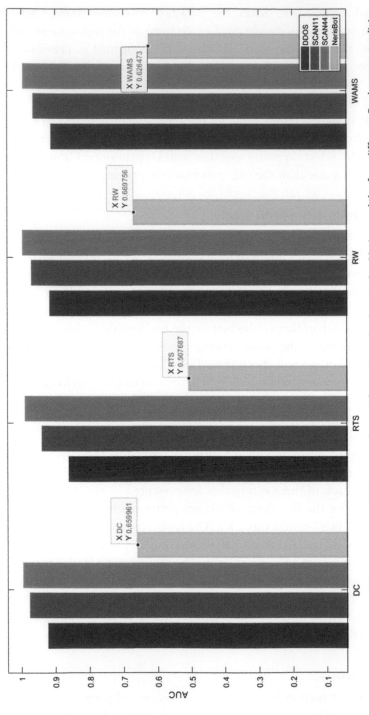

Fig. 9 Results obtained for the Enhanced GA approach considering Negative Assortative Mating and the four different Replacement policies, 10 runs.

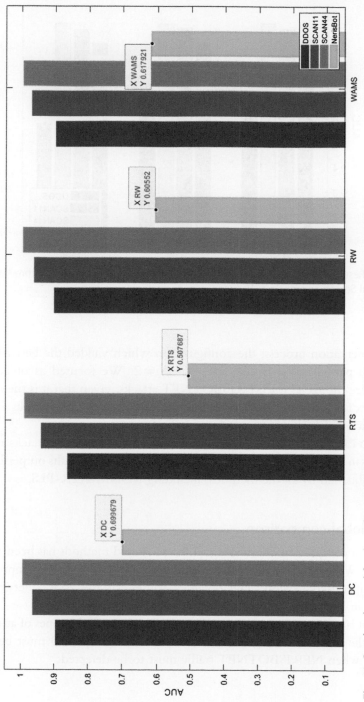

Fig. 10 Results obtained for the Enhanced GA approach considering Linear Ranking and the four different Replacement policies, 10 runs.

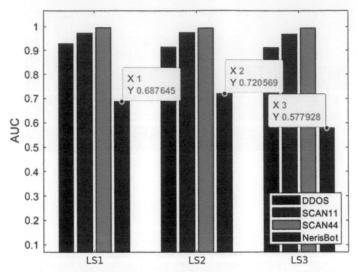

Fig. 11 Results obtained for the Memetic Algorithm approach considering three different Local Search approaches, 10 runs.

experimentation process, the configuration which yielded the best results was: 50 particles; $lr=1$; $\omega=0.01$; ϕ_1, $\phi_2=2$. We focused as objective maximize the AUC for NERISBOTNET attacks, given that it is the most complex for the rest of approaches. Fig. 12 shows the obtained results for all the attacks using this configuration for PSO.

As it can be seen in the graphs, the results for the first three attacks are as good as in the rest of algorithms, however, NERISBOT results outperforms those obtained by all the methods, including MSNM-R2R-PLS.

6.5 Global comparison

Finally, a comparison considering all the proposed methods has been conducted. We have considered as well the original MSNM and its improved semi-supervised version. The obtained values for the AUC metric are shown in Fig. 13.

As it is can be seen there is not a clear winner for all the types of attacks, being the Simple AG a good candidate, but definitely PSO must be the chosen when NERISBOTNET traffic must to be detected.

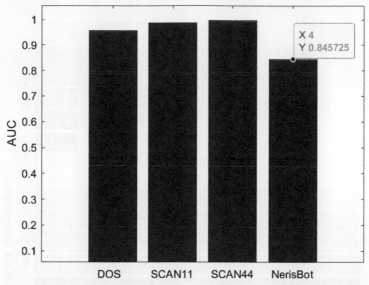

Fig. 12 AUC results obtained for every type of Attack running all the approaches on the same dataset.

7. Conclusions and future work

In this work we have proposed a set of bioinspired techniques, namely Evolutionary Algorithms (EAs) and Particle Swarm Optimization (PSO) methods, aimed to improve the performance of a state-of-the-art Intrusion Detection System, named Semi-Supervised MSNM, based on the recognition of anomalies in data gathered from real network traffic flows. The methods must optimize the set of weights that this algorithm uses for the detection. Four algorithms have been implemented and tested: a Simple Genetic Algorithm, an Enhanced version of it (using different Selection and Replacement policies), a Memetic Algorithm, and a PSO implementation.

Several experiments have been conducted to detect four different types of attacks: DOS, SCAN11, SCAN44 and NERISBOTNET; and some clear conclusions have been reached:

- This is a hard problem to solve, and there is no method better than the rest.

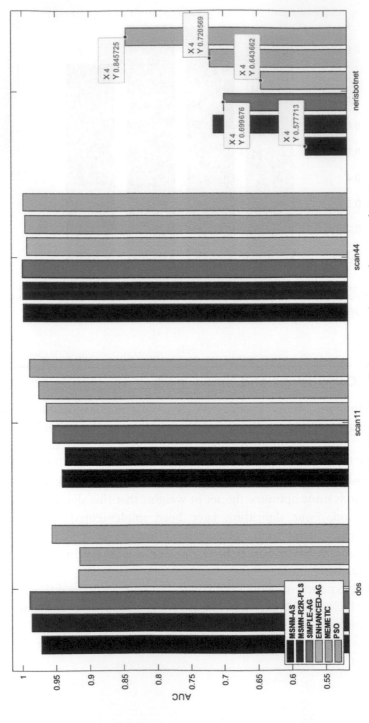

Fig. 13 AUC results obtained for every type of Attack running all the approaches on the same dataset.

- GAs are a very useful tool, able to obtain very good results for all the attacks, except for NERISBOTNET, which is, in turn, the hardest problem.
- The enhanced version of the GA and the Memetic approach also performs very well.
- PSO is the best option to detect NERISBOTNET traffic.

Given these results, we can consider that the proposed methods can be part of a useful IDS.

As future lines of work, we will continue studying variations on the proposed methods, such as other operators that could be better suited for dealing with the most difficult attacks. Different EA models could be also checked, and other metaheuristics could be also tested.

Acknowledgments

This work has been partially funded by Projects PID2020-113462RB-I00, PID2020-115570GB-C22, and PID2020-115570GB-C21, granted by the Spanish Ministry of Economy and Competitiveness. It has been also funded by projects PDC2022-133900-I00, TED2021-129938B-I00 and TED2021-131699B-I00, from the Spanish Ministry of Science and Innovation MCIN/AEI/10.13039/501100011033 and European Union NextGenerationEU/PRTR. As well as project C-ING-027-UGR23 granted by Plan Propio de Investigación UGR.

The authors are very grateful to Professor José Camacho from the Department of TSTC (University of Granada) for his extremely valuable support and help in this work.

References

[1] J.P. Anderson, Computer Security Threat Monitoring and Surveillance, James P. Anderson Co, Fort Washington, Pennsylvania, 1980. Tech. Rep.
[2] O.A. Hathaway, R. Crootof, P. Levitz, H. Nix, A. Nowlan, W. Perdue, J. Spiegel, The law of cyber-attack, Calif. Law Rev. 100 (4) (2012) 817–885.
[3] F. Sabahi, A. Movaghar, Intrusion Detection: A Survey, in: Third International Conference on Systems and Networks Communications, 2008, pp. 23–26.
[4] J. Camacho, A. Pérez-Villegas, P. García-Teodoro, G. Maciá-Fernández, PCA-based multivariate statistical network monitoring for anomaly detection, Comput. Sec. 59 (2016) 118–137.
[5] I.T. Jolliffe, Principal Component Analysis. Springer Series in Statistics, Springer-Verlag, 1986, p. 487.
[6] G. Maciá-Fernández, J. Camacho, R. Magán-Carrión, P. García-Teodoro, R. Therón, UGR'16: a new dataset for the evaluation of cyclostationarity-based network IDSs, Comput. Secur. 73 (2018) 411–424.
[7] J. Camacho, G. Maciá-Fernández, N.M. Fuentes-García, E. Saccenti, Semi-supervised multivariate statistical network monitoring for learning security threats, IEEE Trans. Inf. Forensics Secur. 14 (8) (2019) 2179–2189.
[8] J. Camacho, J. Picó, A. Ferrer, Self-tuning run to run optimization of fed-batch processes using unfold-PLS, Aiche J. 53 (2007) 1789–1804.

[9] A.E. Eiben, J.E. Smith, Introduction to Evolutionary Computing, Springer, Natural Computing Series, 2015.

[10] J. Kennedy, R. Eberhart, Particle swarm optimization, in: Proceedings of ICNN'95 - International Conference on Neural Networks, vol. 4, 1995, pp. 1942–1948.

[11] P. García-Sánchez, A.P. Tonda, A.M. Mora, G. Squillero, J.J. Merelo, Automated playtesting in collectible card games using evolutionary algorithms: a case study in hearthstone, Knowledge-Based Syst. 153 (2018) 133–146.

[12] P. García-Sánchez, G. Romero, J. González, A.M. Mora, M.G. Arenas, P.A. Castillo, C. Fernandes, J.J. Merelo, Studying the effect of population size in distributed evolutionary algorithms on heterogeneous clusters, Appl. Soft Comput. 38 (2016) 530–547.

[13] M. Salem, A.M. Mora, J.J. Merelo, Overtaking uncertainty with evolutionary TORCS controllers: combining BLX with decreasing α operator and grand prix selection, IEEE Trans. Games 14 (2) (2022) 318–327.

[14] M. Salem, A.M. Mora, J.J. Merelo, P. García-Sánchez, Evolving a TORCS modular fuzzy driver using genetic algorithms, in: K. Sim, P. Kaufmann (Eds.), Applications of Evolutionary Computation. Evoapplications 2018, Lecture Notes in Computer Science, vol. 10784, Springer, 2018.

[15] F. Liberatore, A.M. Mora, P.A. Castillo, J.J. Merelo, Comparing heterogeneous and homogeneous flocking strategies for the ghost team in the game of Ms. Pac-Man, in: IEEE Transactions on Computational Intelligence and AI in Games, vol. 8, no. 3, 2016, pp. 278–287.

[16] G. Tambouratzis, Applying PSO to natural language processing tasks: optimizing the identification of syntactic phrases, in: 2016 IEEE Congress on Evolutionary Computation (CEC), IEEE, 2016, pp. 1831–1838.

[17] K. Premalatha, A.M. Natarajan, Hybrid PSO and GA for global maximization, Int. J. Open Problems Compt. Math 2 (4) (2009) 597–608.

[18] I.G. Tsoulos, A. Stavrakoudis, Enhancing PSO methods for global optimization, Appl. Math Comput. 216 (10) (2010) 2988–3001.

[19] D.E. Goldberg, Genetic Algorithms in Search, Optimization and Machine Learning, Addison-Wesley Longman Publishing Co., Inc., Boston, MA, USA, 1989.

[20] Y. He, W. Ma, J. Zhang, The Parameters Selection of PSO Algorithm Influencing on Performance of Fault Diagnosis, MATEC Web of Conferences, 2016.

[21] T.J. Boardman, The statistician who changed the world: W. Edwards Deming, 1900–1993, Am. Stat. 48 (3) (1994) 179–187.

[22] A. Lakhina, M. Crovella, C. Diot, Diagnosing networkwide traffic anomalies, SIGCOMM Comput. Commun. Rev. 34 (4) (2004) 219–230.

[23] J. Camacho, J. Maciá-Fernández, J. Díaz-Verdejo, P. García-Teodoro, Tackling the big data 4 vs for anomaly detection, in: 2014 IEEE Conference on Computer Communications Workshops (INFOCOM WKSHPS), 2014, pp. 500–505.

[24] T. Fawcett, An Introduction to Roc Analysis, Pattern Recogn. Lett. 27 (8) (2006) 861–874. ROC Analysis in Pattern Recognition.

[25] M. Lozano, F. Herrera, J.R. Cano, Replacement strategies to preserve useful diversity in steady-state genetic algorithms, Inform. Sci. 178 (2008) 4421–4433.

[26] C. Fernandes, A. Rosa, A study on non-random matching and varying population size in genetic algorithms using a royal road function, in: Proc. of the 2001 Congress on Evolutionary Computation, 2001.

[27] F. Neri, C. Cotta, P. Moscato, Handbook of Memetic Algorithms. Studies in Computational Intelligence, Vol. 379, Springer Berlin Heidelberg, Berlin, Heidelberg, 2012.

[28] W.E. Hart, Adaptive Global Optimization with Local Search, University of California, 1994.

[29] D. Vrajitoru, Large Population or Many Generations for Genetic Algorithms? Implications in Information Retrieval, in: F. Crestani, G. Pasi (Eds.), Soft Computing in Information Retrieval. Studies in Fuzziness and Soft Computing Physica, Heidelberg, 2000. p. 50.

About the authors

A.M. Mora is professor at Department of Signal Theory, Telematics and Communications. University of Granada, Spain. His current research interests include Artificial Intelligence, Games, Ant Colony Optimization, Data analysis, Prediction, Autonomous players in games and Self-Organizing Map. Author received the PhD degree at the University of Granada (Spain). He has participated in several funded researching projects, and published a number of papers in top-rated international conferences and journals.

P. Merino is a Computer Engineer with a master's degree in Cybersecurity. He currently works as a Cybersecurity Consultant at PwC Spain, where he has gained substantial expertise in the private sector. His experience includes serving as a Blue Teamer in Secure Software Development Life Cycle and Vulnerability Management. In his free time, he is actively pursuing studies to enhance his knowledge in Red Team operations.

Diego Hernández is a seasoned IT professional with a background in Computer Engineering and a master's degree in Computer Science. With a robust foundation in both theoretical knowledge and practical experience, Diego has accumulated significant expertise during his tenure in the private sector. His extensive skill set encompasses a wealth of experience in AWS, where he has excelled as both a Developer and a DevOps specialist. Diego's dedication to staying at the forefront of technological advancements has allowed him to actively contribute to various cutting-edge projects, leveraging his expertise to enhance system capabilities. His proficiency extends to the development and implementation of innovative solutions, pursuing high-level certifications like AWS DevOps.

P. García-Sánchez is associate professor in the Department of Computer Engineering, Automatics, and Robotics at the University of Granada, and the current director of the Free Software Office at the Vice-Rectorate for Digital Transformation. He has participated in 19 research projects, 4 transfer contracts with companies, and has published over 100 scientific articles (25 of them in JCR journals and 60 in international conferences). His interests include service-oriented computing, evolutionary computing, computational intelligence in video games, distributed algorithms, open-source software, and open science.

A.J. Fernández–Ares is assistant professor in the Department of Software Engineering at the University of Granada. His research interests include Explainable Artificial Intelligence, Metaheuristics for Financial Prediction, and Video Game Content Creation. He has actively contributed to cutting-edge R&D projects centered around advancing the capabilities of the Intelligent System for Wireless Communication Capture for Mobility Analysis and Prediction through Soft Computing. His work has not only shaped innovative projects but has also been disseminated through influential articles in academic publications.

CHAPTER FOURTEEN

Modified grey wolf optimization in user scheduling and antenna selection in MU-MIMO uplink system

Swadhin Kumar Mishra[a], Arunanshu Mahapatro[a], and Prabina Pattanayak[b]
[a]Veer Surendra Sai University of Technology, Burla, India
[b]National Institute of Technology Silchar, Silchar, India

Contents

Abstract

The high demand for spectrum and data throughput makes the multiple-input multiple-output (MIMO) system one of the suitable techniques in modern wireless communication networks. The choice of the appropriate set of antenna elements from the base station (BS) and user scheduling are two of the most challenging aspects in a large-scale wireless communication network with a multiuser MIMO (MU-MIMO) system. The exhaustive search algorithm (ESA), in an uplink MIMO system, achieves optimal throughput by choosing the best user and antenna combination after considering

Advances in Computers, Volume 135
ISSN 0065-2458
https://doi.org/10.1016/bs.adcom.2023.12.003

every feasible combination. The optimal throughput achieved by ESA comes with a high level of computational complexity. In a high-speed network with very low latency requirement, the user scheduling and BS antenna selection must be completed within a fraction of coherence time of the channel. By making the scheduling process less computationally complicated, this low latency need can be satisfied. To overcome the difficulties associated with reducing the complexity of the scheduling algorithm, numerous soft-computing solutions are employed. In an uplink MIMO wireless system, the scheduling of users and antenna selection are handled using metaheuristic techniques including ant colony optimization (ACO) and binary grey wolf optimization (BGWO). It is demonstrated that these soft-computing techniques achieve throughput quite close to that of ESA but with a lot less processing load. Out of the discussed metaheuristic methods, BGWO outperforms the binary particle swarm optimization (BPSO) and ACO. A modified variant of the BGWO algorithm is presented and analyzed. The performance of the proposed algorithm is compared with that of the standard BGWO and also with ACO and BPSO.

1. Introduction

In every wireless communication networks developed and commercialized in recent years, one of the integral techniques of the system is the multiple-input multiple-output (MIMO) technology. A MIMO system achieves a channel capacity that is much higher than a conventional single-input single-output system (SISO) [1–3]. Spatial multiplexing, diversity, and beamforming are a few of the specific mechanisms responsible for the performance improvement in MIMO. In a multiuser MIMO (MU-MIMO) wireless network, multiple receive antennas are distributed among multiple users. In this chapter, we examine an uplink wireless communication network where multiple users simultaneously communicate with base-station (BS) antennas. Transceivers present in every pair of transmit-receive antennas consist of a series of hardware components which is referred to as radio frequency (RF) chain. The total hardware complexity of the system increases with the inclusion of more RF chains. Hence, only a few RF chains are used in a MIMO system with huge arrays of antennas in order to simplify the hardware.

One of the primary challenges in a MIMO uplink system with a limited number of RF chains is the selection of a subset of transmit antennas from the BS. The same number of receive antennas is chosen to convey independent and parallel data streams to these transmit antennas. As the number of antennas at the BS and receiver increases, so does the computational processing activity involved in choosing the necessary number of broadcast (BC) and receive antennas from the network. The latency requirement is particularly low for high-speed data networks like 4th generation (4G) and 5th generation (5G) networks. The computational complexity of the selection mechanism in such networks should be kept to a minimum. On the other extreme, the exhaustive search algorithm (ESA) [4] computes and compares every conceivable user and receiving antenna combination, and the optimal combination that produces the highest sum rate capacity is chosen. Since the time complexity grows exponentially with the number of users and receiving antennas, ESA is realistically inefficient.

Many scheduling schemes are proposed with less computational complexity in MIMO system for user selection as well as antenna selection [5]. It has been demonstrated that the hardware cost of MIMO systems can be significantly decreased without sacrificing any of their multiantenna functionality by choosing the ideal user and receiver antenna combinations [6]. In Zhang and Lee [7], a singular value decomposition (SVD)-based scheduling method is put forth that reduces computational complexity without compromising the system throughput. The following users are picked to maximize the upper bound of the sum-rate capacity. A zero-forcing beamforming (ZFBF) methodology is used in Refs. [8, 9] to provide a low-complexity user selection method. Utilizing concurrent orthogonal users prevents interuser interference. However, it is theoretically impossible to choose several orthogonal users. Hence, a set of semi-orthogonal users is chosen to be the scheduled users. As a result, the scheduling technique is known as a semiorthogonal user selection (SUS) algorithm. The SUS approach described in Mao et al. [10] provides a less complicated version with nearly the same throughput performance. The MU-MIMO system uses beamforming techniques to increase signal intensity and range. A limited-feedback-based scheduling algorithm in MIMO-BC system utilizing user grouping is presented in Pattanayak et al. [11].

Reducing the traffic on the reverse channel is another goal of the scheduling plan. [12] outline a quantized technique-based suboptimal user scheduling system [13]. Each user quantizes its signal-to-interference-plus-noise (SINR) quantities and feedbacks the quantized value, there by reducing the feedback burden significantly in the reverse channel. The quantization of channels while implementing various user scheduling algorithms is also presented in Refs. [12, 14, 15]. Choudhary and Mishra [14] implements a two-stage feedback method in the scheduling scheme using the quantization of the channel. For a MIMO-OFDM heterogeneous wireless network, a scheduling scheme based on SINR quantization is presented in Choudhary and Mishra [14]. In each of the scheduling schemes designed using the quantized channel, it is shown that there is a reduction in achieved throughput compared to other low-complex scheduling algorithms. The authors in Ref. [16] propose a scheduling approach in which users are scheduled in order to maximize the sum-rate capacity by picking a group of antennas with the highest channel gains.

In the case of networks with large arrays of antenna elements at the transmitter and a large pool of users, the selection of a combination of a group of transmit antennas and users is necessary. It is suggested by Naeem and Lee [17] to schedule users and transmit antennas concurrently in an effort to maximize the realized sum-rate capacity during each of the time slot [18]. The combined selection of transmit antennas and user scheduling in a massive MIMO-BC communication network is described in Mishra et al. [19]. In Refs. [20–22], both the scheduling of users and selection of antennas are done. The SUS algorithm is utilized for user scheduling, and the strategy of sequentially deleting the worst antennas is employed for antenna selection [23]. Performance is evaluated based on the throughput that is obtained by the planned users using the best set of transmit antennas. A simple low-complex scheduling algorithm for a MIMO-BC system is discussed in Lee et al. [24]. All such scheduling algorithms are called suboptimal scheduling algorithms because most of them are designed to reduce computational complexity as a trade-off with the optimal sum-rate capacity.

Soft-computing methods are frequently used in the real world to address optimal value issues. The authors in Ref. [25] have reviewed and analyzed soft-computing techniques and shown that they are a strong competitor for solving some of the problems associated with optimal allocation of resources in a variety of wireless communications applications. Abedi and Vadgama [26] looked into developing a novel hybrid scheduler for packet networks in high-speed downlink packet access (HSDPA [27]) systems by fusing genetic algorithm (GA) with existing scheduling techniques. Similarly, suboptimal schemes have been proposed for multicarrier systems as well [28, 29]. For wireless sensor networks (WSN), routing and clustering algorithms utilizing particle swarm optimization (PSO) are presented and analyzed in Azharuddin and Jana [30]. Ant colony optimization (ACO) [31–33] technique is among the most effective and successful optimization strategies. In the domain of artificial intelligence, ACO is regarded as one of the most productive subfield of swarm intelligence [34–37]. To enhance the quality of service (QoS) of various applications in cloud-based mobile computing systems, the ACO algorithm is utilized in Wei et al. [38]. The authors in Ref. [39] explained how multipheromone-based ACO is employed to find the most energy-efficient paths possible for transmitting sensing data to the BS in WSNs.

To optimize both spectral and energy efficiency in a multiuser massive MIMO system, a multiobjective optimization technique is discussed in Hei et al. [40]. A multiobjective adaptive GA is utilized to resolve the issue relating to the number of transmit antennas and transmission power. GA is used to perform scheduling in a system with multiantennas using a single carrier in Lau [41]. For a multiantenna system with multiple carriers, GA is used in scheduling in Elliott and Krzymien [42]. For MU-MIMO broadcast channel (BC) systems, authors in Ref. [43] implemented elitism and adaptive mutation (AM) along with binary GA (BGA) and it is demonstrated that BGA with these two features can obtain system sum-rates that are very similar to ESA.

A modern metaheuristic evolutionary optimization method that takes cues from biology is called grey wolf optimization (GWO) [44]. This is influenced by grey wolves, notably by how they hunt. In Nimmagadda [45], a solution is found for the issue of determining the best beam-forming vectors and power distribution in massive MIMO systems. A modified version of the original GWO algorithm along with a modified algorithm is utilized for solving the complex optimization problem. The peak-to-average power ratio (PAPR) contributes in a major way to design the orthogonal frequency division multiplexing (OFDM) system. The authors in Ref. [46] are using a modified GWO scheme to reduce the PAPR in OFDM system. It is also demonstrated that the proposed modified GWO algorithm performs better than the other soft-computing techniques discussed in the chapter in terms of reduction in PAPR.

Soft-computing techniques can be utilized in a MU-MIMO uplink system for choosing the best BS antennas and scheduling a group of users while offering performance that is close to what the ESA scheme achieves. Binary particle swarm optimization (BPSO) is utilized in MIMO uplink system, and the result is compared with that of GA in Refs. [47, 48]. And it is demonstrated that BPSO outperforms in terms of sum-rate capacity compared to GA. To overcome the computational and time complexity involved in implementing ESA for MIMO uplink system without much compromise on the achievable sum-rate capacity of the system, we are motivated to use various soft-computing methods in the scheduling process. In this chapter, we are utilizing BPSO, ACO, and binary GWO (BGWO) soft-computing methods in the selection and schedule of antennas and users, respectively. We analyze the performance of each of these soft-computing methods in terms of achieved throughput and computational complexities and compare these with the values achieved by ESA.

The following is the list of major contributions of the chapter:

1. To design a fitness function for a MU–MIMO uplink system to maximize the achieved sum-rate system capacity. This is achieved by choosing an optimal set of pairs of user and transmit antennas.
2. To utilize the ACO algorithm and other versions of ACO algorithm to optimize the objective function.
3. The objective function is optimized by using the BPSO algorithm.
4. To utilize the original BGWO algorithm along with two different variants of GWO to optimize the objective function.
5. To evaluate the effectiveness of the soft-computing methods mentioned in terms of achieved throughput and computational complexity.
6. To compare the efficiency of the soft-computing methods presented here with the ESA method in the proposed uplink system.

The remaining portions of the essay are structured as follows: Section 2 presents the system model used in this chapter for the MU–MIMO uplink system used in this chapter. The objective function for the proposed soft-computing techniques is also defined in this part. The detailed description and presentation of the soft-computing algorithms ACO and BPSO are done in Section 3. Section 4 discusses the application of the original BGWO algorithm and the new variant of BGWO the algorithm in MU–MIMO uplink network. Different results obtained and the output generated are presented in Section 5. And at last, the conclusions derived from the results and the future work are presented in Section 6.

2. MU-MIMO uplink system model

A MU–MIMO uplink system is considered where multiple users access the antenna elements at the BS. In the model, we assume that there are M antenna elements at the BS. Since we are considering an uplink communication system, the antennas at the BS are considered as receive antennas. It is also assumed that on the user side, there are N independent and geographically dispersed users, and each user consists of C number of transmit antennas. Hence, the wireless system consists of total $N \times C$ number of antennas that act as transmit antennas in the uplink network. We also assume that the network is equipped with N_{rf} number of RF chains with $N_{rf} \leq M$. The objective of the scheduling scheme is to choose the best group of antenna elements S from the BS such that $|S| \leq N_{rf}$ to which the users transmit independent data. The wireless channels between the transmitter antennas at BS and the individual users are assumed to exhibit Rayleigh fading. The multiple access channel (MAC) matrix between the transmit antennas of nth users and the antenna elements at the BS is represented by $\mathbf{H}_n$. The complex signal vector received by the antennas at the BS is given as

$$y = \sum_{n=1}^{N} \mathbf{H}_n \mathbf{s}_n + \mathbf{w} \tag{1}$$

where the dimension of $\mathbf{y}$ is $A \times 1$ and $\mathbf{s}_n$ is the vector consisting of the transmitted signals from the antennas at the nth user and is of size $C \times 1$. The channel matrix $\mathbf{H}_n$ of user n consists of channel coefficients $h_{i,j} = [\mathbf{H}_n]_{i,j}$. The channel coefficients $h_{i,j}$ are complex and symmetric random variables with Gaussian distribution and a mean of zero and variance of one, i.e., $h_{ij} \sim \mathcal{CN}(0, 1)$. The channel elements are identical and independently distributed random variables. Further, it is assumed that the channel exhibits flat fading, i.e., the channel remains quasi-constant over a message duration. The noise vector is represented by $\mathbf{w}$ which is of size $M \times 1$. It is further assumed that the noise vector has unit variance. Assuming P_T is the sum amount of power available for transmission and that all users receive an equal share of power, each user uses an amount of power equal to P_T/N. For a MIMO uplink communication system, the total channel capacity is given as [49]

$$C_{sum} = \log_2 \det \left(\mathbf{I}_{N_{rf}} + \Gamma \sum_{n=1}^{N} \mathbf{H}_n \mathbf{H}_n^H \right) \tag{2}$$

where Γ is the average SNR per transmit antenna at the user side. As noise variance is assumed to be unity, $\Gamma = P_T/NC$. $\mathbf{H}_n^H$ represents the complex conjugate of the channel matrix. We chose a combination of user and receive antennas because picking all the users at once would be impractical. In a symbol duration, maximum N_s users at the transmit side and N_{rf} number of receive antennas from the BS are active. Let ζ_s be the selected set of users such that $|\zeta_s| \leq N_{rf}$, i.e.,

$$\zeta_s = [N_{s_1}, N_{s_2}, N_{s_3}, \dots N_{s_{N_s}}] \tag{3}$$

Such users are chosen which maximizes the achieved sum-rate capacity. Hence, once the users are selected, the achievable uplink capacity is

$$C_{sum}(\zeta_s) = \log_2 \det \left(\mathbf{I}_{N_{rf}} + \Gamma_s \sum_{i \in \zeta_s} \mathbf{H}_i \mathbf{H}_i^H \right) \tag{4}$$

where $\Gamma_s = P_T/N_s C$ is the average SNR per selected users in Γ_s. The objective function for all the soft-computing methods used in this chapter is represented in (4). The aim of the algorithms discussed here is to choose a set of user ζ_s so that the capacity as mentioned in (4) is maximized.

The ESA system considers every potential combination in order to choose the best set in each iteration which is represented as

$$|\zeta_{ESA}| = \binom{A}{N_r} \sum_{i=1}^{N_s} \binom{N}{i} \tag{5}$$

It is computationally inefficient to try every potential combination in search of the best result. We discuss three computationally efficient soft-computing algorithms to solve the optimization problem discussed above. In the next chapter, we discuss the ACO followed by the BPSO algorithms. Then, we present the modified GWO algorithm to find the optimal solution.

3. Proposed optimization algorithms

Various metaheuristic approaches utilized to resolve the optimization problem described above are elaborated upon and thoroughly addressed in this section.

3.1 Ant colony optimization technique

ACO algorithm is based on general-purpose optimization methods [32, 33, 50]. According to artificial intelligence theory, ACO algorithms are one of the most effective subsets of swarm intelligence. It is modeled after how some ant species hunt for food. More specifically, the main inspiration is stigmergy, a special form of communication in insects. Stigmergy is a sort of indirect communication used by insects to affect their environment and transmit information. In their pursuit for food, ants leave behind pheromones along their way. A path with a higher pheromone concentration draws ants. As ants move along certain courses, some of them develop higher pheromone concentrations than others, attracting more ants. In this instance, positive feedback makes it simpler to find the fastest path between the source and the destination. The following three steps are iterated through by the ACO algorithm:

  i. The ants develop several ant solutions.
 ii. A local search on these solutions improves the global solution.
iii. The pheromone is updated.

If F is the objective function that needs to be maximized with the help of any optimization technique, then it is equivalent to minimize the function $L = 1/F$. To update the pheromone, the MAX–MIN ant system [51] is used.

ACO algorithm is described in terms of the pseudocode in Algorithm 1.

The ACO algorithm is implemented by considering the following set of parameters:

1. F_o represents the objective function as mentioned by (4).
2. P denotes the population of ants.
3. Itr denotes the maximum iteration number.
4. Dim is the dimension (length) of path taken by an individual ant.
5. χ_i represents the path traversed by ith ant.
6. C_i represents the cost or value of F_o of the ith ant at χ_i.
7. σ denotes a $2 \times Dim$ pheromone matrix.
8. τ is the evaporating variable used for updating the pheromone.

The ACO algorithm is associated with three phases, which are described in detail.

3.1.1 Create ant solution

The choice of user and reception antennas is made using an artificial ant solution, which also establishes the path that the ant will travel. Let P_{0i} and P_{1i} be the probabilities of ith antenna not being selected and ith antennas

---

ALGORITHM 1 ACO Algorithm.
Require: Parameter set $(F_o, P, Itr, Dim, \chi_i, C_i, \sigma, \tau)$
Ensure: Initialize the pheromone trail
   while Termination requirement not satisfied **do**
      Createantsolutions
      Conductlocalsearch
      Pheromonesupdated
   end while

being selected, respectively. The pheromone matrix provides the basis for these probabilities. The probabilities are derived as

$$P_{ki} = \frac{\sigma_{ki}}{\sigma_{0i} + \sigma_{1i}} \tag{6}$$

where $i = 1, 2, \ldots Dim$ and k takes the values 0 and 1.

3.1.2 Conduct local search
Each iteration ends with the construction of all ant solutions, at which point the iteration's best and overall best solutions are identified. Pheromones are updated using these solutions.

3.1.3 Pheromones updated
Several ACO versions are suggested by the literature, all of which are based on the pheromone update mechanism. The updating of pheromone is done by the participation of all ants in the basic ant system (AS). There exist two effective variants of the basic AS method in which, to update the phero-mone, either the global-best or iteration-best solution for ant is used. Ant colony system (ACS) is one of these method that utilizes the top ants and a second phase of local pheromone updates. And the other is the MAX-MIN ant system (MMAS), which similarly updates using the best ants but with a limited pheromone level. These adjustments do have the ability to enhance the performance of the algorithm dramatically.

In MMAS method, the update of pheromone is performed as mentioned below.

$$\begin{aligned} \sigma_{ij} &\leftarrow (1 - \tau)\sigma_{ij} + \Delta\sigma_{ij}^{best} \\ \sigma_{ij} &= \min(\sigma_{max}, \sigma_{ij}) \\ \sigma_{ij} &= \max(\sigma_{min}, \sigma_{ij}) \end{aligned} \tag{7}$$

where σ_{max} and σ_{min} represent the upper threshold and lower threshold of the pheromone, respectively and $\Delta\sigma_{ij}^{best}$ is denoted as

$$\Delta\sigma_{ij}^{best} = \begin{cases} \dfrac{1}{|L_{best}|} & \text{if } (i,j) \text{ refers to the finest tour iteration} \\ 0 & \text{otherwise} \end{cases} \tag{8}$$

In this chapter for simulation purpose, the values for the parameters are $\tau = 0.3$, $\sigma_{min} = 0.005$, $\sigma_{max} = 1$, and $\sigma_0 = 0.5$

3.2 Particle swarm optimization technique

PSO, a population-based search algorithm, is modeled after the behavior of fish schools and bird flocks. A particle that moves through the search space is taken into consideration as a potential solution. Using the best solution that the particle and the entire swarm have found, each particle's position is updated. Due to the tendency of each particle to migrate in the direction of the best solution, the swarm is guided to the best solution. PSO is useful for optimizing nondifferentiable objective functions because, unlike GA, it does not call for the discovery of the gradients of the objective function. In PSO, for maximizing an objective function F that depends on several variables, such as $y_1, y_2, ..., y_n$, is taken into account. Maximization of $F(y_1, y_2, ..., y_n)$ means determining the vector value $(y_2, ..., y_n)$ using which F achieves the highest fitness value. Therefore, the vector $(y_2, ..., y_n)$ is regarded as a particle's position in an n-dimensional space. The particle moves over the whole n-dimensional search space in search of the optimal position so that the objective function attains the highest fitness value. We consider a binary PSO algorithm here for the optimization algorithm. In BPSO, binary values 0 and 1 are used to represent the position of the particles. The idea of particle movement is to flip the bits.

The following parameters characterize the BPSO algorithm:

1. S_s denotes the search space for all potential solutions.
2. F_o represents the fitness function as mentioned by (4).
3. P stands for the number of particles present in a swarm.
4. Itr denotes the iteration number.
5. Dim denotes the dimension, i.e., the binary string has a length of Dim.
6. χ_j represents the position vector of jth particle.
7. Υ_j represents the velocity of the jth particle.
8. Pb_j denotes the best position attained by the jth particle.
9. G represents the global optimal position of a certain particle within the swarm that has previously been visited.

The search for the global best position is implemented using the BPSO algorithm [47], and the pseudocode for this is presented in Algorithm 2.

In this algorithm, the parameter $U(0, 1)$ is a uniformly distributed random variable ranging from 0 to 1. $U_I(0, 1)$ is another random variable with a uniform distribution that takes values of either 0 or 1. $\psi(r) = (1 + e^{-r})^{-1}$ represents the sigmoid function. The social, cognitive, and inertia characteristics that govern particle motion in the algorithm are c_1, $c_2 0$, and $w > 0$.

ALGORITHM 2 Pseudocode for BPSO.
1: **Input:** F_o, P, Itr, Dim, N, N_s, N_{rf}, G
2: **Initialization:** V_{min}, V_{max}, c_1, c_2, w
3: **for each** particle $j = 1 \rightarrow P$ **do**
4:     **for each** dimension $d = 1 \rightarrow Dim$ **do**
5:         $x_{j,d} \leftarrow U_l(0, 1)$
6:         $v_{j,d} \leftarrow U(V_{min}, V_{max})$
7:     **end for**
8:     $\chi_j =$ CheckBound(χ_j,N, N_s, N_{rf}),
9:     $Pb_j \leftarrow \chi_j$
10:    **if** $F(Pb_j) > F(G)$ **then**
11:        $G \leftarrow Pb_j$
12:    **end if**
13: **end for**
14: **for** $l = 1 \rightarrow Itr$ **do**
15:     **for each** particle $j = 1 \rightarrow P$ **do**
16:         **for each** dimension $d = 1 \rightarrow Dim$ **do**
17:             $v_{j,d} \leftarrow v_{j,d} \times w + (p_{j,d} - x_{j,d}) \times U(0, 1) \times c_1 + (g_d - x_{j,d}) \times U(0, 1) \times c_2$
18:             **if** if $U(0, 1) < \psi(v_{j,d}$ **then**) $x_{j,d} \leftarrow 1$
19:             **else** $x_{j,d} \leftarrow 0$
20:             **end if**
21:             $\chi_j =$ CheckBound(χ_j,N, N_s, N_{rf}),
22:             **if** $F(\chi_j) > F(Pb_j)$ **then**
23:                 $Pb_j \leftarrow \chi_j$
24:             **end if**
25:             **if** $F(Pb_j) > F(\chi_j)$ **then**
26:                 $G \leftarrow Pb_j$
27:             **end if**
28:         **end for**
29:     **end for**
30: **end for**
31: **return** $(G, F(G))$

4. Binary grey wolf optimization

GWO is one of the most well-known biologically inspired meta-heuristic evolutionary optimization strategies [44]. It draws inspiration from grey wolves, specifically from how they hunt. Grey wolves exhibit the most skilled and intelligent behavior, governed by rigorous hierarchy rules. The alpha (α) is the leader of the pack and is in charge of making crucial choices in a wolf pack. The alpha (α) is the leader of the pack and is in charge of making crucial choices in a wolf pack. The beta (β), who is placed second in the hierarchy, assists the alpha in the task. Hierarchically, the omegas (ω), who serve as scapegoats, are at the bottom. The omegas are ruled by the deltas, who submit to the alphas and betas.

In Mirjalili and Mirjalili [44], the authors attempted to define this behavior analytically and address optimization issues. GWO achieves very competitive outcomes when compared to gravitational search algorithm (GSA), differential evolution (DE), fast evolutionary programming (FEP), and PSO. The GWO algorithm is performed by the following stages in sequence.

1. *Encircle the prey*: In this stage, mathematical descriptions are given for the behavior of grey wolves around their prey. The location vector of the wolf and the prey is initialized.
2. *Hunt*: The ability of grey wolves to find their prey and surround it is used during the hunting phase. The position of the wolves is updated in this stage.
3. *Attack the prey*: With the prey cease to move, the hunt is over when the wolves kill the target.
4. *Search the prey*: During this phase of the algorithm, the prey is searched for. Search efforts for grey wolves typically follow the whereabouts of the α, β, and δ wolves. The wolves separate from one another to look for prey and then unite to hunt the predator.

To address a binary optimization issue, Emary et al. [52] expanded the continuous GWO [44, 53]. In this chapter, two BGWO implementations for the aforementioned issue, i.e., to optimize the throughput in a MIMO uplink system, are presented [52]. The BGWO algorithm is compared with the ACO and BPSO algorithms. Also, a comparison with the ESA is also presented. The set of parameters used in the implementation of the BGWO algorithms is presented below.

1. F_o represents the objective function as mentioned by (4).
2. P denotes the population size.
3. Itr represents the maximum iteration number.
4. Dim denotes the dimension (length) of binary string.
5. χ_j represents the position vector of jth particle.
6. a denotes a parameter that takes values between 0 and 2.

4.1 Binary GWO

In this method, the various stages leading to the initial three best solutions are implemented using binary logic. The revised positions of the grey wolves are evaluated with the help of crossover and also represented in binary format. The optimization problems defined in (4) are solved using the GWO method. The pseudocode for the BGWO algorithm [52] is presented below (Algorithm 3).

ALGORITHM 3 Pseudocode for BGWO.

1: **Input:** $[F_o, P, Dim, Itr, N, N_s, N_{rf}]$
2: **Initialization:** $(\alpha, \beta, \delta) \rightarrow -\infty$, and position vectors χ_j for $i \in 1, 2, \ldots P$
3: $l = 1$
4: **while** $(l < Itr)$ **do** $a = 2\left(1 - \frac{l}{Itr}\right)$
5:    **for** $j = 1 \rightarrow P$ **do**
6:        $W_j =$ CheckBound(χ_j, N, N_s, N_{rf}), and
7:        $F_j = F_o(\chi_j)$
8:        **if** $F_j > \alpha$ **then**
9:            $\alpha = F_j$ and $\chi_\alpha = \chi_j$
10:       **else if** $F_j > \beta$ **then**
11:           $\beta = F_j$ and $\chi_\beta = \chi_j$
12:       **else if** $F_j > \delta$ **then**
13:           $\delta = F_j$ and $\chi_\delta = \chi_j$
14:       **end if**
15:       **for each** dimension $i = 1 \rightarrow Dim$ **do**
16:           $A_1 = (2a(\text{rand} - 1)$ and $C_1 = 2\text{rand}$
17:           $D_\alpha = |C_1 \chi_{\alpha,j} - \chi_{i,j}|$
18:           **if** $1/(1 + e^{-10(A_1 D_\alpha - 0.5)}) \geq rand$ **then** $v_1 = 1$
19:           **else** $v_1 = 0$
20:           **end if**
21:           **if** $(\chi_{\alpha,j} + v_1) \geq rand$ **then** $\chi_1 = 1$
22:           **else** $\chi_1 = 0$
23:           **end if**
24:           $A_2 = (2a(\text{rand} - 1)$ and $C_2 = 2\text{rand}$
25:           $D_\beta = |C_2 \chi_{\beta,j} - \chi_{i,j}|$
26:           **if** $1/(1 + e^{-10(A_2 D_\beta - 0.5)}) \geq rand$ **then** $v_2 = 1$
27:           **else** $v_2 = 0$
28:           **end if**
29:           **if** $(\chi_{\beta,j} + v_1) \geq rand$ **then** $\chi_2 = 1$
30:           **else** $\chi_2 = 0$
31:           **end if**
32:           $A_3 = (2a(\text{rand} - 1)$ and $C_3 = 2\text{rand}$
33:           $D_\delta = |C_2 \chi_{\delta,j} - \chi_{i,j}|$
34:           **if** $1/(1 + e^{-10(A_2 D_\delta - 0.5)}) \geq rand$ **then** $v_3 = 1$

ALGORITHM 3 Pseudocode for BGWO.—cont'd

35:         **else** $v_3 = 0$
36:         **end if**
37:         **if** $(\chi_{\delta,j} + v_1) \geq rand$ **then** $\chi_3 = 1$
38:         **else** $\chi_3 = 0$
39:         **end if**

40:
$$\chi_{i,j} = \begin{cases} \chi_1 & \text{if } rand < \dfrac{1}{3} \\ \chi_2 & \text{if } \dfrac{1}{3} < rand < \dfrac{1}{3} \\ \chi_3 & \text{otherwise} \end{cases}$$

41:     **end for**
42:   **end for**
43:   $l = l + 1$
44: **end while**

4.2 Modified BGWO (mBGWO)

To increase performance, a modified variant of the BGWO algorithm [52] is implemented. The employment of a sigmoid function here prevents the process of continually updating positions. The revised positions of the grey wolves are evaluated with the help of stochastic threshold values and represented in binary format. The entire BGWO algorithm is presented in terms of pseudocode as mentioned in Algorithm 4.

ALGORITHM 4 Pseudocode for mBGWO.

1: **Input:** F_o, P, Dim, Itr, N, N_s, N_{rf}
2: **Initialization:** $(\alpha, \beta, \delta) \rightarrow -\infty)$, and position vectors χ_i of each wolves where
    $i \in 1, 2, \dots P$
3: $l = 0$
4: **while** $(l < Itr)$ **do**
5: $a = 2\left(1 - \frac{l}{Itr}\right)$
6:    **for each** position $j = 1 \rightarrow P$ **do**
7:       $\chi_j = \text{CheckBound}(\chi_j, N, N_s, N_{rf})$, and
8:       $F_j = F_o(\chi_j)$
9:       **if** $F_j > \alpha$ **then**
10:         $\alpha = F_j$ and $\chi_\alpha = \chi_j$
11:       **else if** $F_j > \beta$ **then**

Continued

ALGORITHM 4 Pseudocode for mBGWO.—cont'd

```
12:        β = F_j and χ_β = χ_j
13:     else if F_j > δ then
14:        δ = F_j and χ_δ = χ_j
15:     end if
16:     for each dimension i = 1 → Dim do
17:        A_1 = (2a(rand − 1) and C_1 = 2 rand
18:        D_α = |C_1χ_{α,i} − χ_{j,i}|
19:        x_1 = χ_{α,i} − A_1D_α
20:        A_2 = (2a(rand − 1) and C_2 = 2 rand
21:        D_β = |C_1χ_{β,i} − χ_{j,i}|
22:        x_2 = χ_{β,i} − A_1D_β
23:        A_3 = (2a(rand − 1) and C_3 = 2 rand
24:        D_δ = |C_1χ_{δ,i} − χ_{j,i}|
25:        x_3 = χ_{δ,i} − A_1D_δ
26:        x = (x_1+x_2+x_3)/3
27:           if 1/(1 + e^{−10(x−0.5)}) ≥ rand then χ_{i,j} = 1
28:           else χ_{i,j} = 0
29:           end if
30:        end for
31:     end for
32:     l = l + 1
33: end while
```

5. Experimental analysis

5.1 Experimental setup

To perform the experimental analysis, we consider a large-scale MIMO system with 30 antenna elements at the BS and 50 dispersed users. Each user is assumed to consist of three transmit antennas. The system is assumed to use only 3 number of RF chains, i.e., $N_{rf} = 3$. Hence, the three soft-computing techniques described in this chapter are used to select three antenna elements from the BS and three transmit antennas from the user side. The soft-computing techniques are employed to optimize the achievable throughput in a MU–MIMO uplink transmission system. The following steps are performed to implement the algorithms. All the simulations for this part are generated using MATLAB.

- *Step-1:* The channel matrix $\mathbf{H}_n$ of all the users is generated.
- *Step-2:* Choose a specific user and receive antenna pair from all the available choices.
- *Step-3:* The specific fitness value of this selected pair of user and receive antenna is determined in accordance with (4).
- *Step-4:* For each possible pair of user and receive antenna, which are components in the total population of the particular soft-computing technique, itcrate Steps 2 and 3. The particular soft-computing algorithm is executed to find the maximum achievable throughput C_{sum}.
- *Step-5:* Steps 1 to 4 are iterated 1000 times to obtain the expected value of the sum rate.

5.2 Results and analysis for ACO and BPSO

For simulation purpose in the ACO algorithm, the following parameters are considered: $\tau = 0.3$, $\sigma_{min} = 0.005$, $\sigma_{max} = 1$, and $\sigma_0 = 0.5$. The three parameters representing the social, cognitive, and inertia characteristics of PSO for particle motion in the algorithm are c_1, c_2, and w. All these three parameters are chosen to be greater than 0. The simulation results obtained using the proposed ACO and BPSO techniques for uplink communication in MU–MIMO systems are shown in this section. We compare the throughput obtained by the ACO and BPSO with the random scheduling scheme. In the random scheduling process, users and antennas are chosen in a random fashion. In contrast to the ACO method, which sets the pheromone matrix to a fixed default value for every simulation run, the initial set for the BPSO algorithm is produced randomly at every simulation run. For the simulation purposes, the value sets for various parameters are as follows: $N = 30$, $N_s = 3$, $C = 2$, $M = 6$, $N_{rf} = 3$, $P = 30$, and $Itr = 25$. The sum rate achieved in the uplink system using ACO, BPSO, and random scheduling is evaluated for different SNR values. The comparison of the achieved capacity for the above-mentioned techniques is presented in Fig. 1. The performance of these methods is also compared with that of ESA scheme. For characterizing the divergence of results produced by various approaches in comparison to the ESA, a parameter termed percentage deviation in achieve sum rate (PDAS) is proposed. The PDSA in % is evaluated as

$$\text{PDSA}_{Meth} = \frac{C_{Sum}^{Esa} - C_{Sum}^{Meth}}{C_{Sum}^{Meth}} \times 100 \qquad (9)$$

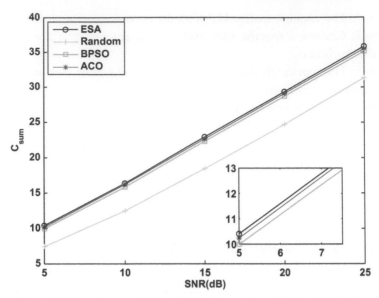

Fig. 1 Achievable throughput vs SNR in dB for ESA, random, BPSO, and ACO algorithms.

where C_{Sum}^{Meth} and C_{Sum}^{Esa} are the achieved sum rate by the soft-computing method and ESA, respectively. Fig. 2 represents the graph showing the percentage deviation in the achieved sum rate using ACO and BPSO from the ESA method. It is observed that the deviation is smaller in the case of ACO in comparison with BPSO, which indicates that ACO performs better than BPSO if the performance metric is achieved sum-rate capacity.

We then present the variation of the throughput when the iteration number in the optimization algorithm is increased. For this, we increase the number of iterations from 2 to 20 in a step of 2, and the corresponding variation in the throughput is presented in Fig. 3. From these curves, it is observed that the ACO scheme converged to the optimal value at a faster rate compared to the BPSO method.

5.3 Results and analysis for BGWO and mBGWO

In order to identify the best option for our practical, real-world problem statement, the performance efficiency of BGWO and the mBGWO is compared in this section. Keeping the values of the parameters used in these algorithms as mentioned in the previous section, we find out the achieved capacity by the MIMO MAC system. The comparison of the achieved throughput versus the

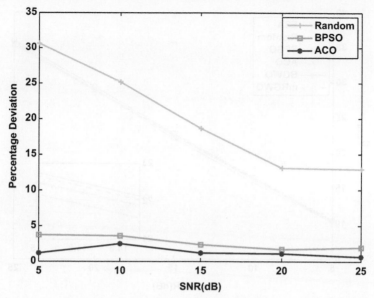

Fig. 2 Percentage deviation in the achievable throughput vs SNR for ESA, BPSO, and ACO.

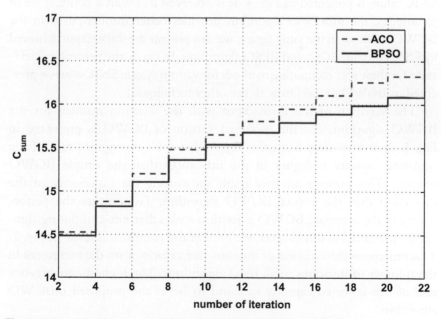

Fig. 3 Sum-rate capacity vs number of iterations for ACO and BPSO.

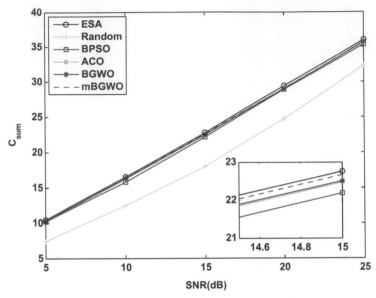

Fig. 4 Achieved capacity vs SNR for different soft-computing schemes.

SNR values is presented in Fig. 4. It is observed that with a constant set of parameters, the mBGWO algorithm provides better throughput than the BGWO method. In the same figure, we also present the throughput achieved by ESA, random, ACO, and BPSO algorithms. In Fig. 5, the percentage deviation for these soft-computing methods for various system SNR values is presented. mBGWO outperforms all the other techniques.

The variation of the throughput with the iteration number for the BGWO algorithm and the modified version of BGWO is presented in Fig. 6. It is very clear from this figure that the convergence rate of the sum-rate capacity is higher in the mBGWO than the simple BGWO scheme. This presents another result showcasing the superiority of the mBGWO than the normal BGWO algorithm. To compare the performance of the proposed BGWO algorithm with other soft-computing algorithms presented in this chapter, we present the comparison result in Fig. 7. This represents the variation of the sum-rate capacity with the increment in the number of iterations in all these algorithms. The highest convergence rate of the achieved capacity is found to be of the proposed mBGWO algorithm.

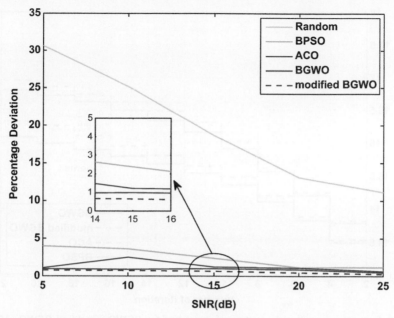

Fig. 5 Percentage deviation in achieved throughput vs SNR for different soft-computing schemes.

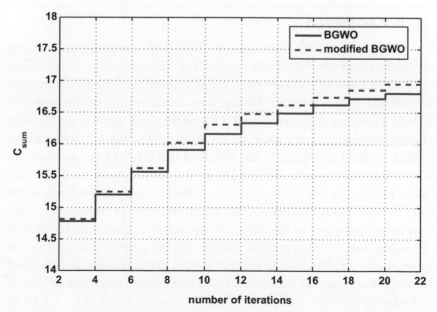

Fig. 6 Sum-rate capacity vs number of iterations for BGWO and modified BGWO.

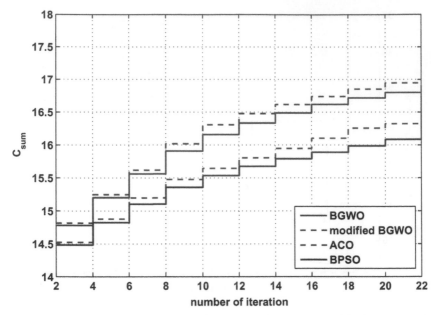

Fig. 7 Sum-rate capacity vs number of iterations for BGWO, modified BGWO, ACO, and BPSO.

6. Conclusion

This chapter makes an attempt to utilize a soft-computing approach in solving the major scheduling issues in wireless communication systems. The ACO, BPSO, and BGWO are the three algorithms used in this chapter in a MU-MIMO uplink scenario. Finally, a modified version of the proposed BGWO algorithm is also presented. Out of the three algorithms, BGWO achieves throughput very close to the optimal throughput achieved by the ESA method. Hence, it is also observed that BGWO outperforms the ACO and BPSO algorithms in terms of archived throughput. From the simulation results, it is also obvious that the mBGWO scheme performs better than the basic BGWO scheme. The other objective to utilize these soft-computing methods is to reduce the computational complexity in the user scheduling cum selection process, thereby lessening the processing load on the system. From all the computational analysis in this chapter, it is observed that the soft-computing methods achieve throughput very close to that of ESA scheme but with a very small fraction of computational load. Out of all the soft-computing algorithms presented here, it is observed that the total computation complexity involved in the mBGWO is the lowest and hence outperforms the other methods. The discussed soft-computing techniques in this chapter can be utilized in solving other optimization problems in MU-MIMO networks. Particularly, the impact of these techniques can be analyzed in a broadcast network and also optimizing the spectral efficiency in wireless networks.

Declaration of conflict of interest

The authors declare that they have no known conflict of interest that could have appeared to influence the work reported in this chapter.

References

[1] G. Caire, S. Shamai, On the achievable throughput of a multiantenna Gaussian broadcast channel, IEEE Trans. Inf. Theory 43 (2003) 1691–1706.

[2] P.P.J. Mohanty, A. Nandi, K.L. Baishnab, D.S. Gurjar, M. Mandloi, MIMO broadcast scheduling using binary spider monkey optimization algorithm, Int. J. Commun. Syst. 34 (17) (2021) e4975. https://onlinelibrary.wiley.com/doi/abs/10.1002/dac.4975.

[3] N. Jindal, W. Rhee, S. Vishwanath, S.A. Jafar, A. Goldsmith, Sum power iterative water-filling for multi-antenna gaussian broadcast channels, IEEE Trans. Inform. Theory. 51 (2005) 1570–1580.

[4] J. Mohanty, P. Pattanayak, A. Nandi, V.K. Trivedi, F.A. Talukdar, Joint antenna and user scheduling for MU MIMO systems using efficient binary artificial bee colony algorithm, IETE J. Res. (2022) 1–12, https://doi.org/10.1080/03772063.2022.2048703.

[5] P. Pattanayak, P. Kumar, Combined user and antenna scheduling scheme for MIMO-OFDM networks, Telecommun. Syst. 70 3–12, https://doi.org/10.1007/s11235-018-0462-0.

[6] A.F. Molisch, M.Z. Win, Y.-S. Choi, J.H. Winters, Capacity of MIMO systems with antenna selection, IEEE Trans. Wirel. Commun. 4 (4) (2005) 1759–1772.

[7] X. Zhang, J. Lee, Low complexity MIMO scheduling with channel decomposition using capacity upperbound, IEEE Trans. Commun. 56 (6) (2008) 871–876.

[8] T. Yoo, A. Goldsmith, On the optimality of multiantenna broadcast scheduling using zero-forcing beamforming, IEEE J. Sel. Areas Commun. 24 (3) (2006) 528–541.

[9] P. Pattanayak, Two-bit SINR quantization based scheduling scheme for MIMO communications, in: 2020 Advanced Communication Technologies and Signal Processing (ACTS), 2020, pp. 1–6.

[10] J. Mao, J. Gao, Y. Liu, G. Xie, Simplified semi-orthogonal user selection for MU-MIMO systems with ZFBF, IEEE Wireless Commun. Lett. 1 (1) (2012) 42–45.

[11] P. Pattanayak, K.M. Roy, P. Kumar, Analysis of a new MIMO broadcast channel limited feedback scheduling algorithm with user grouping, Wirel. Pers. Commun. 80 (2015) 1079–1094.

[12] A. Panda, S.K. Mishra, S.S. Dash, N. Bansal, A suboptimal scheduling scheme for MIMO broadcast channels with quantized SINR, in: International Conference on Advances in Computing, Communications and Informatics (ICACCI), IEEE Press, 2016, pp. 2175–2179.

[13] J. Mohanty, P. Pattanayak, A. Nandi, K.L. Baishnab, F.A. Talukdar, Binary flower pollination algorithm based user scheduling for multiuser MIMO systems, Turk. J. Electr. Eng. Comput. Sci. 30 1317–1336.

[14] D. Choudhary, S.K. Mishra, Sub-optimal user scheduling for MU-MIMO system using channel quantization with two-stage feedback, in: 2nd International Conference on Communication Systems, Computing and IT Applications (CSCITA), IEEE Press, Mumbai, 2017, pp. 169–173.

[15] P. Pattanayak, P. Kumar, Quantized feedback MIMO scheduling for heterogeneous broadcast networks, Wireless Netw. 23 (2016) 1–18.

[16] P. Pattanayak, P. Kumar, Quantized feedback scheduling for MIMO-OFDM broadcast networks with subcarrier clustering, Ad Hoc Netw. 65 (2017) 26–37.

[17] M. Naeem, D. Lee, A joint antenna and user selection scheme for multiuser MIMO system, Appl. Soft. Comput. 23 (2014) 366–374.

[18] P. Pattanayak, P. Kumar, Limited feedback scheduling for MIMO-OFDM broadcast network, in: Proceedings of the 18th IEEE International Symposium on Wireless Personal Multimedia Communications, Hyderabad, India, 13–16 December, 2015.

[19] S.K. Mishra, P. Pattanayak, A.K. Panda, Combined transmit antenna selection and user scheduling in a massive MIMO broadcast system, in: 2020 Advanced Communication Technologies and Signal Processing (ACTS), IEEE Press, 2020, pp. 1–6.

[20] G. Lee, Y. Sung, A new approach to user scheduling in massive multi-user MIMO broadcast channels, IEEE Trans. Commun. 66 (4) (2008) 1481–1495.

[21] P. Pattanayak, D. Pandey, P. Kumar, Error rate performance for multiuser scheduling in MIMO downlink system with imperfect CSI, in: Proceedings of the IEEE 5th International Conference on Wireless Communications, Vehicular Technology, Information Theory and Aerospace & Electronic Systems, December 2015, 2015.

[22] C.W. Chen, S. Jin, K. Wong, A low complexity pilot scheduling algorithm for massive MIMO, IEEE Wireless Commun. Lett. 6 (1) (2017) 18–21.

[23] P. Pattanayak, P. Kumar, SINR based limited feedback scheduling for MIMO-OFDM heterogeneous broadcast networks, in: Proceedings of the IEEE Twenty Second National Conference on Communication (NCC), March 2016, 2016.

[24] B. Lee, L. Ngo, B. Shim, Antenna group selection based user scheduling for massive MIMO systems, in: IEEE Global Communications Conference, 2014, pp. 3302–3307.

[25] P. Pattanayak, P. Kumar, Computationally efficient scheduling schemes for multiple antenna systems using evolutionary algorithms and swarm optimization, in: A.H. Gandomi, A. Emrouznejad, M.M. Jamshidi, K. Deb, I. Rahimi (Eds.), Evolutionary Computation in Scheduling, Wiley, New York, 2020, pp. 105–135.

[26] S. Abedi, S. Vadgama, A genetic approach for downlink packet scheduling in HSDPA system, Soft Comput. 9 (2005) 116–127.

[27] P. Pattanayak, P. Kumar, An efficient scheduling scheme for MIMO-OFDM broadcast networks, AEU Int. J. Electron. Commun. 101 (2019) 15–26. http://www.sciencedirect.com/science/article/pii/S1434841118300359.

[28] P. Pattanayak, Subcarrier wise scheduling methods for multi antenna and multi carrier systems, Wireless Pers. Commun. 114 1485–1500, https://doi.org/10.1007/s11277-020-07434-8.

[29] P. Pattanayak, Proficient user and antenna selection strategies for multi-carrier MIMO communications using adjacent sub-carrier clustering, Wireless Pers. Commun. 125 (2022) 1221–1242.

[30] M. Azharuddin, P.K. Jana, PSO-based approach for energy efficient and energy-balanced routing and clustering in wireless sensor networks, Soft Comput. 21 (2017) 6825–6839.

[31] M. Dorigo, V. Moniezzo, A. Colorni, Ant system: optimization by a colony of cooperating agents, IEEE Trans. Syst. Man Cybern. B 26 (2006) 29–41.

[32] M. Dorigo, L. Gambardella, Ant colony system: a cooperative learning approach to the traveling salesman problem, IEEE Trans. Evol. Comput. 1 (1997) 53–66.

[33] M. Dorigo, L. Gambardella, Ant colonies for the traveling salesman problem, BioSystems 43 (1997) 73–81.

[34] E. Bonabeau, G. Theraulaz, M. Dorigo, Swarm Intelligence: From Natural to Artificial Systems, Oxford University Press, New York, 1999.

[35] P. Pattanayak, D. Sarmah, S. Mishra, A. Panda, Computationally efficient scheduling methods for MIMO uplink networks, Soft Comput. 25 11763–11780, https://doi.org/10.1007/s00500-021-05946-4.

[36] E. Bonabeau, G. Theraulaz, M. Dorigo, Inspiration for optimization from social insect behavior, Nature 406 (2000) 39–42.

[37] A.H. Gandomi, A. Emrouznejad, M.M. Jamshidi, K. Deb, I. Rahimi, Evolutionary Computation in Scheduling: A Scientometric Analysis, Wiley, 2020.

[38] X. Wei, J. Fan, T. Wang, Q. Wang, Efficient application scheduling in mobile cloud computing based on MAXMIN ant system, Soft Comput. 20 (2016) 2611–2625.

[39] V.K. Arora, V. Sharma, M. Sachdeva, A multiple pheromone ant colony optimization scheme for energy-efficient wireless sensor networks, Soft Comput. 24 (2020) 543–553.

[40] Y. Hei, C. Zhang, W. Song, Y. Kou, Energy and spectral efficiency tradeoff in massive MIMO systems with multi-objective adaptive genetic algorithm, Soft Comput. 23 (2019) 7163–7179.

[41] V. Lau, Optimal downlink space-time scheduling design with convex utility functions multiple-antenna systems with orthogonal spatial multiplexing, IEEE Trans. Veh. Technol. 54 (2005) 1322–1333.

[42] R. Elliott, W. Krzymien, Downlink scheduling via genetic algorithms for multiuser single-carrier and multi carrier MIMO systems with dirty paper coding, IEEE Trans. Veh. Technol. 58 (2009) 3247–3262.

[43] P. Pattanayak, P. Kumar, A computationally efficient genetic algorithm for MIMO broadcast scheduling, Appl. Soft. Comput. 37 (2015) 545–553.

[44] S. Mirjalili, S.M. Mirjalili, A. Lewis, Grey wolf optimizer, Adv. Eng. Softw. 69 (2014) 46–61.

[45] S.M. Nimmagadda, Optimal spectral and energy efficiency tradeoff for massive MIMO technology: analysis on modified lion and grey wolf optimization, Soft Comput. 24 (2020) 523–539.

[46] R.S. Suriavel Rao, P. Malathi, A novel PTS: grey wolf optimizer based PAPR reduction technique in OFDM scheme for high-speed wireless applications, Soft Comput. 23 (2019) 2701–2712.

[47] M. Naeem, D. Lee, A joint antenna and user selection scheme for multiuser MIMO system, Appl. Soft Comput. 23 (2014) 366–374.

[48] P. Pattanayak, D. Sarmah, P. Paritosh, Low complexity based scheduling methods for multi-user MIMO systems, Phys. Commun. 43 (2020) 101192. https://www.sciencedirect.com/science/article/pii/S187449072030269X.

[49] S. Vishwanath, N. Jindal, A. Goldsmith, Duality, achievable rates and sum-rate capacity of gaussian MIMO broadcast channel, IEEE Trans. Inf. Theory 49 (2003) 2658–2668.

[50] A. Dey, P. Pattanayak, D.S. Gurjar, Arithmetic/geometric progression based pilot allocation with antenna scheduling for massive MIMO cellular systems, IEEE Netw. Lett. 3 (1) (2021) 1–4.

[51] T.S. Utzle, H. Hoos, Max-min ant system, Futur. Gener. Comput. Syst. 16 (2000) 889–914.

[52] E. Emary, H.M. Zawbaa, A.E. Hassanien, Binary grey wolf optimization approaches for feature selection, Neuro Comput. 172 (2016) 371–381.

[53] D.S. Gurjar, H.H. Nguyen, P. Pattanayak, Performance of wireless powered cognitive radio sensor networks with nonlinear energy harvester, IEEE Sens. Lett. 3 (8) (2019) 1–4.

About the authors

Swadhin Kumar Mishra graduated from Indian Institute of Technology, Madras with an M.Tech in Communication Systems in 2009 after earning a B.E. in Electronics and Telecommunication Engineering from Utkal University, Bhubaneswar, Odisha, in 2000. He is presently employed at the National Institute of Science and Technology in Berhampur, India, as an assistant professor. He is now pursuing his PhD at VSSUT, Burla, India, in the Department of ETC. His current areas of interest in research are optimisation methods, multicarrier communication, and MIMO wireless communication.

Arunanshu Mahapatro received his bachelor's degree in electronics engineering from the University of Pune, his master's degree in electronics and communication engineering from KIIT University, and his doctorate from the National Institute of Technology Rourkela. Presently, he is affiliated to Department of Electronics and Telecommunication Engineering, Veer Surendra Sai University of Technology (VSSUT), Burla. His current research interests include fault tolerant computing, wireless sensor networks, wireless communication and cognitive computing for wireless networks.

Prabina Pattanayak completed his PhD degree from IIT Patna, India, in 2017. He has served as Lead Engineer at HCL Technologies Ltd, after completing his B.Tech from BPUT, Rourkela. He received his B.Tech and M.Tech degrees from Biju Patnaik University of Technology (BPUT) Rourkela, Odisha. He has been working as an Assistant Professor in the department of Electronics and Communication Engineering at NIT Silchar, India since 2018. His current research interests include massive MIMO, multiuser MIMO communications, soft computing techniques, NOMA, Smart Grid Communication Systems and cross-layer scheduling. He is a fellow of IETE India and a senior member of IEEE.

> CHAPTER FIFTEEN

Spectral efficiency optimization by the application of metaheuristic optimization techniques

Jyoti Mohanty[a] and Prabina Pattanayak[b]
[a]Siksha O Anusandhan Deemed to be University, Bhubaneswar, India
[b]National Institute of Technology Silchar, Silchar, India

Contents

Abstract

It is common practice in networks to make use of multiuser multiple-input multiple-output (MU-MIMO) communication systems in order to increase system throughput and spectral efficiency. MIMO systems need careful scheduling of users since the number of users is always much greater than the number of antennas at the base station. To offer optimum system capacity that simultaneously serves a large number of users, dirty paper coding (DPC) is used in a MU-MIMO system. For the combined user and receiver antenna scheduling problems of MU-MIMO systems in the downlink channel, different optimization techniques (i.e., binary flower pollination algorithm (binary FPA), binary spider monkey optimization (binary SMO), binary artificial bee colony optimization (binary ABC)) are utilized in this study. Search space for allocating users to BS receive antennas is huge. Searching across this enormous search area is time-consuming. Therefore, metaheuristic techniques are used to solve this issue.

Advances in Computers, Volume 135
ISSN 0065-2458
https://doi.org/10.1016/bs.adcom.2023.12.004

It is shown in this chapter that binary ABC can handle this massive workload with less time and computational complexity as compared to other methods. The results of simulations that are presented in this research give advantages to the usefulness of binary ABC. Binary ABC has also been shown to have a system sum rate that is almost identical to that of an exhaustive search method. The efficiency of the metaheuristic techniques (binary FPA, binary SMO, binary ABC) has also been evaluated using various statistical parameters.

1. Introduction

Using numerous antennas at both the base station (BS) and the user terminal (UT) improves both system throughput and signal strength stability. These are known as multiple-input multiple-output (MIMO) systems. To get the most out of the available spectrum, MIMO systems with multiple users take use of multiuser diversity (MUD) [1]. For this, multiple users with excellent channel statistics are chosen [2–7]. Parallel data streams may be broadcast at the same time using a multiantenna system, overcoming channel fading and reducing bit error rates. It is possible to dramatically boost spectrum efficiency by placing several antennas on either the transmitting or receiving side of the network. Since several radio frequency (RF) links are required, system hardware costs and complexity increase as a result. Antenna selection is an effective strategy for maximizing antenna advantages while minimizing hardware costs and complexity [8, 9]. In comparison to single-input single-output (SISO) communication systems, the MIMO broadcast channel (BC) is found to have a channel capacity $min(M, N)$ times more than SISO communication systems, where M and N are the number of transmit and receive antennas, respectively [10, 11]. Array gain, spatial diversity, and interference reduction are all advantages of MIMO systems.

With dirty paper coding (DPC), MU-MIMO BC may achieve a near-optimal throughput by scheduling several users at once. The best interference cancellation approach for MIMO broadcast channels is DPC. Because of this, the DPC system requires that BS be fully aware of the channel that UTs are using. Feedback from all users is included in the overall data that is collected as a result of the full feedback feature. As a result, researchers have been working hard in the recent past to find ways to maximize system performance while reducing feedback overhead to the absolute minimum value [12–14].

In this study, it is shown that DPC can attain the highest possible MIMO system sum rate (maximum throughput). User data are encoded and decoded in reverse in the DPC approach. Because each user's encoding is unique, any previous encoding interference is already eliminated. Optimal results can be achieved when $K \gg M$, where K is the number of users present in the network. The UT encoding sequence is essential to the DPC method. To efficiently transmit data from BS to numerous users, a novel transmission technology called DPC has emerged. The optimal utility function value may be calculated after searching through all possible user combinations. In (1), the total number of ordered selections is stated as [15, 16]:

$$N_{OrderedUTs} = \sum_{k=1}^{M} (k!) \binom{K}{k} \tag{1}$$

Exhaustive search algorithms (ESA) [17] are not feasible in a real-world system because of their long scheduling requirements (tentatively a few milliseconds). By applying binary spider monkey optimization (binary SMO) for MU-MIMO downlink wireless communication systems, the maximum sum rate may be maximized. As contrast to the ESA (DPC), here the computational complexity is minimized. Various simulation results are used to demonstrate the performance of binary SMO [18], binary flower pollination algorithm (binary FPA) [11], and binary artificial bee colony optimization (binary ABC) [19]. In prior studies, binary SMO's performance was compared to that of DPC, the random search technique, and inefficient scheduling algorithms. Binary ABC's performance is compared to that of DPC, random search, and other scheduling algorithms in the performance study. The suggested binary ABC provides excellent system sum rate/throughput than the random search method [20] and other metaheuristic algorithms.

Accordingly, Section 2 describes the MU-MIMO system model. Section 3 explains scheduling of DPC in MU-MIMO broadcast channels. Along with it discusses user scheduling using soft-computing techniques; Section 4 explains the simulation results. In terms of throughput and computational complexity, the optimum ESA is compared to the different metaheuristic optimization techniques. Section 5 concludes with an executive summary of this chapter.

2. System model

In this chapter, we will focus on a wireless MIMO system that operates in the downlink manner [21]. In this configuration, the BS is equipped with M sets of antennas. The BS provides support for a set of K users who are located in various places in relation to the BS. Every user is equipped with N antennas. The network makes use of a total of NK receiving antennas. The BS will then choose M of the best receiving antennas from this pool for use in the communication process. Here, it is assumed that $K \gg M$, $M \geq N$. Therefore, the received signal for the Kth user is represented as

$$\mathbf{Y}_k^t = \sqrt{\alpha_k}\mathbf{H}_k^t\mathbf{X}^t + \mathbf{W}_k^t, k = 1, \dots, K \tag{2}$$

For the kth user, the channel gains between receive and transmit antennas are represented by the matrix H_k, which has the dimensions $N \times M$. For this approach, the channel between each user and the BS is represented by a zero-mean, circularly symmetric Gaussian matrix (Rayleigh fading [22]). Single carrier system is being explored. α_k simulates the weakening of a signal caused by the shadowing effect and obstacles in its route. The transmitted signal, denoted by the vector X, is an $M \times 1$ matrix, whereas the noise, denoted by W_k, is a $N \times 1$ matrix with the value $CN (0, 1)$. For user k, the complex gain from transmitter antenna m to receiver antenna n is denoted as $h_k(n, m)$. It is assumed here that the channel matrix is known to the receiver but not the transmitter. According to Refs. [23–26], we assume that the power level of transmission from each antenna is the same. Assume that each antenna has an average transmit power of $1/M$. The signal received by the nth antenna of the kth user looks like:

$$y_k(n) = \alpha_k \sum_{m=1}^{M} h_k(n, m)x(m) + w_k(n) \tag{3}$$

where $x(m)$ is the signal that the user n wants to receive when it is sent via antenna m, with the assumption that $x(m)$ is not dependent on the specific transmit antenna being used. Therefore, the signal-to-interface-plus-noise ratio (SINR) for user k may be stated as:

$$\text{SINR}_{m,n}^k = \frac{|h_k(n, m)|^2}{\left\{ \left(\dfrac{M}{\tau_k}\right) + \sum_{m' \neq m}|h_k(n, m')|^2 \right\}} \tag{4}$$

where τ_k is the average signal-to-noise ratio (SNR) of all users. The average SNR received by each user is assumed to be constant across the network. Consequently, $\tau_k = \tau$, given that $k = 1, \dots, K$. We may set a limit on the system's capacity by generating M random signals and sending them toward the users with a maximum SINR [23]:

$$C_{sum}(H_1, ..., H_K) \leq \mathbb{E}\left[\sum_{m=1}^{M} \log_2 \left(1 + \max_{1 \leq k \leq K, \, 1 \leq n \leq N} SINR_{m,n}^{k} \right) \right] \quad (5)$$

Where $H_1, ..., H_K$ are the channel matrices of all users, and C_{sum} is the maximum capacity of the system that may be achieved. The maximum feedback of the system's SINR [27] for K users with a fixed M is expressed in this way:

$$C_{sum}(H_1, ..., H_K) = \sum_{m=1}^{M} \log_2 \left(1 + \max_{1 \leq k \leq K, \, 1 \leq n \leq N} SINR_{m,n}^{k} \right) \quad (6)$$

3. Proposed algorithms for MIMO broadcast scheduling

3.1 DPC scheduling

Using this approach, the BS gets information from all users $k (1 \leq k \leq K)$ about the highest SINR value and the BS antenna index. Each transmit antenna $m (1 \leq m \leq M)$ chooses the highest SINR from the collection of maximum SINRs received by all users corresponding to that transmit antenna m after receiving all feedbacks from users [28]. It is important that this be repeated for each user.

Eq. (1) shows that for a large number of users K, the number of possible DPC-based scheduling solutions for M users is exponentially large. The authors in Ref. [20] considered limits like these to narrow the variety of possible user combinations down to a more manageable size.
1. Users with the same M combinations will only be considered once during the evaluation.
2. No user should get service from multiple transmit antennas.
3. There should be no duplicate users left in the user combinations database.

Taking these limitations into account, the user and antenna scheduling (UAAS) may calculate the total number of possible user combinations as:

$$N_{UniqueUserSequence} = \binom{K}{M} \quad (7)$$

This figure likewise increases dramatically when more transmit antennas and end users join the network. This causes a large number of instances of the utility function being evaluated. High-speed networks simply cannot handle this amount of computing in real time. Because of this, binary ABC, binary

SMO, and binary FPA have been implemented for the UAAS to lessen both the computational complexity and time complexity [11, 29]. In addition, the BS makes use of high–end digital signal processing (DSP) processors in order to make this process more manageable.

The pseudo-code that describes the execution of DPC Scheduling is provided below:

DPC Scheduling.
1: INITIALIZE $maxSINR(k, m, n) \leftarrow 0$ for $k = 1, ..., K$, $m = 1, ..., M$ and $n = 1, ..., N$.
2: INITIALIZE $Rx_{antennaindex} \leftarrow 0$ for $k = 1, ..., K$, $m = 1, ..., M$ and $n = 1, ..., N$.
3: **for** $K = 1$ to k **do**
4:   **for** $m = 1$ to M **do**
5:     **for** $n = 1$ to N **do**
6:       $SINR \leftarrow 0$, and $R_x \leftarrow 0$
7:       **for** $m = 1$ to M **do**
8:         **for** $n = 1$ TO N **do**
9:           Compute $SINR = SINR_{m,n}^k$ as per (4)
10:            **if** $SINR \geq SINR_{max}$ **then**
11:              $SINR_{max} = SINR$
12:              $R_x = n$
13:            **end if**
14:          **end for**
15:        **end for**
16:        $maxSINR(k, m, n) \leftarrow maxSINR(k, m, n) \cup SINR_{max}$
17:        $Rx_{antennaindex} \leftarrow Rx_{antennaindex} \cup R_x$
18:        Transmit $maxSINR(k, m, n)$ feedback to BS.
19:        The user maintains $Rx_{antennaindex}$ in order to take advantage of the specified receive antenna in data transmission.
20:      **end for**
21:    **end for**
22: **end for**

3.2 MU-MIMO scheduling by using binary ABC

In this section, a new artificial bee colony method is proposed for binary optimization. An alternative binary ABC method is given, in which a binary bitwise operation is utilized in place of the original ABC system's real arithmetic operation [19]. The candidate solution of binary optimization problems serves as the food supply for the artificial bee colony, which can then be used to solve the corresponding optimization issues. A unique binary ABC technique for addressing binary optimization problems is proposed, and it makes use of a surjection mapping, which converts a real vector to a 0–1 vector. A value of 1 for the variable indicates that the employee has been offered the job, whereas a value of 0 indicates that the employee has been passed over. Solutions are generated initially by

$$X_{ij} = \begin{cases} 0, & if \quad U(0,1) \leq 0.5, \\ 1, & if \quad U(0,1) > 0.5, \end{cases} \tag{8}$$

where $U(0, 1)$ is a value that is created uniformly across the whole system. Binary ABC user scheduling parameters are given below:

- Y is the utility function which is shown by Eq. (6).
- *Food_Number* is the number of food sources, and it is equal to half of the colony's total population.
- *max_Cycle* is the maximum number of repetitions allowed.
- D is $M \times [log_2 K]$ long and contains a binary representation of the scheduled users.
- *Limit* refers to the food source limit, which means that the employed bee leaves behind a food supply that could not be extended by limit checks.
- The *Selected_User* row vector holds the M users chosen by BS for service.
- When *Selected_User* is present, the chosen Users' *Selected_RxAntenna* is always a size M vector containing their preferred receive antennas.

◇ **Step 0:**
  1. Initialize the parameters *Food_Number* and *Limit*. Initialize *Selected_User* = 0 and *Selected_RxAntenna* = 0.
  2. The binary space is transformed using the Sigmoid transformation feature. Create a new source of nutrition with the help of the Sigmoid function.
  3. The $m^{th} log_2 K$ contains the binary representation of the user arbitrarily assigned to benefit from the mth transmit antenna.
  4. Every row is a collection of ϕ which is the user sequence, and these ϕs are represented as a binary string γ.

◇ **Step 1:**
  For a near-optimal scheduling solution with minimal computational complexity, the following set of criteria must be followed.
  1. Users with the same M combination will only be counted once in the evaluation.
  2. It is unacceptable to use multiple transmit antennas to serve a single user.

3. It is recommended that duplicate users be deleted from user combinations.

When any given food source violates any of the aforementioned constraints, it is necessary to randomly switch certain components and keep doing so until no such violations occur.

◇ **Step 2**:

1. Perform the computation for ϕ using the most recent population estimation.

2. The mutant solution (new food source) must be unique, and a random supply of food must be utilized in the bee process to create it.

3. If the nectar value is more than that of a food source, then it should be calculated for the surrounding source where the new food source is formed and stored. It is our intention to evaluate (6) in an effort to find the value that minimizes the cost function.

4. The pollens are ranked in descending order of fitness.

5. Update *Selected_User* $= \phi_{opt}$ as well as *Selected_RxAntenna* $= Rx_{opt}$.

◇ **Step 3**:

1. Rotate a roulette wheel to determine the probability of a bee's arrival to a food source, and use that information to direct spectator bees there. Establish fresh communities using probability values.

$$prob = \frac{0.9 \times fitness(X)}{\sum_{\forall X} fitness(X)}. \tag{9}$$

2. To become a scout bee, an employed bee must wait until a certain number of iterations have passed without the corresponding food supply being changed. In order to locate a new food source, the scout bee uses Eq. (8) to perform a random initialization in the solution space. The scout bee will once again be gainfully employed if they locate another food source.

3. If $prob > rand(0, 1)$, update new food source (10).

$$v_{ij} = x_{ij} + \beta_{ij}(x_{ij} - x_{kj}), \tag{10}$$

where the new food source is denoted by v_{ij}. β_{ij} is a random integer between $[-1,1]$, whereas j and k are picked at random as parameters and neighborhoods, respectively.

4. If there is no way to enhance solution, the number of trials must be raised.

◇ **Step 4**:
1. The scout bee finds a new food source if there is an abandoned one.
2. If, after some number of iterations (the *Limit*), no better food source can be located in the area around the current one, the scout bee will choose a new one at random.
3. If *Count* ≤ *max_Cycle*, update *Selected_Consumer* = ϕ_{opt} as well as *Selected_received_antenna* = Rx_{opt}.
4. An additional 1 is added to the *Count*.
5. Fitness, measured by (6), will primarily focus on the system's throughput.
6. Use this information to determine the optimum receiving antenna index and best user index.

3.3 Spider monkey optimization

SMO is developed by observing how spider monkeys forage for food. Spider monkeys' food sources stand in for the solutions to the problem. To determine which food source is best, we compute its fitness value. The interested reader is directed to Ref. [18] for more information.

LocalLeaderLimit, *GlobalLeaderLimit*, *maximumgroup(MG)*, and *perturbation-rate(pr)* are the four control parameters for the SMO algorithm. If the local group leader hasn't been updated in the permitted period of time, the *LocalLeaderLimit* may be used to indicate a change in the group's foraging strategy. The global leader divides the group into subgroups if the *GlobalLeaderLimit* value is reached without an update in global leader. Maximum group (MG) size and perturbation (*pr*) are used to set bounds on the population size and the degree of change to the present position, respectively.

Consider the objective function *F*, which has to be maximized as much as possible (6). The following describes the parameters utilized in the scheduling research using SMO for the specified problems:

- *F* stands for objective function, which means (6).
- The SMO population is *N* in size.
- Number of possible repetitions is limited by *Iter*.
- A route taken by SMO has length (dimension) *D* which is $M \times \lceil log_2 K \rceil$ long and contains a binary representation of the scheduled users.
- A local leader count is referred to as an *LLC*.
- The acronym *GLC* stands for the global leader count.
- The *LocalLeaderLimit* is intended to prevent stagnation.
- *GlobalLeaderLimit* prevents stagnation from occurring.
- The *Selected_User* row vector holds the *M* users chosen by BS for service.

- When *Selected_User* is present, the chosen Users' *Selected_RxAntenna* is always a size M vector containing their preferred receive antennas.

The SMO algorithm that was used for the search problem implementation is explained in the following way.

SMO algorithm.

1: INITIALIZE N and define *LLC*, *GLC*, *LocalLeaderLimit*, *GlobalLeader Limit*, *MG*, and *pr*.

2: Based upon fitness, update local leader and global leader.

3: Use a greedy selection technique to choose the best monkeys from the group of existing and newly created monkeys.

4: Create the monkeys' swarm by assimilating the knowledge of the group's members and the global leader.

5: Apply local leader decision phase for foraging.

6: Apply global leader decision phase for subdivision of groups.

Parameters for the SMO method used in this chapter's design are set as follows according to the recommendations implemented in Ref. [18]: $pr = 0.1$, $GobalLeaderLimit = 50$, $LocalLeaderLimit = 30$, and maximum number of groups in the swarm $(MG) = N/2$, where N is the total number of members in the swarm. All the specific procedures and steps are outlined in Ref. [18]. The fundamental method, which operates in continuous domain, must be adapted for a binary solution. Thus, in this chapter, we also experimented with binary SMO that finds solutions in a search space whose logic values may only be 0 or 1. The binary SMO method is a logical extension of the original SMO algorithm, in which the position-updating equations are altered [18].

3.4 Binary flower pollination algorithm

Yang et al. [30] proposed an FPA algorithm that was inspired by the pollination of flowers. Both biotic and abiotic methods are very important for pollination process. The transfer of pollen from one bloom to another in biotic pollination is accomplished by the movement of the pollen-carrying insects, birds, or bats. With abiotic pollination, bees and other pollinators are unnecessary. We may further classify pollination as either cross-pollination, in which pollen is transferred from one plant's stamen to the stigma of another, or as self-pollination, in which pollen is transferred from one plant's stamen to the stigma of the same species. The following are the four guidelines for pollination as outlined by Yang [30]:

1. Pollen-carrying pollinators are said to engage in Levy flights around the globe as part of cross-pollination and a biotic process.
2. Local pollination may occur via abiotic and self-pollination processes.
3. The likelihood of a successful reproduction between two flowers is related to their degree of resemblance, and this is what we mean by "flower constancy."
4. A switch probability $p \in [0, 1]$ governs the balance between local and global pollination. Due to variables like proximity and wind, local pollination may account for a significant percentage p of global pollination processes.

It is possible to provide a mathematical representation of the global pollination as (11):

$$X_i^{(t+1)} = X_i^{(t)} + \alpha L\left(g_* - X_i^{(t)}\right) \tag{11}$$

where $X_i^{(t)}$ is the ith pollen at iteration t, g_* represents the optimal solution, α represents the scaling factor, s represents the step size, and L represents the Levy flight step size. One way to represent the form of a Levy distribution is as follows (12):

$$L = \frac{\lambda \cdot \Gamma(\lambda) \cdot \sin(\lambda)}{\pi} \cdot \frac{1}{S^{1+\lambda}}, \quad S \gg S_0 > 0 \tag{12}$$

where $\lambda = 3/2$ and $\Gamma(\lambda)$ is the gamma function.

It is possible to provide a mathematical expression for local pollination as,

$$X_i^{(t+1)} = X_i^{(t)} + \epsilon(X_j^{(t)} - X_k^{(t)}) \tag{13}$$

where ϵ is a random number $\in [0, 1]$ and $X_i^{(t)}$ and $X_k^{(t)}$ are pollens from flowers i and k of the same plant species.

For binary FPA, the vector solutions are structured such that each component is a binary value. Binary elements are distributed to the population when it is first generated. On the other hand, after (11) and (13) are applied, the solution vectors transform into vectors that include continuous components. As a result, the formula (14) is used to transform the continuous components back into the binary elements.

$$S(X_i^j(t)) = \frac{1}{1 + e^{-X_i^j(t)}}, \tag{14}$$

$$X_i^j(t) = \begin{cases} 1 & \text{if } S\left(X_i^j(t)\right) > \sigma, \\ 0 & \text{otherwise} \end{cases} \tag{15}$$

where σ is a random integer in the range $[0, 1]$, S is a sigmoid function, and $X_i^j(t)$ represents pollen j from flower X_i at time t.

The method is distinguished by a number of parameters, which are as follows: F, Pop_size, D, $Iter$, K, N, and M where

- F stands for objective function, which means (6).
- The flower population is Pop_Size in size.
- $Iter$ is the maximum number of repetitions allowed.
- D is $M \times \lceil log_2 K \rceil$ long and contains a binary representation of the scheduled users.
- In the interval $[0,1]$, the proximity probability is denoted by P.
- A Levy distribution is used to determine the value of the parameter L, which stands for step size.
- The $Selected\_User$ row vector holds the M users chosen by BS for service.
- When $Selected\_User$ is present, the chosen Users' $Selected\_RxAntenna$ is always a size M vector containing their preferred receive antennas.

This algorithm contains a representation of the binary FPA pseudocode which is shown below:

Flower pollination algorithm.

1: function FPA(F, Pop_Size, D, $Iter$, K, N, M).
2: Input : Switch probability (p), scaling factor (α), size of the population (N).
3: Output : MaxSINR, Best user, Best received antenna index.
4: Initiate a population of flowers, rate their quality, and determine the best one.
5: while the stopping criteria are not met do
6: **for** i =1 TO Pop_Size **do**
7:    **if** *rand* $< p$ **then**
8:      Update (11)
9:    **else**
10:     Go around the neighborhood and pick flowers at random.
11:     Update (13)
12:    **end if**
13:    New solutions should be updated if the fitness function is improved upon (better solutions are identified).
14:    Calculate maximum SINR.
15:    Update (6).
16:    Update the current global best.
17: **end for**
18: Update $Selected\_User$ and $Selected\_RxAntenna$.

---

4. Results and discussions

This section presents and discusses the outcomes of the different soft computing methods covered throughout this chapter. It has been shown how the advantages achieved from using the mentioned binary ABC compare well to other schemes by demonstrating its varied performances. The findings shown in this section are the predicted values obtained after simulating a total of 10,000 random iterations of the channel, given that the channel parameters are random. All of the simulation findings provided and discussed in this research work have been generated by using MATLAB software.

In Fig. 1A and B, we see a comparison of the achievable system through-put in a MU–MIMO system using various methods. In this case, we compare the performance of two different configurations: (a) $M = 5$, $N = 3$, $K = 20$, *Food_Number* $= 20$, and *max_Cycle* $= 20$ and (b) $M = 8$, $N = 6$, $K = 25$, *Food_Number* $= 30$, and *max_Cycle* $= 20$. In Fig. 1A and B, we see a graphical representation of the feasible system sum rate from several approaches vs an SNR range of 0–30 dB. This graph clearly demonstrates that binary ABC achieves system sum rate in close accord with the ESA findings. Unsurprisingly, random search method is underperforming to both binary SMO and binary FPA. Since SNR is proportional to sum rate (6), increasing the SNR results in a greater sum rate and hence a larger channel capacity.

For the purpose of describing the degree to which the results acquired by various approaches deviate from those obtained by the ESA, a metric known as the percentage deviation from the exhaustive search algorithm (PDFTESA) has been established. The PDFTESA may also be written as:

$$\text{PDFTESA} = \left(\frac{\left(C_{sum}^{ESM} \right) - C_{sum}^{Methods}}{C_{sum}^{ESM}} \right) \times 100, \tag{16}$$

where $\left(C_{sum}^{ESM} \right)$ is the sum rate obtained by the exhaustive search strategy that generates the highest system sum rate, and methods may be binary ABC, binary SMO, or binary FPA.

Fig. 2A and B takes into account the MU–MIMO system with the following parameters: (a) $M = 5$, $N = 3$, $K = 20$, *Food_Number* $= 20$, and *max_Cycle* $= 20$ and (b) $M = 8$, $N = 6$, $K = 25$, *Food_Number* $= 30$, and

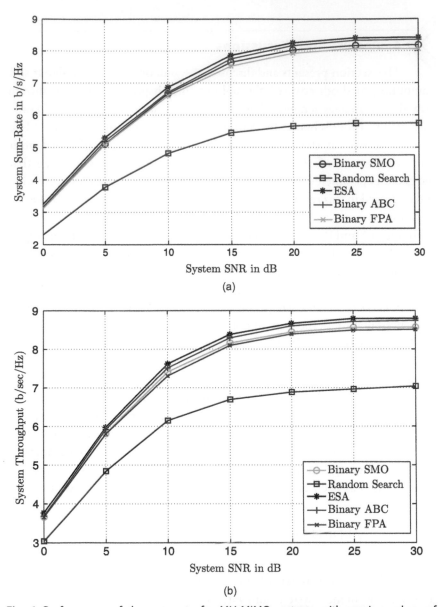

Fig. 1 Performance of the sum rate for MU-MIMO systems with varying values of *max_cycle*, *FoodNumber*, *K, N,* and *M* using ESA (DPC), random search, binary FPA, binary ABC, and binary SMO. Each dot represents an average of well over 1000 iterations of a separate simulation. (A) $M = 5, N = 3, K = 20$, *Food_Number* = 20, and *max_Cycle* = 20. (B) $M = 8, N = 6, K = 25$, *Food_Number* = 30, and *max_Cycle* = 20.

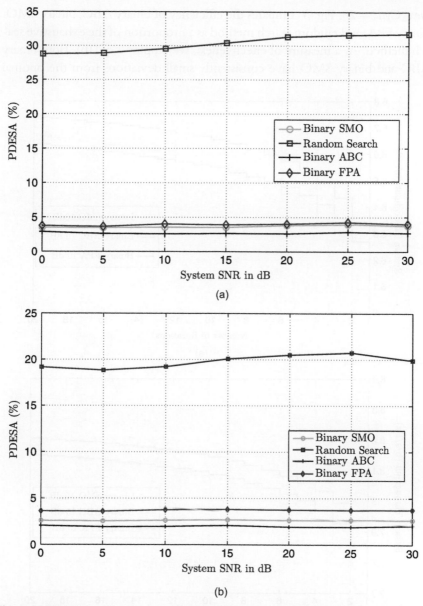

Fig. 2 PDFTESA comparison among binary ABC, binary SMO, binary FPA, and random search with different values of *M, N, K,* and *Food_Number*. (A) *M* = 5, *N* = 3, *K* = 20, *Food_Number* = 20, and *max_Cycle* = 20. (B) *M* = 8, *N* = 6, *K* = 25, *Food_Number* = 30, and *max_Cycle* = 20.

$max\_Cycle = 20$. Fig. 3 compares the efficiency of binary ABC, binary SMO, binary FPA, and random search method as a proportion of the exhaustive search method. As a measure of effectiveness, we employ Eq. (16). Both binary ABC and binary SMO have consistently small deviations from the optimal

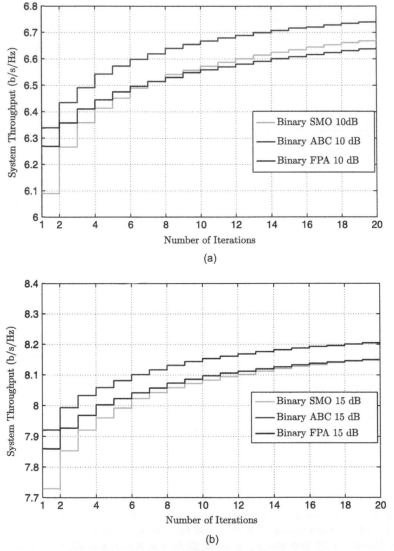

Fig. 3 Performance of binary ABC, binary SMO, binary FPA, and random search method on a per-generation (or per-iteration) basis with different values of *M, N, K,* and *Food_Number* and system SNR = 15 dB. (A) $M = 5, N = 3, K = 20, Food\_Number = 20$, and $max\_Cycle = 20$. (B) $M = 8, N = 6, K = 25, Food\_Number = 30$, and $max\_Cycle = 20$.

value. More so, when maximum number of iterations/generation is held constant, binary ABC consistently yields a lower PDFTESA than binary SMO. The superior performance of binary ABC over binary SMO, binary FPA, and random search method is seen here.

In Fig. 3A and B, we see a relationship between the system's sum rate and the number of generations or iterations. It is possible to calculate the rate of convergence for several algorithms, such as binary ABC, binary SMO, binary FPA, and random search method, using the two figures as shown in Fig. 3A and B. According to the findings, binary ABC performs much better than binary SMO and binary FPA. The results of this study reveal that fitness rises rapidly in the first few iterations; after that, system sum rate steadily reduces and comes to a constant level.

4.1 Complexity analysis

The computational complexity of each method is analyzed and contrasted in this section. As a metric for computational complexity, the number of complex additions and multiplications (NCAAM) is being examined. For each approach (binary ABC, binary SMO, binary FPA, and DPC), we have calculated the NCAAM to demonstrate the computational complexity and approximate run-time. In this case, the objective function is written as (6). As can be observed in (4), $2M$ NCAAMs are needed to calculate SINR. Throughout all, user has $N \times M$ SINR terms. Therefore, a $2M^2N$ number of NCAAMs are needed for a single user (6). It takes $2M^3N$ NCAAMs to support an unpredictable succession of M users. To participate in the ESA approach, $\left[2M^3N\binom{K}{M}\right]$ NCAAMs are necessary (7).

All of the aforementioned soft computing methods, such as binary ABC, binary SMO, and binary FPA, compute (6). Therefore, $[2M^3N \times Food\_Number \times max\_Cycle]$ NCAAMs are required for each soft computing approach. Here, $Food\_Number$ is same as $population\_size$ and max_Cycle is same as $generation\_Number$ for rest of the soft computing techniques. From Table 1, it is clearly observed that the binary ABC outperforms all

Table 1 Computational complexity comparison for different algorithms for the system parameters of $K = 50$, $N = 12$, and $M = 15$.

Algorithm	Population size	No. of generations	NCAAM	Time (ms)
Binary FPA	50	350	14175×10^5	31.640625
Binary SMO	50	320	1296×10^6	28.928571
Binary ABC	50	100	405×10^6	9.040179

other soft computing techniques. For a MIMO system with configuration of $K = 50$, $N = 12$, and $M = 15$, binary ABC requires the least time complexity, which is around 9 ms to reach a near-optimal solution. Other metaheuristic techniques require more number of NCAAMs and accordingly higher computational time to acquire a near-optimal solution.

5. Conclusion

It is shown in this research that the suggested soft-computing-based low-complexity scheduling algorithms have been resulting in throughput in the MU-MIMO system that is very close to that exhaustive search technique. The proposed binary ABC method outperforms binary FPA and binary SMO. Compared to the exhaustive search technique, these low-complexity scheduling algorithms need a much less number of complex multiplications and additions. User and receive antenna scheduling for MU-MIMO downlink communications have been successfully achieved in the time period of emerging data communications by using any of the three suggested low-complexity-based UAAS techniques.

Acknowledgments

This work was supported by the Ministry of Electronics and Information technology (MeitY), Government of India (GoI) under a research grant No. R-23011/3/2020-R&D in CC&BT.

Declaration of conflict of interest

The authors declare that they have no known conflict of interest that could have appeared to influence the work reported in this chapter.

References

[1] P. Pattanayak, K.M. Roy, P. Kumar, Analysis of a new MIMO broadcast channel limited feedback scheduling algorithm with user grouping, Wirel. Pers. Commun. 80 (2015) 1079–1094.

[2] S. Vishwanath, N. Jindal, A. Goldsmith, Duality, achievable rates and sum-rate capacity of gaussian MIMO broadcast channel, IEEE Trans. Inf. Theory 49 (2003) 2658–2668.

[3] P. Pattanayak, P. Kumar, An efficient scheduling scheme for MIMO-OFDM broadcast networks, AEU Int. J. Electron. Commun. 101 (2019) 15–26, https://doi.org/10.1016/j.aeue.2019.01.017.

[4] K.-K. Wong, R. Murch, K. Letaief, A joint-channel diagonalization for multiuser MIMO antenna systems, IEEE Trans. Wirel. Commun. 2 (4) (2003) 773–786, https://doi.org/10.1109/TWC.2003.814347.

[5] P. Pattanayak, Subcarrier wise scheduling methods for multi antenna and multi carrier systems, Wirel. Pers. Commun. 114 1485–1500, https://doi.org/10.1007/s11277-020-07434-8.

[6] C.H. Swannack, E. Uysal-Biyikoglu, G.W. Wornell, MIMO broadcast scheduling with limited channel state information, 2005. https://www.semanticscholar.org/paper/MIMO-Broadcast-Scheduling-with-Limited-Channel-Swannack-Uysal-Biyikoglu/73cfd6851559b6649833aeb4a97d86baafff95a8.

[7] P. Pattanayak, P. Kumar, Limited feedback scheduling for MIMO-OFDM broadcast network, in: Proceedings of the 18th IEEE International Symposium on Wireless Personal Multimedia Communications, Hyderabad, India, 13–16 December, 2015.

[8] H. Tang, Z. Nie, RMV antenna selection algorithm for massive MIMO, IEEE Signal Process. Lett. 25 (2018) 239–242.

[9] P. Pattanayak, D. Pandey, P. Kumar, Error rate performance for multiuser scheduling in MIMO downlink system with imperfect CSI, in: Proceedings of the IEEE 5th International Conference on Wireless Communications Vehicular Technology, Information Theory and Aerospace & Electronic Systems, December, 2015, 2015.

[10] N. Jindal, A. Goldsmith, Dirty-paper coding versus TDMA for MIMO broadcast channels, IEEE Trans. Inf. Theory 51 (2005) 1783–1794.

[11] J. Mohanty, P. Pattanayak, A. Nandi, K.L. Baishnab, F.A. Talukdar, Binary flower pollination algorithm based user scheduling for multiuser MIMO systems, Turk. J. Electr. Eng. Comput. Sci. 30 (2022) 1317–1336.

[12] L.-U. Choi, R. Murch, A transmit preprocessing technique for multiuser MIMO systems using a decomposition approach, IEEE Trans. Wirel. Commun. 3 (1) (2004) 20–24, https://doi.org/10.1109/TWC.2003.821148.

[13] P. Pattanayak, Two-bit SINR quantization based scheduling scheme for MIMO communications, in: 2020 Advanced Communication Technologies and Signal Processing (ACTS), 2020, pp. 1–6, https://doi.org/10.1109/ACTS49415.2020.9350505.

[14] G. Dimic, N. Sidiropoulos, On downlink beamforming with greedy user selection: performance analysis and a simple new algorithm, IEEE Trans. Signal Process. 53 (10) (2005) 3857–3868, https://doi.org/10.1109/TSP.2005.855401.

[15] P. Pattanayak, D. Sarmah, P. Paritosh, Low complexity based scheduling methods for multi-user MIMO systems, Phys. Commun. 43 (2020) 101192, https://doi.org/10.1016/j.phycom.2020.101192.

[16] A.H. Gandomi, A. Emrouznejad, M.M. Jamshidi, K. Deb, I. Rahimi, Evolutionary Computation in Scheduling: A Scientometric Analysis, Wiley, 2020.

[17] P. Pattanayak, D. Sarmah, S. Mishra, A. Panda, Computationally efficient scheduling methods for MIMO uplink networks, Soft Comput. 25 (2021) 11763–11780, https://doi.org/10.1007/s00500-021-05946-4.

[18] J. Mohanty, P. Pattanayak, A. Nandi, K.L. Baishnab, D.S. Gurjar, M. Mandloi, MIMO broadcast scheduling using binary spider monkey optimization algorithm, Int. J. Commun. Syst. 34 (17) (2021) e4975, https://doi.org/10.1002/dac.4975.

[19] J. Mohanty, P. Pattanayak, A. Nandi, V.K. Trivedi, F.A. Talukdar, Joint antenna and user scheduling for MU MIMO systems using efficient binary artificial bee colony algorithm, IETE J. Res. (2022) 1–12, https://doi.org/10.1080/03772063.2022.2048703.

[20] P. Pattanayak, P. Kumar, A computationally efficient genetic algorithm for MIMO broadcast scheduling, Appl. Soft Comput. 37 (C) (2015) 545–553, https://doi.org/10.1016/j.asoc.2015.08.053.

[21] P. Pattanayak, P. Kumar, Quantized feedback MIMO scheduling for heterogeneous broadcast networks, Wirel. Netw. 23 (2016) 1449–1466.

[22] P. Pattanayak, Proficient user and antenna selection strategies for multi-carrier MIMO communications using adjacent sub-carrier clustering, Wirel. Pers. Commun. 125 (2022) 1221–1242, https://doi.org/10.1007/s11277-022-09598-x.

[23] M. Sharif, B. Hassibi, On the capacity of MIMO broadcast channels with partial side information, IEEE Trans. Inf. Theory 51 (2) (2005) 506–522, https://doi.org/10.1109/TIT.2004.840897.

[24] P. Pattanayak, P. Kumar, Computationally efficient scheduling schemes for multiple antenna systems using evolutionary algorithms and swarm optimization, in: A.H. Gandomi, A. Emrouznejad, M.M. Jamshidi, K. Deb, I. Rahimi (Eds.), Evolutionary Computation in Scheduling, Wiley, 2020, pp. 105–135 (Chapter 5).

[25] W. Zhang, K. Letaief, MIMO broadcast scheduling with limited feedback, IEEE J Sel. Areas Commun. 25 (2007) 1457–1467, https://doi.org/10.1109/JSAC.2007.070918.

[26] P. Pattanayak, P. Kumar, SINR based limited feedback scheduling for MIMO-OFDM heterogeneous broadcast networks, in: Proceedings of the IEEE Twenty Second National Conference on Communication (NCC), March 2016, 2016.

[27] P. Pattanayak, P. Kumar, Quantized feedback scheduling for MIMO-OFDM broadcast networks with subcarrier clustering, Ad Hoc Netw. 65 (2017) 26–37.

[28] S.K. Mishra, P. Pattanayak, A.K. Panda, Combined transmit antenna selection and user scheduling in a massive MIMO broadcast system, in: 2020 Advanced Communication Technologies and Signal Processing (ACTS), 2020, pp. 1–6, https://doi.org/10.1109/ACTS49415.2020.9350421.

[29] P. Pattanayak, P. Kumar, Combined user and antenna scheduling scheme for MIMO-OFDM networks, Telecommun. Syst. 70 (2019) 3–12, https://doi.org/10.1007/s11235-018-0462-0.

[30] D. Rodrigues, X.-S. Yang, A.N. de Souza, J.P. Papa, Binary Flower Pollination Algorithm and Its Application to Feature Selection, Springer International Publishing, Cham, 2015, pp. 85–100, https://doi.org/10.1007/978-3-319-13826-8_5.

About the authors

Jyoti Mohanty completed his B.Tech. in electronics and communication engineering from DRIEMS, Cuttack, Odisha in 2007. He also has his M.Tech in electronics and communication engineering from ITER, Siksha O Anusandhan (deemed to be university), Bhubaneswar, Odisha passes on 2011. He received PhD degree in electronics and communication engineering from NIT Silchar, Assam. His varied research interest mostly focuses around wireless Communication, soft computing and advances in MIMO technology.

Prabina Pattanayak completed his PhD degree from IIT Patna, India, in 2017. He has served as Lead Engineer at HCL Technologies Ltd, after completing his B. Tech from BPUT, Rourkela. He received his B.Tech and M.Tech degrees from Biju Patnaik University of Technology (BPUT) Rourkela, Odisha. He has been working as an Assistant Professor in the Department of Electronics and Communication Engineering at NIT Silchar, India since 2018. His current research interests include massive MIMO, multiuser MIMO communications, soft computing techniques, NOMA, Smart Grid Communication Systems and cross-layer scheduling. He is a fellow of IETE India and a senior member of IEEE.

An effective genetic algorithm for solving traveling salesman problem with group theory

Rahul Chandra Kushwaha[a], Dharm Raj Singh[b], and Anubhav Kumar Prasad[c]

[a]Department of Computer Science and Engineering, Rajiv Gandhi University (Central University), Itanagar, Arunachal Pradesh, India
[b]Department of Computer Application, Jagatpur Post Graduate College, Varanasi, Uttar Pradesh, India
[c]Department of Computer Science and Engineering, United Institute of Technology, Naini, Prayagraj, Uttar Pradesh, India

Contents

Abstract

The article proposes a novel algorithm that uses genetic algorithm with group theory for initial population generation, and also proposes a novel crossover for solving traveling salesman problem. In the group tour construction method, each individual/initial tour has distinct start city provided that population size is equal to total number of cities. In the initial population, each individual/tour has a distinct starting city. The distinct starting cites of each tour provide genetic material for exploration for the whole search space. Therefore a heterogeneous starting city of a tour in initial population is generated to have rich diversity. Proposed crossover based on greedy method of subtour connection drives the efficient local search, followed by 2-opt mutation for improvement of tour for enhanced/optimal solution. The result of the proposed algorithm is compared with other standard algorithms followed by conclusion.

Advances in Computers, Volume 135
ISSN 0065-2458
https://doi.org/10.1016/bs.adcom.2024.01.001

1. Introduction

1.1 Genetic algorithm

Genetic algorithm (GA) draws the idea from natural selection and natural genetics principles for searching and optimization algorithm. In this method, we use survival of the fittest rule of natural evolution. Invention of GA was done in 1960 by Holland [1]. It consists of population of chromosomes; with each chromosome representing a solution to the particular problem. Each chromosome is evaluated to obtain its fitness value of the chromosome against some given fitness function. A set of chromosomes are selected for genetic operation(s) (selection, crossover, and mutation) in order to get new chromosomes. Chromosomes are selected according to their fitness values to reproduce/generate the next generation by genetic operations which generate new chromosomes. To achieve this, two transformations namely crossover (generates new chromosomes by overlapping genes of two chromosomes), and mutation (creates a new chromosome by making changes of genes in a single chromosome) are used. After performance of crossover and mutation operation, we generate a new chromosome called child. The process continues by selecting fit chromosomes from parent and child population. Whole process of genetic algorithm is repeated until best individual is obtained or desired number of iterations are completed, providing an optimal/suboptimal solution to the problem [2].

We developed a genetic algorithm for traveling salesman problem (TSP) to provide balance between exploration and exploitation for the search space. For this, all the components of the genetic algorithms were carefully examined.

1.2 Literature review

TSP is a famous combinatorial optimization problem which is still NP-complete [3, 4]. There is no clear evidence for its origin. However, credit goes to Irish mathematician Hamilton and British mathematician Thomas Krikman for its mathematical formulation which is discussed in detail in "Graph Theory 1736-1936" book by Biggs et al. [5]. The general form was firstly studied by mathematician Menger [6]. Schrijver [7] pointed out the connection between the works of Whitney and Menger along with growth of TSP in his paper "On the history of combinatorial optimization (till 1960)". Flood [8] while searching for the solution of school bus routing problem used TSP mathematically for the first time. Whitney [9] of Princeton University introduced the name traveling salesman problem. The popularity of TSP increased considerably during 1950s and 1960s when prizes were offered by RAND Corporation for solving steps of the problem. As a result of it, prime contributions were made by Dantzig et al. [10] from the RAND Corporation who used branch and bound algorithm [11] for solving integer linear program for which they developed the cutting plane method [12] for its solution. Later 532- and 2392-city TSP was solved by Padberg and Rinaldi [13] in 1987, and 1000-city TSP by solved Grötschel and Holland [14]. In 1991, Reinelt [15] introduced TSPLIB, which is still providing problem instances for TSP. In last few decades, many heuristic algorithms have been developed with some of the following notable algorithms: ant colony optimization [16–25], neural network [26–29], self-organizing maps [30], particle swarm optimization techniques [31, 32], simulated annealing [33], weed optimization [34, 35], and genetic algorithm [36–39].

The rest of paper is divided into following sections: details of genetic algorithm is given in Section 1. Section 2 describes our proposed hybrid methods. In Section 3, experimental result is presented followed by conclusion in Section 4.

2. Proposed hybrid method

The methods used here are hybrid because we have used a proposed group theory tour construction algorithm and proposed crossover with 2-opt mutation Croes [40]. The framework of the proposed algorithm is shown in Algorithm 1.

The main idea of the first step is to generate a population of chromosomes (tours) by using proposed group theory approach. Clearly, each chromosome of the population is same but which start city unique providing rich diversity of genetic materials for *exploration*.

Fitness value of each chromosome in the population is calculated in the second step. In the third step, select two parent chromosomes (selected randomly) from the population, and replace the first chromosome with minimum fitness value (tour cost). After that, apply proposed crossover operator on the selected two chromosomes with crossover probability (*pc*). And finally, apply 2-opt mutation operator on selected parent chromosome or

ALGORITHM 1 Proposed algorithm

1: Generate initial population of the tour with population size P using group theory.

2: Gen $= 1$;

3: **while** ($Gen \leq N_{Gen}$) **do**

4:     Calculate the fitness of each tour in P.

5:     $B_s =$ Best tour in P;

6:     Randomly select two parents S_1 and S_2 tour in P;

7:     $S_{1new} = B_s$;

8:     $S_{2new} = S_2$;

9:     Rnd1 $=$ rand(0, 1];

10:     **if** ($Rnd1 <$ crossover probability (pc)) **then**

11:         Perform proposed crossover on selected two parents B_s and S_2 to generate two new children C_1 and C_2;.

12:     **end if**

13:     $S_{1new} = C_1$;

14:     $S_{2new} = C_2$;

15:     Rnd2 $=$ rand(0, 1];

16:     **if** ($Rnd2 <$ mutation probability (pm)) **then**

17:         Perform 2-opt optimal mutation operator on S_{1new} and S_{2new}

18:         update new population P';

19:         $P \leftarrow P'$;

20:     **end if**

21:     Gen $=$ Gen $+ 1$;

22: **end while**

new pair of chromosomes generated after crossover, with mutation probability (pm). Mutation operator helps in generate new population which is then replacing new population with the previous population by the new population. Whole process is repeated until termination condition is satisfied.

2.1 Proposed group theory for population generation

There are various possible methods for generating the initial population [41–43]. One of the simplest ways is generating the initial population randomly using random number generator. Zhang [42, 43] proposed greedy tour construction heuristic with Karp–patching for feasible tour

$$
\begin{array}{cccccccccc}
3 & 4 & 5 & 6 & 7 & 8 & 9 & 10 & 1 & 2 \\
4 & 5 & 6 & 7 & 8 & 9 & 10 & 1 & 2 & 3 \\
5 & 6 & 7 & 8 & 9 & 10 & 1 & 2 & 3 & 4 \\
6 & 7 & 8 & 9 & 10 & 1 & 2 & 3 & 4 & 5 \\
7 & 8 & 9 & 10 & 1 & 2 & 3 & 4 & 5 & 6 \\
8 & 9 & 10 & 1 & 2 & 3 & 4 & 5 & 6 & 7 \\
9 & 10 & 1 & 2 & 3 & 4 & 5 & 6 & 7 & 8 \\
10 & 1 & 2 & 3 & 4 & 5 & 6 & 7 & 8 & 9 \\
1 & 2 & 3 & 4 & 5 & 6 & 7 & 8 & 9 & 10 \\
2 & 3 & 4 & 5 & 6 & 7 & 8 & 9 & 10 & 1 \\
\end{array}
$$

Fig. 1 Population generated using group modulo.

construction and used for solving assignment problem [44]. We proposed the group tour construction heuristic for initial population generation. In this method, nodes of graph are label using group of integers, Z_n with integer modulo n operation.

$$a +_n b = (a + b) mod\ n. \tag{1}$$

where "$+_n$" is operator that represents addition modulo of n, and $(a + b)$ represents the normal addition of integers. This helped in generating the group table shown in Fig. 1. In group table no two row or column elements in the same position are identical. The function used for generating initial population of chromosomes (P) is as follows:

$$P(a) = mod((a + b), n) + 1. \tag{2}$$

where a represents population size whose value is from $a = 1$ to population size, and $b = 1$ to n (as mentioned in earlier, we are taking population size equal to number of cities, therefore population size $= n$). For $n = 10$, the initial population generated using group theory is as follows:

Each chromosome in the initial population being unique (group theory technique) provides a wide diversity of genetic materials for exploring of search space.

2.2 Proposed crossover operator for GA

Sharing information between a pair of chromosomes is called crossover [2]. In this process genes of parent's chromosomes are swapped to generate offspring. The selection of the parent chromosomes is with the possibility that good chromosomes may generate better offspring. Goldberg described

ALGORITHM 2 Algorithm for proposed crossover

```
1: c₁ = zeros(1, n);
2: c₂ = zeros(1, n);
3: for i = 1 : n do
4:     for j = 1 : n do
5:         if (s₂(i) == s₁(j)) then;
6:             c₁(i) = s₂(j);
7:         end if
8:         if (s₁(i) == s₂(j)) then
9:             c₂(i) = s1(j);
10:        end if
11:    end for
12: end for
```

Parent s_1: | 2 | 3 | 4 | 5 | 9 | 6 | 7 | 8 | 1 |
Parent s_2: | 7 | 2 | 4 | 3 | 5 | 6 | 9 | 1 | 8 |

Child c_1: | 9 | 7 | 4 | 2 | 3 | 6 | 5 | 8 | 1 |
Child c_2: | 3 | 5 | 4 | 9 | 7 | 6 | 2 | 1 | 8 |

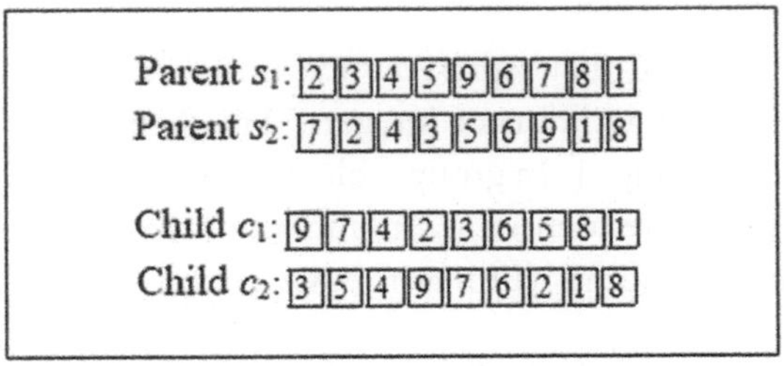

Fig. 2 Proposed crossover.

several order based operators, such as the partially matched crossover (PMX) [45]. The order crossover (OX) was suggested by Syswerda [36]. The position-based crossover (PBX) was introduced by Syswerda [36]. The cycle crossover (CX) was suggested by Oliver et al. [46]. Freisleben and Merz introduced a distance preserving crossover (DPX) [37]. Inspired by DPX crossover, we propose a new crossover in this paper. In the proposed crossover the cities that are identical for the same position in both parents (s_1 and s_2) will not change in child c_1 and c_2. The remaining cities will change accordingly Algorithm 2.

For example, the position 3 and 6 of parent s_1, s_2 are same as in the child c_1 and c_2, respectively, in Fig. 2, remaining positions 1, 2, 4, 5, 7, 8, and 9 cities are swapped as procedure given in Fig. 3.

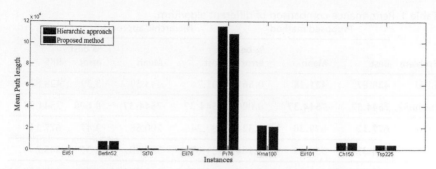

Fig. 3 Average distance (mean) (over 20 trails) for the 9 TSPLIB instances.

3. Experimental results

3.1 Experimental setup

For evaluation purpose, results were generated on 2.20 GHz Intel Core i5 machine with 4 GB RAM.

3.2 Experimental design

The following standard benchmark data set were taken from TSPLIB: pr299, kroA100, berlin52, ch150, pr144, kroB150, pr152, rat195, d198, ts225, kroA200, pr226, lin318, and pcb442 were taken for performance comparison.

For the experiment, the experimental parameters crossover probability ($pc = 0.8$) and mutation probability ($pm = 0.2$) values were set to and, respectively, with number of iteration $= 500$. Population size of each instances equal to number of cities were generated by group tour construction method. For result comparison, Percentage Best Error(% Best Err.) is used. The Percentage Best Error is calculated as follows:

$$PDbest = [(Best\ path\ cost\ from\ n\ trail) - (Best\ Known\ Solution(BKS))]/$$

$$(Best\ Known\ Solution(BKS)) * 100.$$

3.3 Experimental result

For each standard TSP data taken, we performed $n = 20$ trails for first set of comparison shown in Table 1 and Fig. 3 and $n = 30$ trail for second set of

Table 1 Performance comparison of different algorithm.

Problem	Proposed method			Hierarchic approach			BKS
	Best	Mean	% best error	Best	Mean	% best error	
Eil51	**428.87**	**431.28**	**0.561**	431.74	443.39	3.39	428.87
Berlin52	**7544.37**	**7544.37**	**0.000**	**7544.37**	**7544.37**	**0.000**	7544.37
St70	**677.12**	**679.30**	**0.324**	687.24	700.58	3.47	677.11
Eil76	**544.37**	**553.05**	**1.184**	551.07	557.98	2.31	545.39
Pr76	**108,160.00**	**108,232.00**	**0.067**	113,798.56	115,072.29	6.39	108,159.44
Kroa100	**21,285.00**	**21,359.05**	**0.346**	22,122.75	22,435.31	5.40	21,285.44
Eil101	**645.25**	**656.62**	**4.391**	672.71	683.39	6.39	642.31
Ch150	**6588.60**	**6665.29**	**2.103**	6641.69	6677.12	2.21	6532.28
Tsp225	**3878.80**	**3909.04**	**1.297**	4090.54	4157.85	7.74	3859.00

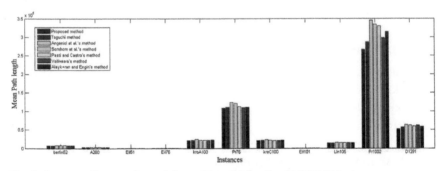

Fig. 4 Average distance (mean) (over 30 trails) for the 11 TSPLIB instances.

comparison shown in Table 2 and Fig. 4. Best results are shown in bold. The reason for two different comparisons is to avoid any biasness which might be thought for comparison. Results are taken from [47] and [48]. It can be clearly seen that proposed method is better than all the methods used for comparison for every dataset taken in to consideration. The reason for different trail (20 and 30) is because comparison was made in and for different trails. Proposed method although not exact but do provide good heuristic solution.

The pictorial presentation for performance comparison in terms of Mean (average) solution for different methods is shown in Figs. 3 and 4.

Table 2 Performance comparison of different algorithm.

TSP instance taken (From TSPLIB)	Proposed method			Taguchi method			Angeniol et al.'s method			Somhom et al.'s method		
	Mean	Best	%best err	Mean	Best	%best err	Mean	Best	%best err	Mean	Best	%best err
berlin52	7542.0	7542.0	0.000	7635.02	7542.0	0.00	8368.60	7778.3	3.13	7984.20	7542.0	0.00
A280	2592.23	2579.0	0.000	2650.01	2583.1	0.16	2672.22	2588.3	0.36	2640.41	2584.8	0.22
Eil51	427.03	426.0	0.000	435.19	426.0	0.00	443.41	432.1	1.43	438.62	433.0	1.62
Eil76	540.53	538.0	0.000	565.28	544	1.12	565.14	554.4	3.05	568.42	545.0	1.28
kroA100	21,287.7	21,282.0	0.000	21,567.34	21,382.3	0.47	24,670.65	23,009.5	8.12	21,835.53	21,641.1	1.66
Pr76	108,159.0	108,159.0	0.000	110,420.00	108,785.4	0.58	123,310.76	116,378.0	7.60	120,830.52	112,372.0	3.75
kroC100	20,773.57	20,749.0	0.000	21,850.18	20,801.2	0.25	23,485.16	22,340.8	7.67	21,408.01	20,924.2	0.84
Eil101	634.37	629.0	0.000	655.27	640.5	1.82	665.87	655.6	4.23	654.22	637.0	1.26
Lin105	14,381.2	14,379.0	0.000	14,475.05	14,380.9	0.01	16,111.44	14,995.5	4.29	14,771.95	14,466.5	0.60
Pr1002	266,645.23	264,216.0	1.996	287,500.11	280,001.1	8.09	345,195.01	332,741.2	28.45	334,591.46	328,451.1	21.13
D1291	51,763.83	51,337.0	1.055	56,904.16	54,915.2	8.10	63,412.43	62,693.2	23.41	61,712.14	59,112.3	14.06

Continued

Table 2 Performance comparison of different algorithm.—cont'd

| | Methods | | | | | | | | | |
TSP instance taken (From TSPLIB)	Pasti and Castro's method			Vallivaara's method			Alaykıran and Engin's method			BKS
	Mean	Best	%best err	Mean	Best	%best err	Mean	Best	%best err	
berlin52	8035.68	7543.9	0.03	7543.33	7543.1	0.01	7650.30	7590.2	0.64	7542
A280	2604.58	2585.9	0.27	2626.44	2591.3	0.48	2800.25	2595.1	0.62	2579
Eil51	438.76	429.0	0.70	434.82	428.2	0.52	458.15	433.3	1.71	426
Eil76	556.16	545.0	1.30	563.12	547.8	1.82	608.02	552.3	2.66	538
kroA100	21,868.41	21,369.0	0.41	21,789.41	21,585.8	1.43	22,560.42	21,670.9	1.83	21,282
Pr76	112,566.42	109,233.2	0.99	110,025.90	108,959.0	0.74	111,050.01	109,382.1	1.13	108,159
kroC100	21,231.63	20,915.6	0.80	21,368.00	20,904.5	0.75	22,109.45	21,035.2	1.38	20,749
Eil101	654.82	641.0	1.91	643.20	640.2	1.78	680.95	653.0	3.82	629
Lin105	14,702.22	14,382.1	0.02	14,745.32	14,389.5	0.07	15,030.39	14,750.2	2.58	14,379
Pr1002	329,857.01	315,678.2	21.86	298,191.82	288,658.3	11.43	314,560.00	306,755.4	18.42	259,045
D1291	59,345.11	58,678.3	15.51	61,712.10	59,112.3	16.36	58,952.44	57,695.2	13.57	50,801

4. Conclusion

This paper proposed a genetic algorithm which works by taking features of group tour construction, proposed crossover and 2-opt mutation. In order to have a heterogeneous population for process initialization with population size equal to number of cities, we applied group tour construction method. After this, the proposed crossover and 2-opt mutation are applied. In order to maintain local optimality, crossover and mutation operators are used. By using crossover operator, new starting points were defined for a local search using information of the current population. The proposed crossover utilizes a greedy method for the duplicated paths in the parents for connecting subtours into the solution. However, 2-opt mutation can easily get stuck in a local optimum to improve the tour quality. The combination of the proposed method is required as 2-opt mutation easily gets stuck in local optimum. From the experimental results, one can easily find that our proposed algorithm gives better performance in comparison of Refs. [16, 30, 47–51].

References

[1] J.H. Holland, Adaptation in Natural and Artificial Systems, University of Michigan Press, Ann Arbor, 1975.
[2] S.N. Sivanandam, S.N. Deepa, Genetic algorithms, in: Introduction to Genetic Algorithms, Springer, 2008, pp. 15–37.
[3] C.H. Papadimitriou, K. Steiglitz, Combinatorial Optimization: Algorithms and Complexity, Prentice-Hall, Inc., New Jersey, 1982.
[4] G.L. Nemhauser, M.W.P. Savelsbergh, G.C. Sigismondi, Constraint Classification for Mixed Integer Programming Formulations, Department of Mathematics and Computing Science, University of Technology, 1991.
[5] N.L. Biggs, E.K. Lloyd, R.J. Wilson, Coloring Maps on Surfaces, Graph Theory, 1736–1936, Clarendon Press, 1976.
[6] K. Menger, Eine neue definition der bogenlänge, in: Ergebnisse eines Mathematischen Kolloquiums, vol. 2, 1932, pp. 11–12.
[7] A. Schrijver, On the history of combinatorial optimization (till 1960), Handbooks Oper. Res. Manag. Sci. 12 (2005) 1–68.
[8] M.M. Flood, The traveling-salesman problem, Oper. Res. 4 (1) (1956) 61–75.
[9] H. Whitney, The mathematics of physical quantities: Part I: Mathematical models for measurement, Am. Math. Mon. 75 (2) (1968) 115–138.
[10] G. Dantzig, R. Fulkerson, S. Johnson, Solution of a large-scale traveling-salesman problem, J. Oper. Res. Soc. Am. 2 (4) (1954) 393–410.
[11] G. Laporte, The traveling salesman problem: an overview of exact and approximate algorithms, Eur. J. Oper. Res. 59 (2) (1992) 231–247.
[12] G. Reinelt, The Traveling Salesman: Computational Solutions for TSP Applications, Springer-Verlag, 1994.

[13] M. Padberg, G. Rinaldi, Optimization of a 532-city symmetric traveling salesman problem by branch and cut, Oper. Res. Lett. 6 (1) (1987) 1–7.

[14] M. Grötschel, O. Holland, Solution of large-scale symmetric travelling salesman problems, Math. Program. 51 (1-3) (1991) 141–202.

[15] G. Reinelt, TSPLIB—a traveling salesman problem library, ORSA J. Comput. 3 (4) (1991) 376–384.

[16] R. Pasti, L.N. de Castro, A neuro-immune network for solving the traveling salesman problem, in: The 2006 IEEE International Joint Conference on Neural Network Proceedings, IEEE, 2006, pp. 3760–3766.

[17] S.-M. Chen, C.-Y. Chien, Solving the traveling salesman problem based on the genetic simulated annealing ant colony system with particle swarm optimization techniques, Expert Syst. Appl. 38 (12) (2011) 14439–14450.

[18] W. Deng, R. Chen, B. He, Y. Liu, L. Yin, J. Guo, A novel two-stage hybrid swarm intelligence optimization algorithm and application, Soft Comput. 16 (10) (2012) 1707–1722.

[19] K. Jun-man, Z. Yi, Application of an improved ant colony optimization on generalized traveling salesman problem, Energy Procedia 17 (2012) 319–325.

[20] J.M. Cecilia, J.M. García, A. Nisbet, M. Amos, M. Ujaldón, Enhancing data parallelism for ant colony optimization on GPUs, J. Parallel Distrib. Comput. 73 (1) (2013) 42–51.

[21] M. Mavrovouniotis, S. Yang, Ant colony optimization with immigrants schemes for the dynamic travelling salesman problem with traffic factors, Appl. Soft Comput. 13 (10) (2013) 4023–4037.

[22] M. Tuba, R. Jovanovic, Improved ACO algorithm with pheromone correction strategy for the traveling salesman problem, Int. J. Comput. Commun. Control 8 (3) (2013) 477–485.

[23] H.-Y. Yun, S.-J. Jeong, K.-S. Kim, Advanced harmony search with ant colony optimization for solving the traveling salesman problem, J. Appl. Math. 2013 (2013).

[24] T. Saenphon, S. Phimoltares, C. Lursinsap, Combining new fast opposite gradient search with ant colony optimization for solving travelling salesman problem, Eng. Appl. Artif. Intell. 35 (2014) 324–334.

[25] M. Mahi, Ö.K. Baykan, H. Kodaz, A new hybrid method based on particle swarm optimization, ant colony optimization and 3-opt algorithms for traveling salesman problem, Appl. Soft Comput. 30 (2015) 484–490.

[26] C.H. Papadimitriou, The adjacency relation on the traveling salesman polytope is NP-complete, Math. Program. 14 (1) (1978) 312–324.

[27] M.K.M. Ali, F. Kamoun, Neural networks for shortest path computation and routing in computer networks, IEEE Trans. Neural Netwo. 4 (6) (1993) 941–954.

[28] J.-C. Créput, A. Koukam, A memetic neural network for the Euclidean traveling salesman problem, Neurocomputing 72 (4-6) (2009) 1250–1264.

[29] T.A.S. Masutti, L.N. de Castro, A self-organizing neural network using ideas from the immune system to solve the traveling salesman problem, Inf. Sci. 179 (10) (2009) 1454–1468.

[30] S. Somhom, A. Modares, T. Enkawa, A self-organising model for the travelling salesman problem, J. Oper. Res. Soc. 48 (9) (1997) 919–928.

[31] X.H. Shi, Y.C. Liang, H.P. Lee, C. Lu, Q.X. Wang, Particle swarm optimization-based algorithms for TSP and generalized TSP, Inf. Process. Lett. 103 (5) (2007) 169–176.

[32] Y. Marinakis, M. Marinaki, A hybrid multi-swarm particle swarm optimization algorithm for the probabilistic traveling salesman problem, Comput. Oper. Res. 37 (3) (2010) 432–442.

[33] Y. Chen, P. Zhang, Optimized annealing of traveling salesman problem from the nth-nearest-neighbor distribution, Physica A: Stat. Mech. Appl. 371 (2) (2006) 627–632.

[34] G.G. Roy, S. Das, P. Chakraborty, P.N. Suganthan, Design of non-uniform circular antenna arrays using a modified invasive weed optimization algorithm, IEEE Trans. Antennas Propag. 59 (1) (2010) 110–118.

[35] A. Sengupta, T. Chakraborti, A. Konar, A. Nagar, Energy efficient trajectory planning by a robot arm using invasive weed optimization technique, in: 2011 Third World Congress on Nature and Biologically Inspired Computing, IEEE, 2011, pp. 311–316.

[36] G. Syswerda, Scheduling optimization using genetic algorithms, in: Handbook of Genetic Algorithms, 1991.

[37] B. Freisleben, P. Merz, A genetic local search algorithm for solving symmetric and asymmetric traveling salesman problems, in: Proceedings of IEEE International Conference on Evolutionary Computation, IEEE, 1996, pp. 616–621.

[38] B. Freisleben, P. Merz, New genetic local search operators for the traveling salesman problem, in: International Conference on Parallel Problem Solving from Nature, Springer, 1996, pp. 890–899.

[39] M. Albayrak, N. Allahverdi, Development a new mutation operator to solve the traveling salesman problem by aid of genetic algorithms, Expert Syst. Appl. 38 (3) (2011) 1313–1320.

[40] G.A. Croes, A method for solving traveling-salesman problems, Oper. Res. 6 (6) (1958) 791–812.

[41] G. Gutin, A.P. Punnen, The Traveling Salesman Problem and Its Variations, 12, Springer Science & Business Media, 2002. vol.

[42] W. Zhang, Truncated branch-and-bound: a case study on the asymmetric TSP, in: Proc. of AAAI 1993 Spring Symposium on AI and NP-Hard Problems, vol. 160166, 1993.

[43] W. Zhang, Depth-first branch-and-bound versus local search: a case study, in: AAAI/IAAI, 2000, pp. 930–935.

[44] R.M. Karp, A patching algorithm for the nonsymmetric traveling-salesman problem, SIAM J. Comput. 8 (4) (1979) 561–573.

[45] D.E. Goldberg, Genetic Algorithms in Search, Optimization, and Machine Learning, Addison Wesley Publishing Co. Inc., 1989.

[46] I.M. Oliver, D.J.d. Smith, J.R.C. Holland, Study of permutation crossover operators on the traveling salesman problem, in: Genetic Algorithms and their Applications: Proceedings of the Second International Conference on Genetic Algorithms: July 28–31, 1987 at the Massachusetts Institute of Technology, Cambridge, MA, L. Erlhaum Associates, Hillsdale, NJ, 1987.

[47] M. Gündüz, M.S. Kiran, E. Özceylan, A hierarchic approach based on swarm intelligence to solve the traveling salesman problem, Turk. J. Electr. Eng. Comput. Sci. 23 (1) (2015) 103–117.

[48] M. Peker, B. ŞEN, P.Y. Kumru, An efficient solving of the traveling salesman problem: the ant colony system having parameters optimized by the Taguchi method, Turk. J. Electr. Eng. Comput. Sci. 21 (Sup. 1) (2013) 2015–2036.

[49] B. Angeniol, G.D.L.C. Vaubois, J.-Y. Le Texier, Self-organizing feature maps and the travelling salesman problem, Neural Netw. 1 (4) (1988) 289–293.

[50] I. Vallivaara, A team ant colony optimization algorithm for the multiple travelling salesmen problem with minmax objective, in: Proceedings of the 27th IASTED International Conference on Modelling, Identification and Control, ACTA Press, 2008, pp. 387–392.

[51] K. Alaykıran, O. Engin, Ant colony meta heuristic and an application on TSP, J. Fac. Eng. Archit. Gazi Univ. 1 (2005) 69–76.

About the authors

Rahul Chandra Kushwaha is currently working as assistant professor with the Department of Computer Science and Engineering, Rajiv Gandhi University (Central University), Doimukh (Itanagar), Arunachal Pradesh, India. Additionally, he is the Coordinator of the Technology Enabling Centre at Rajiv Gandhi University, funded by the Department of Science & Technology, Government of India. Earlier he has worked as assistant professor at Banaras Hindu University, India and as senior associate (SA/SA-SD) at National Council of Educational Research and Training (NCERT), New Delhi, India. He has received PhD degree in Computer Science from Banaras Hindu University, Varanasi, India. His research interests include machine learning, deep learning, soft computing, learning science and technologies.

Dharm Raj Singh is currently working as assistant professor with the Department of Computer Application at Jagatpur Post Graduate College, Varanasi affiliated to Mahatma Gandhi Kashi Vidyapith, Varanasi, India. He has received PhD degree in Computer Science from Banaras Hindu University, Varanasi, India. His research interests include machine learning, soft computing, optimization and graph theory.

Anubhav Kumar Prasad is currently working as assistant professor with the Department of Computer Science and Engineering at United Institute of Technology, Naini, Prayagraj, India. He has received PhD degree in Computer Science from Banaras Hindu University, Varanasi, India. His research interests include machine learning, soft computing and optimization techniques.

PART V

Deep learning and neural networking systems

> CHAPTER SEVENTEEN

Adaptation of nature inspired optimization algorithms for deep learning

Yeshwant Singh[a] and Anupam Biswas[b]
[a]School of Computer Science, UPES, Dehradun, Uttarakhand, India
[b]Department of Computer Science and Engineering, National Institute of Technology Silchar, Silchar, Assam, India

Contents

Abstract

Deep Learning (DL) models have found widespread application across diverse domains. With the prevalent use of DL models, there is a growing need to develop optimal architectures and enhance their performance. In recent years, Nature-Inspired Optimization Techniques (NIOTs) have gained attention as effective tools for optimizing DL models. These techniques draw inspiration from natural phenomena and have proven their prowess in solving intricate optimization problems. This chapter delves into the exploration of various NIOTs and their applicability to DL models. We investigate their roles in the realms of DL architecture design and weight adaptation. The chapter offers an in-depth analysis of the approaches used for representing solutions, designing objective functions, and handling constraints specific to DL models. Additionally, it sheds light on the primary challenges and open questions that arise when employing NIOTs in the context of DL models.

Advances in Computers, Volume 135
ISSN 0065-2458
https://doi.org/10.1016/bs.adcom.2023.12.005

1. Introduction

In recent years, the utilization of nature-inspired optimization techniques (NIOTs) to tackle highly intricate and non-linear optimization problems has witnessed a remarkable surge. These algorithms have exhibited superior performance compared to traditional calculus-based optimization methods when it comes to seeking near-optimal solutions for complex optimization challenges. Unlike conventional methods that revolve around identifying critical points, which yield minimum or maximum values based on the desired objective function, NIOTs offer a paradigm shift. Traditional approaches rely on the computation of higher-order derivatives of the objective function, followed by equating the results to zero. However, this process demands significant computational power and is susceptible to errors, rendering these calculus-based optimization methods less favorable compared to their nature-inspired counterparts.

NIOTs employ stochastic and probabilistic techniques to ascertain optimal solutions for intricate non-linear optimization problems, offering efficiency and reduced computational complexity. Furthermore, NIOTs have proven particularly effective in architectural design and weight adaptation for diverse Deep Learning (DL) models. While traditional stochastic-based methods are commonly employed for determining DL model weights, NIOTs excel in automating architectural design processes, minimizing manual intervention, and enhancing the precision of solutions.

Researchers have harnessed various NIOTs, including the Whale Optimization Algorithm, Artificial Immune System Algorithm, and Paddy Field Algorithm, in conjunction with a variety of DL models such as Deep Belief networks and Restricted Boltzmann machines. These hybrid NIOT-DL models have been applied across a spectrum of fields, from medical image analysis to speech recognition and cybersecurity. With the proliferation of DL models, the study of the approaches and concepts applied to architectural design has become increasingly pivotal.

NIOTs, rooted in natural phenomena, have proven to be robust tools for addressing complex optimization problems. This chapter is dedicated to exploring the realm of NIOTs and their applicability within the context of DL models. It delves into the methodologies associated with solution representation, objective function design, and constraint handling when applied to DL models. Additionally, the chapter addresses the primary challenges and outstanding issues that arise when adopting NIOTs for DL models.

The chapter is structured as follows: In Section 2, a comprehensive introduction to NIOTs and their various types is presented, along with insights into the applications of commonly used algorithms in diverse DL architectures. Section 3 delves into the intricate concepts underpinning these algorithms as they relate to deep learning, including solution representation, objective function design, and constraint handling. In Section 4, we tackle the primary challenges and open questions that emerge when integrating NIOTs into deep learning models. Finally, Section 5 provides a succinct conclusion derived from the insights presented throughout the chapter.

2. Nature-inspired optimization techniques

Nature-inspired optimization techniques (NIOTs) are some efficient optimization approaches that are derived from various natural processes. These algorithms are broadly classified into three types: bio-inspired, physics-inspired, and chemistry-inspired optimization algorithms. This section briefly explains the concepts involved with these three algorithms and their applications in deep learning weight and architecture optimization.

2.1 Bio-inspired optimization algorithms

Bio-inspired optimization algorithms, also known as nature-inspired optimization techniques, are a class of computational methods that draw inspiration from natural processes, ecosystems, and the behavior of living organisms to solve complex optimization problems. These algorithms have gained immense popularity and have been applied across various domains due to their ability to find near-optimal solutions efficiently. They emulate the adaptability, efficiency, and resilience observed in the natural world, making them valuable tools for tackling real-world challenges.

One prominent category of bio-inspired optimization algorithms is genetic algorithms (GAs). GAs are inspired by the process of natural selection and genetics. They operate by evolving a population of potential solutions over multiple generations, mimicking the way genetic traits are passed down and refined over time. By incorporating selection, crossover, and mutation operators, GAs explore a wide solution space and gradually converge toward the best possible solution. These algorithms have been employed in a multitude of applications, from machine learning and parameter tuning to financial modeling and engineering design. Another noteworthy class of bio-inspired algorithms is the particle swarm optimization (PSO).

Inspired by the flocking behavior of birds and fish, PSO models individual solutions as particles that move through a multi-dimensional search space. Each particle adjusts its position and velocity based on its own experience and the experiences of its neighbors, ultimately converging toward the optimal solution. PSO has found applications in optimizing neural network architectures, image processing, and even in solving complex, multi-objective problems.

Ant colony optimization (ACO) is yet another bio-inspired algorithm that takes inspiration from the foraging behavior of ants. In ACO, a population of artificial ants cooperates to discover the optimal path or solution in a search space. As ants leave pheromone trails to communicate with each other, ACO algorithms employ pheromone-based mechanisms to guide the search process. ACO has demonstrated its prowess in solving complex combinatorial problems such as the traveling salesman problem and network routing.

Bio-inspired optimization algorithms continue to evolve and adapt to a wide range of applications, offering powerful tools for addressing complex optimization challenges in fields as diverse as engineering, finance, and artificial intelligence. Their ability to harness the wisdom of nature's mechanisms makes them a valuable asset for researchers and practitioners seeking innovative solutions to complex problems.

2.1.1 Application of bio inspired algorithms in deep learning

The application of bio-inspired optimization algorithms in the realm of deep learning has been a transformative development in recent years. These algorithms draw inspiration from the complex and efficient problem-solving mechanisms observed in biological and natural systems, such as genetic evolution, animal behavior, and ecological processes. When applied to deep learning, they have proven to be invaluable tools for optimizing neural networks, enhancing their performance, and automating various aspects of the deep learning pipeline.

One of the key areas where bio-inspired optimization algorithms have made a significant impact is in training deep neural networks. Traditional gradient-based optimization methods, like stochastic gradient descent, while effective, often get stuck in local optima and require careful hyperparameter tuning. Bio-inspired algorithms, on the other hand, bring a level of adaptability and robustness to the optimization process. Genetic algorithms, for instance, can explore a wide range of network architectures and hyperparameters, evolving and selecting the most promising combinations to improve model performance. This capability enables the automatic design of neural network architectures tailored to specific tasks, reducing the need for manual architecture engineering.

Another fascinating application lies in the domain of hyperparameter tuning. Deep learning models are notoriously sensitive to hyperparameter settings, and finding the right configuration can be a time-consuming and challenging task. Bio-inspired optimization algorithms, such as particle swarm optimization and ant colony optimization, can efficiently explore the hyperparameter space, seeking the optimal combination that leads to improved model performance. This not only reduces the burden on data scientists and machine learning engineers but also often results in better-performing models.

Furthermore, bio-inspired optimization algorithms have found their place in transfer learning and feature selection for deep learning. Techniques like the Particle Swarm Optimization (PSO) can identify the most relevant features from vast datasets, reducing computational overhead and potentially improving the generalization of deep learning models. In transfer learning, where models are fine-tuned for specific tasks, these algorithms can assist in selecting the most suitable layers and parameters to adapt pre-trained models effectively.

In conclusion, the application of bio-inspired optimization algorithms in deep learning is a promising frontier that offers a range of benefits, from automating the architectural design of neural networks to optimizing hyperparameters and improving model generalization. These algorithms harness the power of nature-inspired processes to enhance the efficiency and effectiveness of deep learning, paving the way for advancements in a variety of fields, from computer vision to natural language processing and beyond. As research in this area continues to evolve, we can expect bio-inspired optimization algorithms to play an increasingly vital role in the advancement of deep learning technology.

2.1.2 Classification of bio-inspired optimization algorithms

Bio-inspired optimization algorithms are a diverse and growing family of computational techniques that draw inspiration from biological, ecological, and natural systems to solve optimization problems. These algorithms can be classified into several categories based on their underlying principles and inspiration sources. Below is a classification of bio-inspired optimization algorithms:

1. Evolutionary Algorithms (EAs):
   a. Genetic Algorithm (GA)
   b. Genetic Programming (GP)
   c. Evolutionary Strategies (ES)
   d. Differential Evolution (DE)
2. Swarm Intelligence Algorithms:
   a. Particle Swarm Optimization (PSO)
   b. Ant Colony Optimization (ACO)
   c. Bee Colony Optimization (BCO)
   d. Firefly Algorithm
3. Natural Selection Algorithms:
   a. Darwinian Algorithms
   b. Memetic Algorithms (MA)
   c. Estimation of Distribution Algorithms (EDA)
4. Physical and Chemical–Based Algorithms:
   a. Simulated Annealing
   b. Quantum Computing-Inspired Algorithms
   c. Harmony Search (HS)
   d. Artificial Bee Colony (ABC)
5. Membrane Computing Algorithms:
   a. P Systems
   b. Tissue-Like P Systems (TPS)
   c. Cell-like P Systems (Cellular Automata)
6. Brain-Inspired Algorithms:
   a. Neural Network–Based Optimization
   b. Spiking Neural Networks
   c. Brainstorm Optimization (BSO)
7. Plant Growth and Life Cycle Algorithms:
   a. Root Growth Algorithm
   b. Fruit Fly Optimization Algorithm (FOA)
8. Animal Behavior Algorithms:
   a. Cuckoo Search (CS)
   b. Bat Algorithm
   c. Gray Wolf Optimizer (GWO)
9. Biological and Physiological Systems Algorithms:
   a. Immune System-Based Algorithms
   b. Hormone-Inspired Algorithms
   c. Bacterial Foraging Optimization (BFO)

10. Hybrid and Multi-disciplinary Algorithms:
   a. Nature-inspired optimization combined with Machine Learning
   b. Nature-Inspired Optimization in Conjunction with Deep Learning
   c. Metaheuristic Hybridization
11. Other Novel Bio-inspired Algorithms:
   a. Social Spider Optimization (SSO)
   b. Elephant Herding Optimization (EHO)
   c. Big Bang–Big Crunch Algorithm (BB–BC)
   d. Krill Herd Algorithm (KHA)

These categories encompass a wide range of bio-inspired optimization algorithms, each with its unique approach to problem-solving. Researchers and practitioners often choose an algorithm from one of these categories based on the nature of the optimization problem and the algorithm's compatibility with the problem's characteristics. The choice of algorithm can significantly impact the efficiency and effectiveness of optimization processes in various domains, including engineering, finance, healthcare, and more.

2.1.3 Evolutionary algorithms and their applications in deep learning

Evolutionary algorithms are inspired by the concepts of Darwin's theory, "survival of the fittest". These algorithms imitate various strategies living organisms use to interact with one another and are developed based on the biological evolution of nature. Paddy Field Algorithm (PFA), Evolutionary Strategy (ES), Genetic Programming (GP), Differential Evolution (DE), and Genetic Algorithm (GA) together constitute the family of successful Evolutionary algorithms [1]. A wide variety of applications, simplicity in implementation, and parallelism are some of the advantages of Evolutionary Algorithms.

Each of the Evolutionary Algorithms involves the following steps:

- **Initialization**: At the beginning, an initial population of feasible solutions is created randomly within the constraints of the given problem. Each member of the population is referred to as an individual.
- **Calculation of Fitness Score**: An appropriate fitness function suitable for the problem is used for the evaluation of fitness scores of each individual.
- **Selection of Parents**: After calculating the fitness scores of every individual, the individuals having better fitness scores are selected as parents, thereby giving them more chance to reproduce than others.

- **Crossover (Breeding Children)**: After the selection of parents, these individuals are used to create the next generation of the algorithm. The characteristics of the selected parents are utilized to create these new individuals (children).
- **Mutation**: New random genes are inserted during the creation of the new generation for maintaining the population diversity. It is done by changing a small portion of children with individuals having different genes.
- **Termination**: The algorithm is repeated until the satisfaction of a termination criterion. This termination criterion is generally some threshold of performance. If the termination criteria are satisfied, the final solution is selected and returned.

The following flowchart in Fig. 1 best explains the steps of Evolutionary Algorithms.

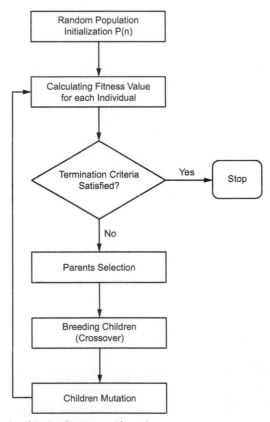

Fig. 1 Steps involved in Evolutionary Algorithms.

The various Evolutionary optimization algorithms and their applications in DL are described in the following sections:

2.1.3.1 Genetic algorithm for deep belief network

The Genetic Algorithm is a stochastic optimization algorithm based on evolutionary principles.[2] This algorithm has been widely used in optimizing parameters of deep learning architectures, like Deep Belief Networks (GA-DBN). For example, the author in [3] has used an improved GA for the optimization of deep belief network (DBN) parameters and developed an intrusion detection model needed for the security of the network layer in IoT devices. The authors in have also proposed a traffic detection model for improvement in intelligent transportation systems by using a GA-DBN model. Moreover, deep learning models for flood susceptibility mapping [4] and paper-making wastewater re-searchers in the past few years also developed [5] treatment based on GA-DBN models.

2.1.3.2 Genetic algorithm for restricted Boltzman machine

The researchers have extensively used the GA for optimizing the parameters of the Restricted Boltzmann Machine. It is used for weight determination and automatic evolution of RBM architecture, as proposed in the paper [6] (called GA-RBM). Additionally, GA can also be incorporated [7] with DeRBM (GADeRBM) and weighted nearest neighbor (GAWNN).

2.1.3.3 Evolution strategies for convolutional neural network

The technique of Evolution Strategies (ES) takes its inspiration from the species-level or macro-level process of evolution (phenotype, hereditary, variation) [1] unlike GA, which deals with micro or genomic level (chromosomes, genes). The currently used variant of Evolution Strategy, (μ, λ)-(ES), is applied by the re-searchers to optimize CNN parameters. For example, in the paper [8], the authors proposed a hybrid method that uses both backpropagation and evolutionary strategies for the training purpose of CNNs, where the evolutionary strategies are used for avoiding local minima and ne-tune the weights so that the network achieves higher accuracy results. This ES and CNN hybrid model is found to significantly improve image classification performance in CIFAR-10.

2.1.3.4 Genetic programming for deep belief network

Genetic Programming (GP) is an extension of the Genetic Algorithm. However, the solution representation in GP is different from the solution representation in GA. It is used extensively in deep learning architectures like Deep Belief Networks. Researchers have efficiently used GP to optimize the Deep Belief Networks (GP-DBN). These GP-DBN models and their variations are applied by the researchers in Parkinson's disease progression tracking [9], for time-series forecasting as well as drought forecasting [10]. A framework of Prior Formula Knowledge and GP (PFK-GP) is also proposed in the paper [11] to reduce the space of GP Searching. The PFK architecture is built on the Deep Belief Networks.

2.1.3.5 Paddy field algorithm for RBF neural network

Paddy Field Algorithm (PFA) is inspired by the reproductive principle of plant populations and was proposed in 2009 [12]. PFA does not involve crossover between individuals as in other evolutionary algorithms. It instead involves the use of pollination and dispersion steps. The PFA is used for the optimization of Radial Basis Function (RBF) Neural Network centre parameters because of its good convergence speed and strong global search capacity, as shown in the paper [13]. The Paddy Field Algorithm, when compared with PSO for the optimization of RBF Neural Network parameters, is found to perform significantly better than the latter and produce lower predictive errors.

EAs have found numerous applications in the eld of DL, offering unique advantages for optimization and model development. Here are some of the key applications of EAs in DL:

1. Neural Architecture Search (NAS):
   - EAs are used for automating the design of neural network architectures. They explore various architectural configurations, including the number of layers, types of layers, and their connections.
   - Algorithms like Genetic Algorithms and Evolution Strategies help optimize neural architectures for specific tasks, improving model performance.
2. Hyperparameter Tuning:
   - EAs are employed to optimize hyperparameters in DL models. This includes tuning parameters like learning rates, batch sizes, dropout rates, and weight initialization strategies.
   - Evolutionary methods can efficiently search the hyperparameter space to find configurations that result in better training and generalization.

3. Feature Selection:
   - In scenarios where feature engineering is essential, EAs can assist in feature selection by evaluating and evolving subsets of features that contribute most to model performance.
   - This is especially valuable when dealing with high-dimensional data.
4. Optimizing Convolutional Neural Networks (CNNs):
   - EAs are used to optimize the architecture and hyperparameters of CNNs, which are widely used in computer vision tasks.
   - They help in finding the optimal filter sizes, number of filters, and layers for specific image classification tasks.
5. Reinforcement Learning:
   - EAs can be applied to optimize policies in reinforcement learning problems. This includes finding optimal sets of actions in complex environments.
   - Evolutionary methods have been used to evolve neural network policies for game playing and robotics.
6. Transfer Learning:
   - EAs can adapt pre-trained models to new tasks or domains. They ne-tune existing models by evolving specific layers or connections.
   - This approach reduces the need for training models from scratch in many cases.
7. Generative Adversarial Networks (GANs):
   - EAs have been used to optimize the architecture of GANs for generating high-quality synthetic data. They help generate more realistic images, audio, or other data types.
   - EAs can be used to evolve the generator and discriminator networks in GANs.
8. Optimization of Recurrent Neural Networks (RNNs):
   - EAs are applied to optimize the architecture of RNNs, which are used in sequential data processing tasks.
   - They help find the most suitable architectures for tasks like natural language processing, speech recognition, and time series analysis.
9. Multi-objective Optimization:
   - EAs are useful for optimizing DL models with multiple objectives, such as minimizing error while maximizing interpretability.
   - Multi-objective evolutionary algorithms help in finding trade-off solutions.

10. Custom DL Model Design:
* EAs can be used to design custom DL models tailored to specific tasks and datasets, thereby improving efficiency and performance.

In summary, Evolutionary Algorithms have a wide range of applications in Deep Learning, spanning neural architecture search, hyperparameter tuning, feature selection, and more. Their ability to efficiently explore complex solution spaces makes them valuable tools for optimizing and designing DL models. Researchers and practitioners continue to explore novel ways to harness EAs to enhance the capabilities of deep learning systems in various domains.

2.1.4 Swarm intelligence based algorithms and their applications in deep learning

Swarm Intelligence (SI) Based Algorithms takes their inspiration from the self-organizing behavior of several interacting agents [14]. The word "swarm" comes from the irregular movements of particles in the problem space [1]. Each of these algorithms uses multiple agents that imitate the collective social behavior of birds, fishes, insects, and other animals [14]. Each of these algorithms uses multi-agents inspired by the collective behavior of social insects, like ants, termites, bees, and wasps, as well as from other animal societies like flocks of birds or sh. The SI-based algorithms are based on the collective social behavior of organisms, unlike Evolutionary Algorithms, which are based on the genetic adaptation of organisms [1].

The SI-based algorithms are further classified based on the SI system they are inspired from. Therefore, the SI based algorithms can be classified into three types:
* Natural River System
* Convergent Social Phenomenon in Animal and Microbes
* Human Immune System

2.1.4.1 Natural river system
The Natural River System algorithm class includes the Intelligent Water Drops Algorithm (IWD). Hamed Shah-Hosseini proposed the IWD algorithm in 2007. It takes its inspiration from the behavior of water droplets in natural river systems and the changes that take place within the river environment [15].

2.1.4.2 Convergent social phenomenon in animal and microbes
Many SI-based algorithms fall in the class of algorithms derived from Convergent Social Phenomenon in Animal and Microbes. Bacterial

Foraging Optimization Algorithm (BFOA), Cuckoo Search Algorithm (CSA), Flower Pollination Algorithm (FPA), Ant Colony Optimization (ACO), Particle Swarm Optimization (PSO) are some of them.

2.1.4.3 Human immune system
The Artificial Immune System algorithm falls in the Human Immune System algorithms class. It was proposed by Gupta [16] in 1999 and is an effective optimization algorithm that possesses various strengths like robustness, feature extraction, reinforcement learning, and immune recognition.

The popular SI-based optimization algorithms and their applications in Deep Learning are discussed in the following sections.

2.1.4.4 Particle swarm optimization for deep feed-forward neural network
The Particle Swarm Optimization (PSO) algorithm has been widely used for optimizing the parameters of various deep learning architectures. In the paper [17], a canonical PSO algorithm is found to be used for optimizing the weights of DFNN classifiers. The main idea of this variant is like the movement of the flock of birds. In this algorithm, the particles in the search space are initialized first (done by random generation of positions and velocities of particles). The position and velocity of each particle are then updated in every iteration. More formally, the population of PSO, i.e., the group of particles (each denoting a bird), is represented by a position vector P_n and velocity vector V_n. x_i, which is the position of each particle, represents the weights of the DFNN, while the particle fitness value fi represents the DFNN loss value. In every iteration, the position of each particle is updated by adding the velocity vector as shown in Eq. (1):

$$x_{id}^{k+1} = x_{id}^k + v_{id}^{k+1} \tag{1}$$

where vid is the velocity of ith particle at $(k+1)^{th}$ iteration.

The velocity vector of a particle is also updated using the Eq. (2):

$$v_{id}^{k+1} = \left(v_{id}^k + c_1 \text{ x } r_1 \left[P_{id}^k - x_{id}^k \right] + c_2 \text{ x } r_2 \left[P_{gd}^k - x_{id}^k \right] \right) \tag{2}$$

where v_{id}^k is the velocity of ith particle at k^{th} iteration, P_{id}^k is the local best position of particle and P_{gd}^k represents the global best position of particle at kth iteration. On implementation, this variation of PSO is found to have better performance than the DFNN-MTO algorithm in DFNN parameter optimization [17].

2.1.4.5 Ant colony optimization for deep recurrent neural networks

The Ant Colony Optimization (ACO) algorithm, besides being used for optimizing the parameters of various deep learning architectures, is also used for the evolution of the structure of deep recurrent neural networks (RNNs) as proposed by the authors in [18]. This ACO-based strategy for RNN structure evolution involves a connection between RNN neurons as the potential path for ants and the generation of neural network design by having the ants choose a path biased with the amount of pheromone on each connection [18].

2.1.4.6 Cuckoo search algorithm for deep belief network

Researchers have used the Cuckoo Search Algorithm (CSA) for ne-tuning the parameters of the widely used deep learning architecture viz. Deep Belief Networks (DBN). Rodrigues et al., [19] in addition to proposing the CSA-DBN model, compared its performance with other algorithms like PSO and Harmony Search. Another hybrid model Rider-CSA-DBN was proposed by Cristin et al., for the detection of plant diseases. The Rider Optimization Algorithm and Cuckoo Search Algorithm, Rider-CSA, were employed to train the Deep Belief Network used in the Rider-CSA-DBN model.

2.1.4.7 Whale optimization algorithm in deep learning architectures

The Whale Optimization Algorithm (WOA) mimics the behavior of humpback whales. It is inspired by the bubble-net hunting strategy of humpback whales and is a recently developed SI-based optimization algorithm [20]. Since its proposal in 2016, the WOA has been applied to several Deep Learning Architectures because of its high performance. A WOA-based training algorithm was put forward in the paper [21] for the optimization of the parameters of Feed-forward Neural Network (DFNN). WOA has also been used along with Deep Neural Network (DNN) models for the optimization of y rocks induced by blasting [22].

2.1.4.8 Intelligent water drops for convolutional neural network

The IWD algorithm takes its inspiration from the behavior of water droplets in natural river systems. It considers an artificial water drop having two important properties, viz. amount of soil carried by the water droplets and current velocity of the water drop [15]. IWD is widely used to optimize CNN parameters. The authors in the paper [23] have used IWD for the segmentation of MRI images, which in turn was blocked for training by CNN.

The proposed IWD-CNN model is found to perform better than the existing FCM clustering algorithm and has an improved pixel-segmentation based classification accuracy [23].

2.1.4.9 Artificial immune system algorithm for hopfield neural network

The Artificial Immune System (AIS) algorithm is a population-based optimization algorithm inspired by the highly evolved human immune system, and clonal selection principle [16]. The authors [24] have used this algorithm for optimizing the parameters of the Hop eld Neural Network (HNN). This HNN-AIS model has been used for solving the Maximum k-satisfiability (MAX k-SAT) problem, and the results showed that this methodology comprehensively outperforms the conventional method viz. Brute force search algorithm integrated with HNN.

2.1.5 Ecology based algorithms and their applications in deep learning

Ecology based optimization algorithms are one of the recently invented bio-inspired optimization algorithms. These algorithms include those algorithms inspired by the mechanisms of natural ecosystems. The natural ecosystems and the various phenomena involved can provide researchers with a wide variety of sources and techniques for solving complex real-life computer science and engineering problems. This class of optimization algorithms takes its inspiration from both inter-species and intra-species interaction, thereby providing a unique technique for developing the optimization problem solutions [1] [25].

The Ecology based bio-inspired optimization algorithms have been further divided into three classes depending on the inter-species and intra-species inter-actions:

- Symbiosis-Inspired Optimization These algorithms are inspired by the heterogeneous interaction/cooperation between species. It may include mutualism, commensalism, and parasitism among members of two different species. PS2O is a Symbiosis-Inspired Ecology-based optimization algorithm [26].
- Weed Colony Inspired Optimization The colonization patterns inspire these algorithms in the weed population. They imitate the adaptation and randomness of colonizing weeds. Invasive Weed Colony Optimization (IWCO) is one of the Weed Colony-based bio-inspired algorithms [27].
- Biogeography Inspired Optimization These algorithms are based on the concept of biogeography. Biogeography is the study of the distribution

of species in nature over time and space, i.e., the immigration and emigration of species between habitats. Over the past few years, they have been widely used to optimize Feed-Forward neural networks successfully [25].

2.2 Physics-inspired optimization algorithms

Not all the meta-Heuristic algorithms are bio-inspired; they draw inspiration from physics and chemistry. Most algorithms that do not appear to be bio-inspired are grown by imitating Physics laws, electrical charges, the force of attraction, stream order, etc. As various standard systems exist, having to do with the current class. [14].

2.2.1 Application of physics inspired algorithms in deep learning

Physics-inspired models in the generative adversarial network are an emerging area of exploration, e.g., the work on a feasible model of GANs [28] happen an inference of the earlier mathematical physics everything on connected to the internet learning fashionable perceptron. Skilled is a diversified collaboration between deep learning and physics, e.g., the preparation period in the life of many machine learning algorithms happen done by way of a theory of prob-ability gradient assault which bears direct similarity in the study of complex generated power landscapes [29].

Ideas fabricating in physics have informed progress in DL for many years. However, the Computer Science community neglects the genealogy of many such ideas. Here algorithms focus on the present and past ideas from physics that are useful in advanced DL. Recent fostering in physics-inspired concepts in Deep Learning have also searched physics visions that may ensure the contract of opening the black box of deep learning [30].

2.2.2 Newtons gravitational based algorithms

These algorithms exist to establish the Newtonian gravity; all pieces in outer space draw attention to every other bit accompanying the force that is straight-forwardly equivalent to the product of the public and conversely corresponding to the square of the distance middle from two-point bureaucracy. Mutual attraction happens the tendency of the public to increase speed toward each other. Individuals of the four fundamental interactions are electromagnetic force, gravity, weak force, and strong nuclear force. All particle chic outer space attracts the broken piece. Gravity occurs in all places. The inescapability of gravity creates it various from all other organic points. The habit Newton's gravitational force behaves is named "something done at a distance." This way, importance acts between divided pieces without someone who negotiates and without delay. Fashionable by the Newton law of force of attraction, each atom attracts a broken part with a gravitational pull. The gravitational force middle from two points two-piece is the shortest route equivalent to the product of their crowd and with the order reversed, corresponding to the square of the distance middle from two-point bureaucracy.

$$F = G\frac{M_1.M_2}{r^2} \tag{3}$$

where F is the gravitational force magnitude, G is the gravitational constant, M_1 and M_2 are the mass of the first and second particles, and r is the distance between the two particles. Newton's second law says that when a force, F, is applied to a particle, its acceleration, a, depends only on the force and mass.

$$a = \frac{F}{m} \tag{4}$$

2.2.3 Electro magnet based algorithms

An electromagnet happens a type of bait in which an energetic current produces a magnetic eld of currents. In contrast to the constant temptation, an electromagnet bears single opposition, which happens contingent upon the direction of the energetic wind and may be transformed by changing the management of the active current. Moreover, electromagnets have two forces: the ability to draw attention and disgust. Electromagnets with identical opposition repulse each other, and those accompanying opposite oppositions attract each one. The ability to draw attention to the force among electromagnets (5{10 allotments) is more forceful than the repulsion force. Our treasure uses this idea and replaces the ratio middle of two points ability to draw attention and hatred forces with the good percentage. This area helps particles thoroughly check the question search space and find a forthcoming optimum solution.

In classical physics, a charge of a piece generally remains part loyal. Still, in this place, curious about the amount of each point that exists that is not constant. (q) changes from redundancy to redundancy. The amount of each point I decide the capacity of attraction or repulsion.

$$q^i = \frac{e^{-n\left(f(x^i)-f\left(x^{best}\right)\right)}}{\sum\limits_{k=1}^{m} f\left(x^i\right) - f\left(x^{best}\right)} \qquad (5)$$

where the mass exists, the total number of points and n happens the number in the range. This rule shows that facts with better objective principles will have or obtain higher charges this curious act empty signs to display actual or negative cost as for fear of an energetic leader. Thus, the route of force (either drawing attention or very repulsive force) exists contingent upon the objective function principles (appropriateness) of two points.

2.2.3.1 Electromagnetism-like heuristic for deep belief network

Shadow detection is proper in the application of image analysis, as it can make or become better scene understanding. Recent shadow discovery methods use near–infrared (NIR) cameras and deep learning to support the enhanced separation of the shadow eld in representation. The comparative advantage concerning this arrangement over the state-of-the-creation meant to communicate or appeal to the senses or mind happens that allure performance happens without needing some special gear for activity, to a degree NIR cameras, while it is natural [31].

2.2.4 Electrostatics-based algorithms

2.2.4.1 Charged system search algorithm for skeletal structures

CSS uses a few charged molecules (CP) which influences each other considering their division distances and wellness esteems, thinking about the administering laws of Coulomb and Gauss from electrical material science and the overseeing rules of movement from the Newtonian mechanics [32]. The CSS calculation advances some bracket and edge structures. Examination of the CSS results with other meta-heuristic calculations shows the strength of the new calculation.

2.2.4.2 Adaptive charged system search algorithm for optimal tuning

This paper is for the ideal tuning of Takagi-Sugeno relative basic fluffy regulators (T-S PI-FCs). The five phases are commitment, clarification, elaboration, investigation, and assessment, including adjusting the speed, acceleration, and distance separation parameters to the iteration index [33]. The ACSS calculation takes care of the advancement issues intends to limit the true capacities communicated as the amount of outright control mistake in addition to squared result awareness work, bringing about ideal u y control frameworks with diminished parametric responsiveness. The ACSS-based tuning of T-S PI-FCs is applied to second-arrange servo frameworks with a necessary part. The ACSS-found brings into the amicability of T-S PI-FCs exist used to second-arrange servo arrangement with an important part.

2.2.5 Quantum-mechanics-based algorithms

Following the branch of quantum physics, electrons affect or in the general area the nucleus fashionable a curve path, famous as orbits. Contingent upon the momentum and strength level, electrons happen located in different trajectories. A power fashionable lower-level orbit can jump to the taller level circuit by absorbing a certain amount of strength; similarly, more elevated level power can jump to a lower strength level by releasing a certain amount of force.

This kind of vaulting exists considered as an individual. Skilled happen no middle state fashionable middle from two points two energy levels. The position places an energized matter lies ahead of the orbit exist changeable; it may lie at some position fashionable rotation at the occasion. The instability of the electron's position is referred to as a superposition of energized matter.

Fashionable classical estimate, some exist represented either by 0 or 1, but stylish quantity computing this happen name something as a qubit. State a qubit, maybe 0 or 1, though fashionable superposition state. This superposition of qubit mimics the superposition of electrons or particles. State of qubit at some opportunity is delimited stylish agreement of probabilistic amplitudes. The position of an electron happens writing in conditions of qubits by a heading named quantum state heading. A quantity state vector may be explained in speech with the equating likely beneath.

2.2.5.1 Quantum-inspired evolutionary algorithm for deep belief network

Convolutional neural networks (CNN) exist widely used and productive deep learning methods for concept categorization. But the architecture of CNN in the way that Lett and AlexNet happen devise elaborately by experts' cause plotting the neural networks happen late and demand expert knowledge. This treasure quantity-inspired about evolution or development invention to search the neural architectures. They encrypt CNNs into quantity chromosomes and identify these chromosomes from the Convolutional Layer, Combine Tier, fully affiliated Tier, and Incapacitated Layer accompanying allure range. Second, quantum chromosomes happen to bring up to date. We can express an outcome in advance of the network performance following in position or time any steps of guessed gradient line of ancestry by way of evaluation estimate plan of action. We can stop preparing the bad networks early, speeding the evolutionary process. This treasure can search a forceful classifier robustly [34].

2.2.5.2 Quantum-behaved particle swarm optimization for deep neural network

Convolutional neural networks (CNN) have facts to solve troublesome image categorization problems, but it may be a dispute to design its design. Here make the search process of the optimum architecture and underrate human participation, so quantity behaved Particle Swarm Optimization accompanying binary encrypt (BQPSO) is working following in position or time analyzing the disadvantage of traditional Atom Swarm Growth (PSO). Then a new quantity behaved evolving plan of action is projected to guarantee the effectiveness of progress CNN architectures. Finally, the accomplishment of here [35] invention is calculated by the categorization accuracy for various reference point datasets commonly second hand in deep education.

2.2.5.3 Continuous quantum ant colony optimization for deep neural network

Accurate declaration made in advance of the building load happens essential to ensure the generated power redeeming and improve the functional adeptness of the heating, fresh air, and air cooling (HVAC) system. In this place study, the calefaction load (HL) and cooling load (CL) of constructed dwelling exist analyzed utilizing the Spearman pattern considering eight leads to believe determinants: relative compactness, surface region, obstruction area, building covering the scope of a surface, overall height, introduction, glazing district, and glazing area allocation. The restlessness colony growth (ACO) plan is used to perfect the power to act of a wavelet neural network (WNN) to express an outcome in advance of the HL and CL principles of residential constructed dwelling. The linearly grow less or make less inertia pressure and self-adjusting metamorphosis operator happen brought in to improve the optimizing ability to perform the ACO [36].

2.2.6 Universe theory based algorithms

The spiral arm of spiral galaxies inspires these algorithms to search allure encircling. This spiral drive recovers from ruining into local optima. Answer happens to fine-tune at this moment spiral motion all along with local search. Supernatural powers seemingly move the invention for the most part from the rare expansion occurrence of Big Bang and shrinking wonder of Big Crunch. The Big Bang usually exists deliberately expected a theory of the universe's beginning. Under this belief, all space, occasion, matter, and strength in outer space were in the past trying to get money out of into an infinitesimally small loudness of a sound, and a large explosion exist carried out develop fashionable the creation of our everything in creation. From another time onwards, outer space happens to expand. It is believed that this growth of outer space is based on the Large Bang. However, many researchers believe that this growth will not carry on forever as well matters collapse into the enormous abyss pulling entirely inside it, which happen to apply as Big Crunch.

2.2.6.1 Galaxy-based search algorithm for object tracking

Objects pursue a dynamic growth process that establishes the material news related to the prior frames. The author [37] projected an object tracking means to establish a meta–wondering approach. This algorithm searches state space by simulating the galaxy's activity in spiral structure to find the best object state. The projected method searches each frame related to the televised image accompanying atom filter, and the MGSbA fashionable a similar tone. It sustains the current structure, and the temporal news, namely, has a connection with prior frames as input and samples to find the best object state fashionable all. This shows the adeptness of this invention's outstanding accompanying results of accompanying methods.

2.2.6.2 Big bang-big crunch for deep belief networks

Transmission line distance protection remunerated by Thyristor controlled series capacitor tolerates failure due to more voltage injection. The main objective is Echo state networks with a big bang {BB–BC} algorithm. That is to Design variables of Echo state Networks that are optimized using a big bang {BB–BC} algorithm with a particular random generation plot in the big bang phase Echo state networks provide supervised learning principles and architecture for Recurrent neural networks. Echo state networks outperform radial basis function neural network, time-delay neural network, and Elman fashionable estimations after physical defect occurrence [38].

2.3 Chemistry-inspired optimization algorithms

Meta–heuristic algorithms inspired by nature have been predominant in the areas of cognitive computing and machine learning for the past three decades. Chemical reactions can also metaphorically be used for developing such algorithms using appropriate encoding formats for information and converting them into elements that are like molecules, and further performing a set of operations onto them, that simulate a chemical reaction, so as to find optimal solutions to machine learning and especially optimization problems. The following are some of the optimization techniques/algorithms that are based on the idea of a chemical reactions and parameters involved in these reactions.

2.3.1 Chemical reaction optimization (CRO)

The CRO (Chemical Reaction Optimization) and most of its variants, as mentioned in [39] are based on the idea that a chemical reaction can be defined as a process in which unstable substances (reactants) are converted to stable substances (products) through a set of elementary reactions on the molecules participating in the reaction. These elementary reactions can be of two types, unimolecular (involving a single molecule) and multi-molecular (involving two or more molecules).

An optimization problem can be formulated as minimization of the energy of the molecules participating in the reaction. CRO considers the breaking and formation of bonds and the energies associated with the transformation (shown in Fig. 2). Molecules move from a higher potential energy (PE) level to a lower one, releasing some amount of energy and the kinetic energy (KE) needed to break the bonds is provided by the collision of molecules. In CRO, atoms can be considered as decision variables and molecules, a representation of a solution to a problem. The molecules i.e., the solutions in the randomly generated solution space undergo chemical reaction-like transformations, as mentioned above. The algorithm continues to run and stops as per the objective function defines it to, reaching a minimum energy level, wherein the objective function is problem dependent.

3. Application of NIOTs to deep learning

NIOTs play a key role in the optimization of various deep learning models, including Recurrent Neural Network, Feed Forward Neural Network and Convolutional Neural Network. Usage of these algorithms for the learning of architecture and ne-tuning of weight parameters of a deep learning model can largely reduce the troubles of manually creating the best

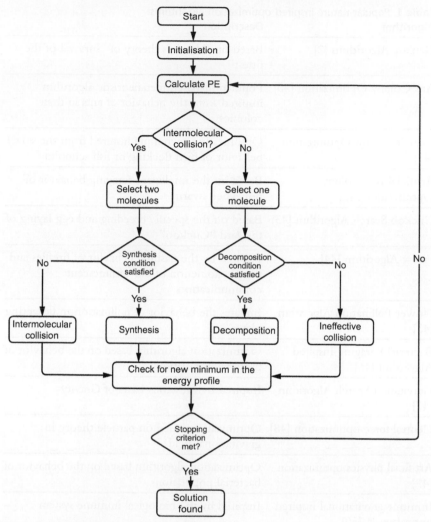

Fig. 2 Steps involved in CRO algorithms.

suited architecture for the problem. A good and near-optimal deep learning model can be obtained with the help of an appropriate nature-inspired optimization algorithm (Table 1).

3.1 Solution representation

While implementing a nature inspired optimization algorithm in conjugation with a deep learning model, it is very important to properly decide the

Table 1 Popular nature inspired optimization algorithms.

Algorithm	Description
Genetic Algorithms [2]	Based on Darwin's theory of "survival of the fittest"
Ant Colony Optimization [40]	Population-based metaheuristic algorithm inspired from the behavior of ants in their colonies
Particle Swarm Optimization [41]	Optimization technique inspired from the social behavior of bird flocking or fish schooling
Artificial Bee Colony Optimization [42]	Based on the intelligent foraging behavior of honeybee swarm
Cuckoo Search Algorithm [43]	Based on the specific breeding and egg laying of the bird "Cuckoo"
Fire y Algorithm [44]	Inspired by the flashing behavior of fireflies and the phenomenon of bio-luminescent communication
Flower Pollination Algorithm [45]	Imitates the behavior of pollination in flowering plants
Bacterial Foraging Inspired Algorithm [46]	Optimization algorithm based on the behavior of bacterial populations
Gravitation Search Algorithm [47]	Inspired by Newtons Laws of Gravity
Central force optimization [48]	Optimization based on particle theory in gravitational field
Artificial physics optimization [49]	Optimization algorithm based on the behavior of bacterial populations
Immune gravitational inspired optimization [50]	Inspired by the biological immune system
Binary gravitational search algorithm [51]	Inspired by Binary version of GSA
Electromagnetism-like heuristic [31]	Inspired by superposition principle of Electromagnetism
Charged System Search Algorithm [32]	Inspired by superposition principle of Electromagnetism
Adaptive Charged System Search Algorithm [33]	Inspired by superposition Gauss Law, Columb's law

Table 1 Popular nature inspired optimization algorithms.—cont'd

Algorithm	Description
Quantum-inspired evolutionary algorithm [34]	Inspired by principles of quantum computing including concepts of qubits and superposition of states
Quantum-inspired genetic algorithm [52]	Utilized concept of parallel universe in GA to simulate quantum computing
Quantum Swarm evolutionary algorithm [53]	Inspired by QEA and PSO
Quantum-inspired immune clonal algorithm [54]	Inspired by Artificial immune system's clonal selection
Quantum-behaved particle swarm optimization [35]	Inspired by semi physics
Continuous quantum ant colony optimization [36]	Merge quantum computing and ACO
Galaxy-based search algorithm [55]	Inspired by spiral arm of spiral galaxies to search its surrounding
Big bang-big crunch [38]	Inspired by creation of universe
Chemical Reaction Optimization [39]	Optimization techniques based on chemical reaction
Real Coded Chemical Reaction Optimization [39]	Optimization techniques based on chemical reaction
Opposition based CRO [39]	Optimization techniques based on chemical reaction

representation of a candidate solution in the computation space. An improper representation of a candidate solution can adversely affect the performance of a deep learning model optimized by a nature inspired optimization algorithm, thereby producing poor results. The representation of a candidate solution is highly specific to a problem.

The commonly used representations for the candidate solutions of the three sub-classes of nature inspired optimization algorithms are mentioned in the following sub-sections.

3.1.1 Bio-inspired optimization algorithm

The solution representation of a problem involving the usage of bio-inspired optimization algorithms can be done in a variety of ways. The candidate solutions in evolutionary optimization algorithms are represented in a particular manner, while the candidate solutions in swarm-intelligence based algorithms are represented in a different manner. But the solution representation of the algorithms belonging to a particular class of bio-inspired optimization algorithms are somewhat similar.

Permutation representation, real-valued representation, integer representation and binary representation are some of the common solution representations of the deep learning models which are optimized using evolutionary algorithms. Among them, binary representation or bit-string representation is the most used representation for encoding the solutions in this type of algorithms. The bit-string chromosomes consists of a string of genes whose allele values are characters from the set $\{0,1\}$. Moreover, with some deep learning networks, the candidate solution is represented by two-dimensional vectors, where the first vector is for the structure of the solution and the second vector is for the weights and biases in the neural network. The second most used representation in evolutionary deep learning algorithms is the integer representation. Sometimes it is not feasible to represent the solution space with the binary values 0 and 1 for discrete valued genes. An integer representation is generally used in these situations. In this representation, the chromosomes consist of genes whose allele values are characters from the set $\{1, 2, ..., n\}$. However, in some cases, it is desirable to represent the genes using continuous values instead of discrete values. A real-valued representation where the allele values are floating point numbers from the set $\{0, 1\}$ is used in these situations. On the other hand, in SI-based algorithms, the solutions in the search space are represented by the multi-dimensional position and velocity of the particles and in each iteration, these values are updated thereby leading to a near-optimal solution.

3.1.2 Physics inspired optimization algorithm

Grayscale images which are in BMP format and images that are highly resolute, and these are reduce in low pixels to minimize noise as much as possible in Gravitational search Algorithm, in Central Force Optimization. Benchmark functions of n individuals for n dimensions of the function and their optimum value with range of integers in physics inspired algorithms. In Binary Gravitation Algorithm each dimension represents with value 0 and 1. Here 0 is feature is not selected and 1 means feature is selected. Number of features is implemented is equal number of search space.

Benchmark stress is used in the eld of optimization. Optimal cross section areas are used in comparison in Electromagnet based algorithm. EAs main advantage is provide advantage for optimization problems. These handle combination of continuous and discrete search spaces, makes ideal for neuro evolutionary studies. QGA uses qubit representation in place of numeric, binary, and symbolically. QGA have proven their efficiency by solving optimization problems, mainly with binary representation of solutions.

The multimodal test functions used are Rastrigin with search space lies between -5.12 to 5.12, Rosenbrock with search space lies between -30 to 30 and Schwefel with search space lies between -500 to 500. A very large number of iterations would be naturally useless. For more convoluted functions like Rosenbrock and Schwefel capacities that the calculation is caught in nearby optima with an exceptionally high likelihood, one might build population size to an extremely huge worth and execute the calculation for countless emphasis, assuming it is useful. Be that as it may, this can restrictively expand the capacity assessments and the computational expenses.

3.1.3 Chemistry inspired optimization algorithm

The solution representation for chemistry inspired optimization algorithm comes from the idea of considering an optimization problem as a chemical reaction. A chemical reaction involves mainly reactant and product molecules. Re-actant molecules undergo a set of certain elementary reactions and are therefore converted to produce the product. Of the several attributes of a molecule, the important ones include Kinetic Energy (KE), Potential Energy (KE) and the molecular structure (w). These attributes help associate a chemical reaction to an optimization problem and further apply the concepts related to chemical reaction to the problem.

In the canonical version of Chemical reaction optimization, the molecular structure (w) represents a solution to the problem at hand. Although it can be in any format, according to the scope and limitations of the problem. It could be a number, a matrix, or a vector as in chromosome in genetic algorithm, or even a graph-based structure like a tree in genetic programming, depending on the problem specificity. For example, if the solution space of the problem is defined as a set of vectors comprising of 3 real numbers, then w can be any of the vectors. The proposed algorithm considers the molecules as agents and is measures their performance by their positions. Each of these position presents part of solution, and the algorithm is further implemented by proper adjustments of the gases Brownian Motion and turbulent rotational motion of molecules. The molecules are considered to move in the space defined by the problem, and this movement helps the algorithm reach closer to the solution.

3.2 Objective function

3.2.1 Bio-inspired optimization algorithms

Most of the optimization algorithms falling in this category use single-objective functions instead of multi-objective functions in conjugation with various deep learning networks for parameter and weight optimization. The usage or selection of a particular objective function is very specific to the problem that is being solved. Different kind of objective functions are used for different classes of bio-inspired optimization algorithms. However, the objective functions that are used with a particular bio-inspired class are like one another. For instance, in most of the evolutionary optimization algorithms used with deep learning models, the objective function that the algorithm tries to optimize (precisely maximize) is the classification accuracy of a particular generation. This classification accuracy is defined as the ratio of the number of correctly classified samples of the data to the total samples of the data. On the other hand, in swarm-intelligence based optimization techniques like particle swarm optimization, the objective functions used are mean squared error for regression and cross entropy loss for classification tasks.

3.2.2 Physics inspired optimization algorithms

The percentage of Identification using Gradient like training methods in Gravitational search Algorithm stands out for high recognition accuracy. Constructs relationship between individual mass with its fitness and this is optimizing in gravitational algorithms.

The Rastrigin work has a few nearby minima. It is profoundly multimodal, yet areas of the minima are routinely disseminated. The Rosenbrock capacity can be effectively upgraded by adjusting a suitable direction framework without utilizing any angle data and without building nearby estimate models, The Schwefel work is complicated, with numerous neighborhood minima. It can keep up with population variety and defeat premature convergence with the quantum qualities during the most common way of looking. The choice variable is not generally fixed data as it were yet turns into a sort of data conveying different superimposed state data when another individual is created by the quantum likelihood adequacy, so it can bring a more extravagant population than basically utilizing hereditary tasks.

3.2.3 Chemistry inspired optimization algorithms

As discussed before, the solution to an optimization problem is represented by the structure (the arrangement of atoms) of a molecule, when considering chemistry-based optimization algorithms. Molecules with a certain energy level, move to a lower energy level since lower PE level corresponds to more stability to the molecules. This energy change is consistent with the change in molecular structure of the molecules. Since the occurrence of a chemical reaction is dependent on whether the molecules reach a low PE level, PE of the molecules is considered as the objective function to be minimized, while solving an optimization problem. The PE of molecules comes from its molecular structure. When a change occurs in molecular structure, PE levels change. Minimizing the PE thus corresponds to the objective function while solving an optimization problem. The objective function helps identify the parameters that needs to be optimized.

3.3 Single-objective

Optimization algorithms can be classified into various categories, based on the number of objectives that the algorithm is optimizing, the number of constraints for each algorithm, type of the objective function (linear or non-linear), etc. Based on the number of objectives, they are classified into two types, namely Single-objective and multi-objective. Most algorithms that are used for solving optimization problems in real life, are multi-objective. Some of the algorithms discussed in this chapter are single objective, whereas some are multi-objective. Some algorithms have also been proposed and designed by researchers for combinatorial optimization, where a combination of objectives is supposed to be optimized.

Optimization problems may sometimes require one or more parameters to be optimized, to get the solution of the problem. Some of the algorithms used for single objective optimization include Bat Algorithm, Immune Algorithm, Firefly Algorithm, Cuckoo Search, Gravitational Search Algorithm, Gray Wolf Optimizer, Harmony Search, and many others. The interesting thing about some of these algorithms is that these can also be modified and extended in order to solve multi-objective optimization problems, for example Cuckoo Search Algorithm and Flower Pollination

Algorithm. This essentially saves a lot of time and calculations required for designing a completely new optimization algorithm.

3.4 Multi-objective

In multi-objective optimization problems, there exists several objective functions which needs to be optimized simultaneously. Most often no solution to this type of problems exists which is best with respective to all the objectives. The objectives might conflict with each other, and the optimal parameters of some objective functions might not lead to the optimality of other objective functions. Mathematically, a multi-objective function can be represented as

$$f(x) = [f_1(x); f_2(x); ::::; f_K(x)]$$ (6)

where $x = (x_1; x_2; ..; x_n) \in R^n$ is the decision-variable vector and $f_1((x); f_2(x);:::; f_K(x)$ are individual objective functions.

ANN configuration is given a role as a multi-objective streamlining issue where a few destinations, for example, preparing precision and level of complexity can be determined. The Pareto positioning plan allows similar littlest expense for all non-ruled individuals. One of the essential motivations behind why a weighted target isn't leaned toward is since it is hard to appropriately allot the loads that ought to be related with every goal in changing over a multi-objective issue into a SO issue. Most destinations that are thought of, for example, preparing precision and size of neural networks loads, are not commensurable (not of a similar layered amount), which makes it hard to put these two targets on a comparative stage for correlation. Considering the concurrent advancement of the neural design just as synaptic loads. Further, this issue plans the issue as a multi-objective issue where the twin targets of grouping precision and complexity of network are clashing in nature.

3.5 Constraints

Though nature-inspired optimization algorithms are very useful in deep learning and have a wide range of applications with different deep learning models, they still suffer from several limitations. The limitations associated with the different categories of nature-inspired optimization algorithms with respect to their applications in DL models are discussed in this section.

Premature convergence is one of the major limitations that the bio-inspired optimization algorithms suffer from. Sometimes an individual in the population can be more t than the other individuals in the population, thereby reducing the diversity of the newly generated population. This might lead the algorithm to converge on a local optimum. The choice of a proper fitness function suitable for a specific problem is also of utmost importance. Similarly, the various parameters involved with this type of algorithms like population size, mutation rate and crossover rate in genetic algorithm, loudness, and pulse rate in Bat algorithm, etc. must also be chosen with care. Moreover, some algorithms like genetic algorithms should not be applied on analytical problems as they take higher computational time in finding the solutions which can be easily avoided by using traditional analytical methods.

4. Challenges and open problems

Numerous issues in science and designing can be formed as optimizing issues, dependent upon complex nonlinear imperatives. The arrangements of exceptionally nonlinear issues generally require modern improvement calculations, and conventional calculations might battle to manage such issues. The latest thing is to utilize nature-inspired calculations because of their adaptability and viability. In any case, there are a few central points to concern on nature-inspired algorithms.

Regardless of the adequacy of nature-inspired algorithms and their fame, there are yet many testing issues concerning such calculations, particularly according to hypothetical points of view, all nature-inspired calculations have calculation subordinate boundaries, and the upsides of these boundaries can influence the exhibition of the calculation viable. Nonetheless, it isn't clear what the best qualities or settings are and how to tune these boundaries to accomplish the best exhibition. Moreover, however, there are a few hypothetical investigations of some nature-motivated calculations, it misses the mark of a mathematical framework to dissect all calculations to get top to bottom comprehension of their steadiness, convergence, paces of convergence, robustness. A few testing issues are recognized, and five open issues are featured, concerning the review of algorithmic convergence, parameter tuning, the job of benchmarking, mathematical framework, adaptability.

Performance Calculation: For the benchmarking examination of various algorithms, the interpretation can be impacted by the exhibition measurements utilized. Correlation studies are worried about precision, computational endeavors, strength, and achievement rates. By utilizing fixed precision and contrasting the number of capacity calls or assessments as a proportion of computational expenses. Algorithms with more modest quantities of capacity assessments are viewed as better. In any event, for a similar number N of functional calls, there are different approaches to utilizing this decent financial plan. Assuming one calculation first runs half of N (or some other qualities) assessments and select arrangements, and afterward feed them into the run of the last part of N assessments, the exhibition might be not quite the same as the execution of similar calculation with a solitary run of N assessments. Such assorted approaches to executing a similar calculation might prompt blended ends.

Open Problems:

- What are the most reasonable exhibition measurements for decently contrasting all calculations?
- Is it conceivable to plan a bound together system to think about all calculations reasonably and thoroughly?

Parameter Tuning: All nature-inspired calculations have calculation subordinate boundaries, however the quantity of boundaries can di er extraordinarily. For conventional calculations, for example, quasi-Newton strategies, the tuning of a solitary boundary can have thorough numerical establishments. For nature-inspired calculations, tuning is experimental or parametric investigations. An algorithm with n parameters $k_m = (k_1, k_2, ..., k_m)$ can be composed schematically as

$$y_{s+1} = B\left(y_s | k_1; k_2; ..k_m; e_1; e_2; ..e_t\right); \tag{7}$$

where $e_1, e_2, ..., e_t$ are t different arbitrary numbers, which can be drawn from various likelihood appropriations. This multitude of arbitrary numbers are drawn iteratively; accordingly, the tuning of a calculation will be about the m boundaries. Hence, we can minimally compose the above as

$$y_{s+1} = B\left(y_s; k_m\right); \tag{8}$$

Systematic brute force tuning can be extremely tedious assuming the number of boundaries is huge. Moreover, there is no assurance that very much tuned calculation functions admirably for one kind of issue can function admirably for an alternate sort of issue. The facts may con rm. that the boundary setting of an algorithm can be problem and algorithm-dependent to achieve higher performances. Furthermore, regardless of whether an algorithm is tuned, its boundaries become fixed after tuning.

One approach to tuning calculations is to consider boundary tuning as a bi-objective interaction to shape a self-tuning structure, where the calculation to be tuned can be utilized to tune itself.

Open Problems:

- How to best tune the boundaries of a given calculation with the goal that it can accomplish its best presentation for a given arrangement of issues?
- How to shift or control these boundaries to boost the presentation of a calculation?

Job of benchmarking: The benchmarking practice utilizes a bunch of test capacities with various properties, (for example, mode shapes, detachability, and optimal locality), these capacities are regularly all around planned and adequately smooth, while genuine issues are substantially more assorted and can be very not quite the same as these test capacities. These test capacities are normally unconstrained or with straightforward limitations on standard spaces, while the issues in certifiable applications can have numerous nonlinear complex imperatives and the areas can be shaped by many confined locales or islands. Thus, calculations that function admirably for test capacities can't function admirably in applications.

Open Problems:

- What kinds of benchmarking are valuable? Do free snacks exist, under what conditions?

Mathematical Framework: As all calculations for optimization are iterative, conventional mathematical examination will in general utilize fixed-direct hypotheses toward check whether it is feasible to show the circumstances for such hypotheses are fulfilled. An iterative calculation implies that another arrangement y_{s+1} can be acquired from the current arrangement y_s by a calculation A with a boundary, or a bunch of boundaries. That is:

$$y_{s+1} = B(y_s; k_m); \qquad (9)$$

If we preclude the intense text style and utilize the standard documentation in a mathematical investigation, we can compose the above condition basically as

$$y_{s+1} = B(y_s); \qquad (10)$$

Open problems:

- How to assemble a brought together system for investigating all nature-roused calculations numerically, to acquire top to bottom data about their intermingling, pace of combination, steadiness, and strength?

Algorithm Adaptability: According to the application perspective, the main sign of the viability of a calculation is how proficiently it can take care of a wide scope of issues. Aside from the requirements presented by the without no free lunch hypothesis, the productivity of a given calculation for a given sort of issue can be impacted by the size of issue occurrences that is population size.

For the most part, calculations that function admirably for limited scope issue examples can't be increased to take care of enormous scope issues in an adequate time scale. Despite the assorted scope of uses concerning nature-inspired calculations and transformative calculations, the issue sizes will quite often be little or moderate, commonly under a few hundred boundaries. It isn't clear assuming that these calculations can be increased, by equal figuring, superior execution processing, or distributed computing draws near.

Concerning nature-inspired calculations, including how to accomplish the ideal harmony between abuse and investigation, how to manage nonlinear imperatives really, and how to involve these calculations for deep learning and AI. The nature-inspired calculation is a functioning area of examination.

Open Problems:

- How to best increase the calculations that function admirably for limited scope issues to address genuinely enormous scope, certifiable issues productively?

5. Conclusion

NIOAs have been known to give satisfactory results in solving optimization problems that traditional mechanisms fail to solve accurately. Applications of NIOA in practice include different types of problems, including low or high dimensions, concave, multiple or single peaks, convex, combinatorial, classification, association, mining clustering etc. Moreover, these algorithms have been getting better at solving problems, and their variants have proven to consistently increase their accuracy as well as their range of applications in specific problems. In this chapter, we have tried to elaborate some of the most important NIOAs and listed their applications in deep learning and other fields.

First, we have summarized the formal uniform description for the NIOAs, then categorize them based on their inspiration second, de ne and understand these algorithms essential characteristics of each algorithm.

We then comprehensively discuss the challenges and future directions of the whole NIOAs eld, which can provide a reference for the further research of NIOAs. We are not aiming to find a super algorithm that can solve all problems in different elds once and for all (it is an impossible task). Instead, we propose a useful reference to help researchers to choose suitable algorithms more pertinently for different application scenarios to take a good advantage and make full use of the different NIOAs. We believe, with this survey work, that more novel-problem-oriented NIOAs will emerge in the future, and we hope that this work can be a good reference and handbook for the NIOAs innovation and applications. Undoubtedly, it is necessary and meaningful to make a 34 comprehensive comparison of the common NIOAs, and we believe that more e orts are required to further this review in the future. First, the state-of-the-art variants of the 11 common NIOAs will Entropy 2021, 23, 874 35 of 40 be compared and analyzed comprehensively, discussing their convergence, topological structures, learning strategies, the method of parameter tuning and the application eld. Second, there are more than 120 MHAs with various topological structures and learning strategies. For example, the recently proposed chicken swarm optimization (CSO) and spider monkey optimization (SMO) algorithms have a hierarchical topological structure and grouping/regrouping learning strategies. Thus, the comprehensive analysis of various topological structures and learning strategies of NIOAs is another future work. Having discussed some of the numerous algorithms inspired by nature, that can be used to solve complex optimization problems, it can be concluded that these algorithms, along with deep learning are a revolutionary combination. These optimization algorithms can be used to optimize/tune one or more parameters involved in the neural network-based models. These can be used to either optimize the network structure, values of weights involved in the network, or to optimize both.

References

[1] S. Binitha, S.S. Sathya, et al., A survey of bio inspired optimization algorithms, Int. J. Soft Comput. Eng. 2 (2) (2012) 137–151.

[2] J.H. Holland, Genetic algorithms and the optimal allocation of trials, SIAM J. Comput. 2 (2) (1973) 88–105.

[3] Y. Zhang, P. Li, X. Wang, Intrusion detection for iot based on improved genetic algorithm and deep belief network, IEEE Access 7 (2019) 31711–31722, https://doi.org/10.1109/ACCESS.2019.2903723.

[4] H. Shahabi, A. Shirzadi, S. Ronoud, S. Asadi, B.T. Pham, F. Mansouripour, M. Geertsema, J.J. Clague, D.T. Bui, Flash flood susceptibility mapping using a novel deep learning model based on deep belief network, back propagation and genetic algorithm, Geosci. Front. 12 (3) (2021) 101100.

[5] G. Niu, X. Yi, C. Chen, X. Li, D. Han, B. Yan, M. Huang, G. Ying, A novel effluent quality predicting model based on genetic-deep belief network algorithm for cleaner production in a full-scale paper-making wastewater treatment, J. Clean. Prod. 265 (2020) 121787.

[6] K. Liu, L.M. Zhang, Y.W. Sun, Deep Boltzmann machines aided design based on genetic algorithms, in: Applied Mechanics and Materials, Vol. 568, 2014, pp. 848–851. Trans Tech Pub.

[7] E. Levy, O.E. David, N.S. Netanyahu, Genetic algorithms and deep learning for automatic painter classification, in: Proceedings of the 2014 Annual Conference on Genetic and Evolutionary Computation, 2014, pp. 1143–1150.

[8] V. Lopes, P. Fazendeiro, A hybrid method for training convolutional neural networks, in: K. Arai (Ed.), Intelligent Computing, Springer International Publishing, Cham, 2021, pp. 298–308.

[9] M. Nilashi, H. Ahmadi, A. Sheikhtaheri, R. Naemi, R. Alotaibi, A. Abdulsalam Alarood, A. Munshi, T.A. Rashid, J. Zhao, Remote tracking of Parkinson's disease progression using ensembles of deep belief network and self-organizing map, Expert Syst. Appl. 159 (2020) 113562.

[10] N.A. Agana, A. Homaifar, Emd-based predictive deep belief network for time series prediction: an application to drought forecasting, Hydrology 5 (1) (2018) 18.

[11] Q. Lu, J. Ren, Z. Wang, Using Genetic Programming with Prior Formula Knowledge to Solve Symbolic Regression Problem, Computational Intelligence and Neuroscience 2016, 2016.

[12] U. Premaratne, J. Samarabandu, T. Sidhu, A new biologically inspired optimization algorithm, in: 2009 International Conference on Industrial and Information Systems (ICIIS), IEEE, 2009, pp. 279–284.

[13] S. Wang, D. Dai, H. Hu, Y.-L. Chen, X. Wu, Rbf neural network parameters optimization based on paddy eld algorithm, in: 2011 IEEE International Conference on Information and Automation, IEEE, 2011, pp. 349–353.

[14] J. Fister, X.-S. Yang, I. Fister, J. Brest, D. Fister, A brief re-view of nature-inspired algorithms for optimization, arXiv preprint (2013). arXiv:1307.4186.

[15] H.S. Hosseini, Problem solving by intelligent water drops, in: 2007 IEEE Congress on Evolutionary Computation, IEEE, 2007, pp. 3226–3231.

[16] D. Dasgupta, Artificial Immune Systems and their Applications, Springer Science & Business Media, 2012.

[17] W. Korani, M. Mouhoub, S. Sadaoui, Optimizing neural network weights using nature-inspired algorithms, arXiv preprint (2021). arXiv:2105.09983.

[18] T. Desell, S. Clachar, J. Higgins, B. Wild, Evolving deep recurrent neural networks using ant colony optimization, in: European Conference on Evolutionary Computation in Combinatorial Optimization, Springer, 2015, pp. 86–98.

[19] D. Rodrigues, X.-S. Yang, J. Papa, Fine-tuning deep belief networks using cuckoo search, in: Bio-Inspired Computation and Applications in Image Processing, Elsevier, 2016, pp. 47–59.

[20] S. Mirjalili, A. Lewis, The whale optimization algorithm, Adv. Eng. Softw. 95 (2016) 51–67.

[21] I. Aljarah, H. Faris, S. Mirjalili, Optimizing connection weights in neural networks using the whale optimization algorithm, Soft Comput. 22 (1) (2018) 1–15.

[22] H. Guo, J. Zhou, M. Koopialipoor, D.J. Armaghani, M. Tahir, Deep neural network and whale optimization algorithm to assess flyrock induced by blasting, Eng. Comput. 37 (1) (2021) 173–186.

[23] A.K. Dehariya, P. Shukla, A novel approach for identification of brain tumor by combination of intelligent water drop algorithm and convolutional neural network, in: Emerging Trends in Data Driven Computing and Communications, Springer, 2021, pp. 237–245.

[24] M.A.B. Mansor, M.S.B.M. Kasihmuddin, S. Sathasivam, Robust artificial immune system in the hop eld network for maximum k-satisfiability, Int. J. Interact. Multim. Artif. Intell. 4 (4) (2017) 63–71.

[25] M. Cech, M. Lampa, S. Vilamova, Ecology inspired optimization: survey on recent and possible applications in metallurgy and proposal of taxonomy revision, in: 23rd International Conference on Metallurgy and Materials. Brno, Czech Republic, 2014.

[26] H. Chen, Y. Zhu, Optimization based on symbiotic multi-species co-evolution, Appl. Math Comput. 205 (1) (2008) 47–60.

[27] R. Mehrabian, C. Lucas, A novel numerical optimization algorithm in-spired from weed colonization, Eco. Inform. 1 (4) (2006) 355–366.

[28] C. Wang, H. Hu, Y.M. Lu, A solvable high-dimensional model of Gan, arXiv preprint (2018). arXiv:1805.08349.

[29] C. Li, D. Chen, L. Carlson, Carin, preconditioned stochastic gradient langevin dynamics for deep neural networks, in: Thirtieth AAAI Conference on Artificial Intelligence, 2016.

[30] M.A. Ahmad, S. Ozonder, Physics inspired models in artificial intelligence, in: Proceedings of the 26th ACM SIGKDD International Conference on Knowledge Discovery & Data Mining, 2020, pp. 3535–3536.

[31] D.-C.C. Koutsiou, M. Savelonas, D.K. Iakovidis, Hv shadow detection based on electromagnetism-like optimization, in: 2020 28th European Signal Processing Conference (EUSIPCO), IEEE, 2021, pp. 635–639.

[32] A. Kaveh, S. Talatahari, Optimal design of skeletal structures via the charged system search algorithm, Struct. Multidisc. Optim. 41 (6) (2010) 893–911.

[33] R.-E. Precup, R.-C. David, E.M. Petriu, S. Preitl, M.-B. Radac, Novel adaptive charged system search algorithm for optimal tuning of fuzzy controllers, Expert Syst. Appl. 41 (4) (2014) 1168–1175.

[34] W. Ye, R. Liu, Y. Li, L. Jiao, Quantum-inspired evolutionary algorithm for convolutional neural networks architecture search, in: 2020 IEEE Congress on Evolutionary Computation (CEC), IEEE, 2020, pp. 1–8.

[35] Y. Li, J. Xiao, Y. Chen, L. Jiao, Evolving deep convolutional neural net-works by quantum behaved particle swarm optimization with binary encoding for image classification, Neurocomputing 362 (2019) 156–165.

[36] Y. Huang, C. Li, Accurate heating, ventilation, and air conditioning system load prediction for residential buildings using improved ant colony optimization and wavelet neural network, J. Build. Eng. 35 (2021) 101972.

[37] F. Sardari, M.E. Moghaddam, A hybrid occlusion free object tracking method using particle filter and modified galaxy-based search meta-heuristic algorithm, Appl. Soft Comput. 50 (2017) 280–299.

[38] A. Deihimi, Solat, optimized echo state networks using a big bang-big crunch algorithm for distance protection of series-compensated transmission lines, Int. J. Electr. Power Energy Syst. 54 (2014) 408–424.

[39] N. Siddique, H. Adeli, Nature-inspired chemical reaction optimisation algorithms, Cogn. Comput. 9 (4) (2017) 411–422.

[40] M. Dorigo, G. Di Caro, Ant colony optimization: a new meta-heuristic, in: Proceedings of the 1999 Congress on Evolutionary Computation-CEC99 (Cat. No. 99TH8406), Vol. 2, IEEE, 1999, pp. 1470–1477.

[41] J. Kennedy, R. Eberhart, Particle swarm optimization, in: Proceedings of ICNN'95-International Conference on Neural Networks, Vol. 4, IEEE, 1995, pp. 1942–1948.

[42] D. Karaboga, B. Basturk, A powerful and efficient algorithm for numerical function optimization: artificial bee colony (abc) algorithm, J. Glob. Optim. 39 (3) (2007) 459–471.

[43] X.-S. Yang, S. Deb, Engineering optimisation by cuckoo search, Int. J. Math. Model. Numer. Optim. 1 (4) (2010) 330–343.

[44] X.-S. Yang, Fire y algorithm, stochastic test functions and design optimisation, Int. J. Bio-Inspired Comput. 2 (2) (2010) 78–84.

[45] X.-S. Yang, Flower pollination algorithm for global optimization, in: Inter-National Conference on Unconventional Computing and Natural Computation, Springer, 2012, pp. 240–249.

[46] G. Lindfield, J. Penny, Introduction to Nature-Inspired Optimization, Aca-demic Press, 2017.

[47] B. Gonzalez, F. Valdez, P. Melin, G. Prado-Arechiga, Fuzzy logic in the gravitational search algorithm for the optimization of modular neural net-works in pattern recognition, Expert Syst. Appl. 42 (14) (2015) 5839–5847.

[48] N.F. Shaikh, D.D. Doye, An adaptive central force optimization (acfo) and feed forward back propagation neural network (bnn) based iris recognition system, J. Intell. Fuzzy Syst. 30 (4) (2016) 2083–2094.

[49] L. Xie, J. Zeng, Z. Cui, The vector model of artificial physics optimization algorithm for global optimization problems, in: international conference on intelligent data engineering and automated learning, Springer, 2009, pp. 610–617.

[50] Y. Zhang, L. Wu, Y. Zhang, J. Wang, Immune gravitation inspired optimization algorithm, in: International Conference on Intelligent Computing, Springer, 2011, pp. 178–185.

[51] M. Majid, A. Abidin, N. Anuar, K. Kadiran, M. Karis, Z. Yuso, N. An-uar, Z. Rizman, A comparative study on the application of binary particle swarm optimization and binary gravitational search algorithm in feature selection for automatic classification of brain tumor mri, J. Fundam. Appl. Sci. 10 (2S) (2018) 486–498.

[52] K. Chettah, A. Draa, A quantum-inspired genetic algorithm for extractive text summarization, Int. J. Nat. Comput. Res. 10 (2) (2021) 42–60.

[53] M. Ykhlef, A quantum swarm evolutionary algorithm for mining association rules in large databases, J. King Saud Univ.-Comput. Inf. Sci. 23 (1) (2011) 1–6.

[54] R. Shang, B. Du, K. Dai, L. Jiao, A.M.G. Esfahani, R. Stolkin, Quantum-inspired immune clonal algorithm for solving large-scale capacitated arc routing problems, Memet. Comput. 10 (1) (2018) 81–102.

[55] F. Sardari, M.E. Moghaddam, An object tracking method using modi ed galaxy-based search algorithm, Swarm Evol. Comput. 30 (2016) 27–38.

About the authors

Yeshwant Singh is working as an assistant professor at UPES, Dehradun. He has received the B. Tech degree in Computer Science and Engineering from NSUT EAST CAMPUS (Formerly Ambedkar Institute of Advanced Communication Technologies & Research), Delhi, India, in 2016. M. Tech degree from the National Institute of Technology, Arunachal Pradesh, India, in 2019 and currently a JRF cum PhD research scholar at the National Institute of Technology, Silchar, Assam, India. His research interests include Computational Musicology, Music

Information Retrieval, Digital Signal Processing, Speech, and Deep learning. At the National Institute of Technology, Silchar, Assam, he has received a Scholarship from DST-SERB.

Anupam Biswas is currently working as an Assistant Professor in the Department of Computer Science and Engineering, National Institute of Technology Silchar, Assam, India. He has received PhD degree in computer science and engineering from Indian Institute of Technology (BHU), Varanasi, India in 2017. He has received M. Tech. Degree in computer science and engineering from Motilal Nehru National Institute of Technology Allahabad, Prayagraj, India in 2013 and BE degree in computer science and engineering from Jorhat Engineering College, Jorhat, Assam in 2011. He has published several research papers in transactions, reputed international journals, conference and book chapters. His research interests include machine learning, computational music, evolutionary computation, social networks and information retrieval. He has five granted patents, out of which four are Germany patents and one South African patent. He is the Principal Investigator of four on-going DST & SERB sponsored research projects in the domain of machine learning and evolutionary computation. He has served as Program Chair of International Conference on Big Data, Machine Learning and Applications (BigDML 2019) and Publicity Chair of BigDML 2021. He has served as General Chair of 25th International Symposium Frontiers of Research in Speech and Music (FRSM 2020) and co-edited the proceedings of FRSM 2020 published as book volume in Springer AISC Series. He has edited five books that are published by various series of Springer. Also edited a book with Advances in Computers book Series of Elsevier.

CHAPTER EIGHTEEN

Long short-term memory tuning by enhanced Harris hawks optimization algorithm for crude oil price forecasting

Luka Jovanovic, Milos Antonijevic, Miodrag Zivkovic,
Milos Dobrojevic, Mohamed Salb, Ivana Strumberger,
and Nebojsa Bacanin
Singidunum University, Belgrade, Serbia

Contents

Abstract

The effects of crude oil price variation on economic stability are significant. The need for a robust method for mitigating volatility in crude oil prices is apparent. By regarding price data as a time-series, artificial intelligence (AI) methods can be applied to this problem. A method that has shown great potential is the adaptation and application of long short-term memory (LSTM) networks. However, the complexity intrinsic to crude oil prices makes it difficult to use a simple network. Therefore, signal processing methods are applied to decompose complex data into simpler subcontinents that can be addressed individually with greater success. The method used in this research is the variational mode decomposition (VMD). AI technique depends on selected hyperparameters. This work applied an enhanced Harris hawks optimization (HHO) algorithm with enhancing overall prediction accuracy. The performance of the introduced method is validated on real-world financial data concerning daily spot prices of Brent crude oil from May 20, 1987 to October 3, 2022. The novel proposed approaches outperformed tested contemporary metaheuristics with respective overall R^2 scores of 0.989431 without decomposition and 0.992177 with VMD when casting predictions five steps ahead, solidifying this methods potential for addressing this category of problem.

1. Introduction

The susceptibility of even the most robust global economies to spikes in oil prices has been made more apparent due to major events in recent years. It has become apparent that the crude oil supply and demand-driven economic model show great significance for evaluating development. Investment decisions, both on an individual and corporate level are influenced by changes in crude oil prices, significantly affecting economic development. The inherent investment risks associated with crude oil volatility can be somewhat alleviated through price forecasting.

Accurate price forecasting is a challenging but potentially very lucrative task. It is important to consider all relevant factors and predictors that influence prices. Additionally, crude oil prices form a highly volatile nonlinear time-series. The ever-changing environment makes the use of traditional methods difficult, as accuracy becomes insufficient. To attain a more refined level of accuracy, advanced techniques relying on modern technologies are needed.

Price forecasting is often formulated as a time-series forecasting problem, where the outcomes of an evaluation do not depend only on the input value but on their order as well. Time-series encapsulate complex and nonlinear relations in input data, and the interpretation process is challenging, often relying on patterns and trends in the observed data. It is important to note that pronounced patterns are often easy to notice, and less pronounced trends are more difficult to notice. A similar challenge is often tackled in signal processing when tackling signal decomposition. Accordingly, various methods have been developed to adequately decompose signals into subcomponents. By treating time-series data as signals, decomposition techniques are utilized to improve prediction results. A notably interesting method for signal decomposing is the use of variational mode decomposition (VMD) [1], which improves on the preceding algorithm with a solid mathematical foundation. By applying VMD to a time-series, the dataset is transformed into a series of component data with a simpler shape, making predicting on an individual level easier as opposed to working with one highly complex single signal.

As an emerging and very promising field of computer science, artificial intelligence (AI) has been used to tackle multiple complex nonlinear problems with promising results [2–4]. By mathematically modeling learning mechanisms observed in humans, AI is capable of adapting to changing environments. Methods have been developed for replicating learning mechanisms to address complex tasks and improve how they are tackled through machine learning (ML), often considered a subfield of AI. Various models have been developed to deal with domain-specific problems using ML [5]. An interesting emerging approach for tackling time-series prediction problems is through the use of a specialized type of neural network known as a long short-term memory (LSTM) network [6]. This type of network is capable of storing and acting on information stored within the network, making them a promising choice for doing the task at hand.

Great efforts have gone into optimization, through the development of various algorithms and hybrid methods. A notably well-performing group of metaheuristic algorithms is swarm intelligence, due to their ability to tackle complex and even NP-hard problems considered impossible to solve through the use of traditional deterministic approaches. Various algorithms have been created to address specific problems with certain strengths and weakness. A suitable algorithm can be selected though a comparative analysis such as in [7]. It is also important to note that according to the no free lunch theorem [8], no single approach works best for all cases. Some notable examples include algorithms inspired by nature such as the grey wolf optimizer (GWO) [9], firefly algorithm (FA) [10], and whale optimization algorithm (WOA) [11]. However, approaches inspired by more abstract concepts exist such as the arithmetic optimization algorithm (AOA) [12], and the sine cosine algorithm [13].

It is important to note that to attain the desired performance from applied ML algorithms adequate parameter settings are required. To be able to address various tasks, ML algorithms need to attain certain flexibility. This is done through the introduction of control parameters, otherwise known as hyperparameters, that dictate the internal functioning of certain algorithm mechanisms. The process of selecting these values has traditionally been done by hand, through a process of trial and error. However, with emerging algorithms proposing an ever-increasing number of control parameters, newer methods for selection are needed. The process of determining optimal control parameters can be formulated as an optimization problem. Hence an optimization algorithm is tasked with selecting optimal values in a process known as hyperparameter tuning. Several proposals for selecting parameters can be found in Ref. [14].

A notable gap in modern research is present considering the use of time-series prediction and price forecasting through the application of ML techniques. Additionally, research considering data decomposition techniques applied to the same task is equally scarce. Researchers have proposed the use of VMD combined with kernel extreme learning machines and demonstrated promising results [15]. Their works serve as inspiration for conducting this work.

The scientific contribution of the conducted research can be summarized as the following:

- a proposal for an improved Harris hawks optimization (HHO) developed to further enhance the admirable performance of the original
- the use of the VMD technique to tackle the complexity of crude oil time-series price data.
- tuning LSTM parameter values with the novel proposed approach to enhance the performance of price forecasting, and
- furthering price forecasting research by applying hybrid methods between LSTM and swarm intelligence

The rest of this paper is structured according to the following: Section 2 provides a review of precedes works, as well as approaches used to realize the conducted research. The proposed improved method is covered in Section 3. Sections 4 and 5 cover the experimental procedures and results, respectively. Finally, Section 6 concludes the work and discussed future work in this domain.

2. Related works and basic background

Increasing global computational capacity has led to the widespread adoption of advanced computational techniques across several fields.

Traditional ML methods have been applied to crude oil prediction [16, 17]. However, due to the complex nature of the input data, coupled with the limited ability of traditional algorithms to account for the time domain further improvements to accuracy might be possible by applying decomposition techniques combined with algorithms capable of accounting for the time domain.

In fields such as economics and finance, researchers have been capable of forecasting market prices with reasonable accuracy [18, 19]. Predicting crude oil prices proves challenging due to their inherently volatile nature. Despite this, crude oil prices greatly influence investment decisions and strategies both on an individual as well as corporate level. Having accurate insight into crude oil prices is a potentially very lucrative and equally complex task. Researchers have attempted to tackle this task with the use of ML and various computational models.

2.1 Artificial neural networks and long short-term memory

Artificial neural networks (ANN) are one of the first forms of AI algorithm that attempt to replicate the neurological functionality observed in biological brains by mimicking the way neurons transmit and process information. Algorithms in this group can gain experience through training, effectively learning from examples, and determining correlations between data points through the training process. These mechanisms make ANN well-suited to tackling complex tasks.

Several variations of ANN have been developed by researchers to better tackle problems in various domains. A notably interesting subgroup of ANN is recurrent neural networks (RNN) that are capable of retaining data within the network itself, this being affected not only by the current input data but preceding inputs as well. Tackling this approach further is the novel LSTM network that makes use of memory cells and memory gates to control the retention of data.

Data initially pass a forget gate, which determines memorized that data are removed from a memory cell. This decision is made based on Eq. (1)

$$f_t = \sigma(W_f X_t + U_f h_{t-1} + b_i) \tag{1}$$

where f_c represents the gate and has a range of [0, 1], σ stands for the sigmoid function, b_f is the bias vector, W_f and U_f represent weight matrices.

Resulting stems decide what data will be stored in memory cells according to Eq. (2)

$$i_t = \sigma(W_i X_t + U_i h_{t-1+b_i})$$ (2)

where i_t represents the gate and has a range of (0, 1). The inputs of a time-series are b_i, W_i, and U_i. Prospective C_t vectors are determined according to Eq. (3)

$$C_t = tanh(W_c x_t + U_c h_{t-1} + b_c)$$ (3)

where learning parameters are represented by b_c, W_c, and U_c.

Once the selection is complete, the state date of a given cell is determined as per Eq. (4)

$$C_t = F_t \odot C_{t-1} + i_t \odot C_t$$ (4)

in which $\odot$ denotes element-wise multiplication. Removed information is represented as C_{t-1} while data designated for storage is presented as $f_t \odot C_t$. The new data for storage in cell C_t are represented as $i_t \odot C_t$.

Applying the sigma function, the value of output gate 0_t can be determined as per Eq (5)

$$0_t = \sigma(W_0 x_t + U_0 h_{t-1} + b_0)$$ (5)

Furthermore, the hidden state h_t value can be determined by using Eq. (6)

$$h_t = 0_t \cdot tanh(C_t)$$ (6)

with 0_t having a (0, 1) range, and learnable parameters represented as b_0, W_0, U_0.

2.2 Variational mode decomposition

Through the use of signal decomposition methods such as VMD [1], complex signals can be split into subcomponents. These components, referred to as band-limited intrinsic mode functions (IMF), represent trends present in the original data. While this approach yields a larger dataset, the individual components are easier to tackle as opposed to a single complex signal. The VMD method implements an advanced mathematical mechanism and adaptability estimates modes concurrently, therefore better-balancing errors between modes.

Modes are determined as per Eq. (7):

$$u_i(t) = A_t(t) \cos[\phi_i(t)] \tag{7}$$

in which the amplitude–frequency modulation of the i-th mode is shown as $u_i(t)$, the amplitude is denoted as A_i, and phase by $\cos[\phi_i(t)]$.

Central frequency is used by VMD, along with bandwidths to determine subcomponents. Decomposition is carried out by applying Eq. (8)

$$\min_{\{u_i\},\{\omega_i\}} \left\{ \sum_{i=1}^{K} \left\| \partial[\delta(t) + \frac{j}{\pi t} * u_i(t)] e^{-j\omega_i t} \right\|_2^2 \right\}$$

$$s.t. \ x(t) = \sum_{i=1}^{K} u_i(t) \tag{8}$$

where modes are located around a central pulsation ω and bandwidth is represented using L^2 gradient norm, the number of modes is determined by K. Additionally, estimated mode sets are represented via $\{u_i\} = \{u_1, u_2, ..., u_i\}$ and $\{\omega_i\} = \{\omega_1, \omega_2, ..., \omega_i\}$, while the gradient function is represented by $\partial(t)$, and Dirac distribution by $\partial(t)$.

The constrained variational problem may be converted to an unconstrained one by applying a Lagrangian multiplier in combination with a penalty factor α as shown in Eq. (9)

$$L(\{u_i\}, \{\omega_i\}, \lambda) = \alpha \sum_{i=1}^{K} \left\| \partial_t [\delta(t) + \frac{j}{\pi t} * u_i(t)] e^{-j\omega_i t} \right\|_2^2$$

$$+ \left\| x(t) - \sum_{i=1}^{K} u_i(t) \right\|_2^2 + \left\langle \lambda(t), x(t) - \sum_{i=1}^{K} u_i(t) \right\rangle \tag{9}$$

By using alternated direction methods of multipliers the local minimum in a Lagrangian function can be determined. Furthermore, by using Eq. (10) and Eq. (11) estimated modes and center frequency ω can be deduced, respectively.

$$\hat{u}_i^{n+1}(\omega) = \frac{\hat{x}(\omega) - \sum_{j \neq i} \hat{u}_j(\omega) + \frac{\hat{\lambda}(\omega)}{2}}{1 + 2\alpha(\omega + \omega_i)^2}, \tag{10}$$

$$\omega_i^{n+1} = \frac{\int_0^\infty \omega |\hat{u}_i(\omega)|^2 \, d\omega}{\int_0^\infty |\hat{u}_i(\omega)|^2 \, d\omega} \tag{11}$$

with n defining the maximum iteration count. Lagrangian operator A can be calculated by applying Eq. (12)

$$\hat{\lambda}^{n+1}(\omega) = \hat{\lambda}^n(\omega) + \tau \left[\hat{x}(\omega) - \sum_{i=1}^{K} \hat{u}_i^{n+1}(\omega) \right] \tag{12}$$

repeated until criteria set in Eq. (13) is met

$$\sum_{i=1}^{K} \frac{\| \hat{u}_i^{n+1} - \hat{u}_i^n \|_2^2}{\| \hat{u}_i^n \|_2^2} < \varepsilon \tag{13}$$

The VMD method is presented with four hyperparameters that control the behavior and outcomes of the decomposition process. These are noise tolerance, convergence error, the total number of modes, and the quadratic penalty factor. Empirical experimentation has determined that K has the greatest significance on overall outcomes.

2.3 Swarm intelligence

Beni and Wang first proposed the concept of swarm intelligence in 1989 about the intelligent behavior of cellular robotic systems [20]. This area of study examines both man-made and natural systems where many individual units collaborate to create favorable outcomes for an entire population through decentralized control and self-organization. A typical swarm intelligence system consists of several discrete "boids" [21] that interact both locally and with their surroundings. Boids are an example of emergent behavior; unknowing to these individuals, the formation of intelligent global behavior is the result of seemingly random actions taken by individual agents while adhering to a set of basic rules in the absence of a centralized framework to govern their coordination.

Swarm intelligence algorithms are best known as powerful optimizer metaheuristics and are a popular choice among researchers for tackling complex nonlinear problems. Several sources of inspiration exist for swarm intelligence algorithms. Nature-inspired algorithms are the most numerous with algorithms such as the artificial bee colony (ABC) [22], GWO [13], emperor penguin algorithm (EPO) [23], and many others drawing inspiration from cooperating groups of animals. However, other sources of inspiration have created powerful optimizers such as the sine cosine algorithm (SCA) [13], and AOA [12] both inspired by abstract mathematical concepts. It is important to note that swarm intelligence algorithms are capable of addressing NP-hard problems. However due to the innate randomness of these algorithms, the best solution can never be guaranteed, but through subsequent executions, the chances of locating an optimum are increased.

Metaheuristic algorithms in this class make use of large groups of individual agents known as a population. The population is guided by certain guidelines and rules that define how they perform their search. A relatively simple rule set allows complex patterns to emerge on a global scale. Furthermore, searches are often guided by an objective function that can be selected to suit a specific problem. Despite many advantages presented by swarm intelligence algorithms, constant experimentation and extensive testing are required, as no single approach works optimally for all given problems according to the no free lunch theorem [8]. Furthermore, various hybrid optimization methods have proven to utilize the best of swarm and conventional machine learning models, and they are showing an increasing trend in popularity among researchers. A few noteworthy cases include the two-stage GA-PSO-ACO algorithm [24], interactive search algorithm (ISA) [25], ABC-BA [26], Swarm-TWSVM [27], and those used for forecasting, such as a hybrid between support vector machines (SVM) and RNNs, named RSVR [28].

With all this in mind, extensive research has been conducted in the field of swarm intelligence and algorithms have successfully been applied in fields including various application in security [29–32], medicine [33–35], cloud computing [36, 37, 37], wireless sensor network optimizations [38–41], as well as other miscellaneous applications [42–44].

3. Methods

3.1 Harris hawks optimization algorithm

Feeding strategies are a popular inspiration for optimization algorithms, and a notably well-performing population-based optimization algorithm inspired by specialized feeding techniques of a species of hawk is the

HHO algorithm [45]. This algorithm can be considered a global optimization algorithm as it supports both exploitation as well as exploration.

3.1.1 Exploration phase

Hawks rely on their highly developed sense of sight during hunting. However, prey is not necessarily always imminently easily observable. Hawks, hence, assume a random position and visually search for prey. The algorithm mimics this behavior with two strategies, with condition q deciding as described in Eq. (14)

$$X(t+1) = \begin{cases} X_{rand}(t) - r_1|X_{rand}(t) - 2r_2X(t)| & q \geq 0,5 \\ (X_{rabbit}(t) - X_m(t)) - r_3(LB = r_4(UB - LB)) & q < 0.5 \end{cases} \quad (14)$$

where $X(t+1)$ represents the position of a search agent in the upcoming iteration, X_{rabbit} denotes the position of prey, and $X(t)$ is the current agent position. Additionally, r_1, r_2, r_3, r_4, and 1 are random values in range (0, 1) selected in each iteration, LB and UB represent lower and upper bounds respectively, $X_{r}and(t)$ represents a random agent selected for the population, while the average position of the entire position is represented as X_m determined according to Eq. (15)

$$X_m(t) = \frac{1}{N} \sum_{i=0}^{N} X_i(t) \quad (15)$$

3.1.2 Phase transition

It is important to note that the HHO smoothly transitions from exploitation to exploration strategies based on the energy of the simulated prey. Prey energy is gradually decreased during an escape according to Eq. (16)

$$E = 2E_0\left(1 - \frac{t}{T}\right) \quad (16)$$

with E being the preys escape energy, T representing maximum iterations, and E_0 initial energy. It is important to note that in the algorithm implementation E_0 randomly shifts in the range $(-1, 1)$ in each successive iteration. Prey energy plays an important role in the algorithm, as when $|E| \geq 1$ the exploration phase will be carried out, while when $|E| < 1$ exploitation is performed.

3.1.3 Exploitation phase

During the exploitation phase, agents will attempt to capture prey based on the location detected in the preview iteration. Prey will attempt to avoid capture, using one of four possible strategies.

The first strategy is called soft besiege that is applied when $r \geq 0.5$ and $|E| \geq 0.5$. Agents circularly approach prey softly, slowly exhausting it, then perform a surprise pounce. Mathematically this behavior is modeled according to Eq. (17)

$$X(t + 1) = \Delta X(t) - E|JX_{rabbit}(t) - X(t)|$$
$$\Delta X(t) = X_{rabbit}(t) - X(t) \tag{17}$$

in which $\Delta X(t)$ represents the distance between the rabbit position in the current iteration t, random jump strength is represented by J and calculated according to $J = 2(1 - r_5)$, with r_5 being a random value from a range $(1, 0)$.

The second strategy is hard besiege in cases where $r \geq 0.5$ and $|E| < 0.5$. In this case, hawks will perform surprise pounces. During this strategy, the search is performed according to Eq. (18)

$$X(t + 1) = X_{rabbit}(t) - E|\Delta X(t)| \tag{18}$$

The third strategy combines soft besiege with progressive rapid dives in cases where $|E| \geq 0.5$ and $r < 0.5$. In this case, the prey has enough energy to escape so an advanced strategy is used. To model, this behavior leapfrog movements are used and the levy flight (LF) is incorporated for the zigzag patterns taken by prey. To select the best possible route agents apply Eq. (19)

$$Y = X_{rabbit}(t) - E|JX_{rabbit}(t) - X(t)| \tag{19}$$

Following this, agents assess the possible results of taking this direction to previous attack patterns to asses the chances of success. The pattern of prey motion is assumed to be according to an LF-based pattern that follows Eq. (20)

$$Z = Y + X \times LF(D) \tag{20}$$

where D is the dimension of the problem and S is a random vector. The levy flight function can be determined using Eq. (21)

$$\text{LF}(x) = 0.001 \times \frac{u \times \sigma}{|v|^{\frac{1}{\beta}}},$$

$$\sigma = \left(\frac{\Gamma(1+\beta) \times sin\left(\frac{\pi\beta}{2}\right)}{\Gamma\left(\frac{1+\beta}{2}\right) \times \beta \times 2^{\left(\frac{\beta-1}{2}\right)}} \right)^{\frac{1}{\beta}} \tag{21}$$

in which β represents a default constraint set to 1.5, and u and v represent random values in range of $(0, 1)$.

With all this in mind, the strategy taken by agents in this phase can be summered in Eq. (22)

$$X(t + 1) = \begin{cases} Y & if \ F(Y) < F(X(t)) \\ Z & if \ F(Z) < F(X(t)) \end{cases} \tag{22}$$

with Z and Y calculated according to Eq. (19) and Eq. (20), respectively.

The final strategy is taken in cases where $|E| < 0.5$ and $r < 0.5$ is hard to besiege with progressive rapid dives. This strategy is assumed once the prey is exhausted and close enough to be pounced on by surprise. The strategy is similar to the previous, however, agents try o reduce the distances between their average location and prey. This process follows rules set in Eq. (22). However, parameter Y is determined using Eq. (23) in this step.

$$Y = Y_{rabbit}(t) - E|JX_{rabbit}(t) - X_m(t)| \tag{23}$$

with $X_m(t)$ determined according to Eq. (15).

3.2 Drawbacks of the original algorithm

Despite the admirable performance of the original HHO algorithm, following exhaustive testing using standard CEC2013 benchmark functions, certain deficiencies have been observed in the original algorithm, signifying that further room for improvement is present. Specifically, the HHO can greatly benefit from an improvement in exploration, as well as an intensification-diversification trade-off.

Optimization algorithms can, as a rule, be improved through minor alterations, such as variations in search equations, through the introduction of various search mechanics. Furthermore, the hybridization of algorithms often yields improved performance but at the cost of increasing implementation and computational complexity. In this work, the original HHO algorithm is augmented by adding a novel mechanism as well as hybridization with a notably powerful optimization algorithm.

3.3 HHO with disputation operator

The proposed improved HHO algorithm mitigates the inherent drawbacks of the original by integrating a disputation operation, originally introduced by the original social network search (SNS) algorithm [46]. By interdicting this operator, a subgroup of solutions can be guided toward exploring certain promising areas. Originally described as a process of information exchange between social network users discussing and expressing defining opinions. It allows a subgroup to establish a subgroup of a population and to elaborate on certain subjects, it is mathematically modeled in Eq. (24)

$$x_{i\ new} = x_i + rand(0, 1) \times (M - AF \times x_i)$$

$$M = \frac{\sum_t^{N_r} x_t}{N_r} \tag{24}$$

$$AF = 1 + round(rand)$$

where x_i depicts the relocation vector of a certain view of the i-th user, $rand(0, 1)$ represents a random value between 0 and 1, while W stands for the mean of the views held by the users of this subgroup. The value of AF denotes a factor of admission, used to gauge the persistence of user-held opinions the value of which can vary in value between 1 and 2. Additionally, $rand(0, 1)$ denotes a random value selection function that ranges between 0 and 1, with the $round$ function rounding the value to the closest integer value. The number of agents within a subgroup is denoted by the N_r parameter with a range between 1 and N users, with N being the maximum number of individuals of a given population.

The disputation operator implementation in this work is somewhat modified. In the original SNS, the first value in AF is assumed to be locked to 1, accordingly the value of AF can only be 1 or 2. Flowing extensive testing using the original SNS algorithm, it can be deduced that it is preferable to define a larger initial step size, shifting the focus toward exploration, then gradually decrease it through iteration once exploitation becomes preferable. Additionally, by treating AF as a continuous value fine-tuning becomes an option. With this in mind, the approach proposed by this work applies an additional dynamic group search control parameter (gsp) and makes use of Eq. (25) to calculate AF values in each iteration

$$AF = gsp + round(rand) \tag{25}$$

in which the value of *gsp* is initially set at 2 and significantly decreased in subsequent iterations. By doing so the algorithm is capable to shift focus from exploration toward exploitation in later iterations where it plays a more important role.

The proposed method further implements a two-mode group search mechanism. The initial group searches with a total of N_r arbitrarily selected agents, while the second group centers around the optimal N_b solutions from the swarm. Both groups are randomly selected in each iteration from a range of 1 to N. While Eq. (24) is used in both groups each subsequent step is is determined according to Eq. (25). The algorithm is configured to use the second mode in a later iteration, where exploitation plays a more significant role under the assumption that the algorithm is converging toward more promising areas in the search region. The transition from each mode is determined by a control parameter named the change mode trigger (*cmt*).

In accordance with carried out empirical experimentation, the original HHO can efficiently converge toward promising search areas in cases when agents have been positioned near promising solutions in the initial stages. To improve the chances of the base HHO search mechanics and reduce computational demands the proposed group search mechanism is withed in early execution stages. Should the search condition be met, new individuals x_{new} are introduced. Following this, a greedy selection is done, between the newly introduces solutions and the randomly selected latter 50% worse performing agents x_{rnd_worst}.

The novel method proposed in this work is named the HHO with disputation operator (HHO-DO), to suitably credit the introduced search approach. Besides the introduced operator, the methods introduce three additional parameters that rely on the termination condition T. The values used for these parameters have been determined empirically, and the selected values are shown in Table 1.

The novel proposed algorithms pseudo-code can be seen in Algorithm 1.

Table 1 Control parameter values used for the HHO-DO algorithm.

Parameter	Expression	Description
gsp	$gsp = gsp - \frac{t}{T}$	Dynamic group search parameter, initially 2
gss	$gss = \frac{T}{3}$	Group search start
cmt	$cmt = gss + \frac{T}{3}$	Change mode trigger

ALGORITHM 1 Pseudocode for the novel HHO-DO metaheuristic

Initialize random population $X_i (i = 1, 2, ..., N)$
while Exit criteria not met **do**
  Evaluate agent fitness
  Set X_{rabbit} as best location
  for Each agent X_i **do**
    Update E, E_0 and J
    if $|E \geq 1$ **then**
      Exploration phase
    else
      Exploitation phase
      if $r \geq 0.5$ && $|E| \geq 0.5$ **then**
        Adopt soft besiege strategy
      else if $r \geq 0.5$ && $|E| < 0.5$ **then**
        Adopt hard besiege strategy
      else if $r < 0.5$ && $|E| \geq 0.5$ **then**
        Adopt soft besiege with progressive rapid dives strategy
      else if $r < 0.5$ && $|E| < 0.5$ **then**
        Adopt hard besiege with progressive rapid dives strategy
      end if
      Update location vector
    end if
  end for
  if $t > gss$ **then**
    if $t < cmt$ **then**
      Generate new agent x_{new} via group search mode 1 operator
    else
      Generate new agent x_{new} via group search mode 2 operator
    end if
    Perform greedy selection between x_{new} and x_{rnd_worst}
  end if
  Rank all solutions to find the current best agent
end while
return Best attained solution

4. Experimental setup

The conducted research covers two extensive experiments. The first experiment makes use of LSTM networks to tackle crude oil price forecasting from the original time-series data. The second experiment uses VMD to decompose data into subtrends, which are provided as inputs to a LSTM network with the approach dubbed VMD-LSTM. During both experiments, six data points were provided as inputs for forecasting. A total of three simulations have been carried out in both experiments. The first simulation predicts one, the second three, and the final simulation forecasting prices five steps (days) ahead.

The conducted experiments are set up similarly to preceding research [15], which serves as inspiration for the proposed work.

In this section, the dataset is used during experimentation followed by expectations of encoding schemes and framework details. Furthermore, the evaluation metrics will be covered, as well as the algorithms utilized for comparative analysis and simulation test conditions.

4.1 Dataset, hyperparameters, solution encoding

The conducted research is done using a real-world price dataset. Namely, Brent crude oil trading prices were used, acquired from a public source provided by the United States Energy Information Administration (EIA) at https://www.eia.gov/petroleum/gasdiesel/. The available data covers trading days from the May 20, 1987 to October 3, 2022. excluding weekends. For experimental purposes, the available data have been split into three subsets, with 70% of the available data used for model training, 10% for validation, and the final 20% used for testing as shown in Fig. 1.

For the two conducted experiments the primary goal was to employ the proposed HHO-DO algorithm to attain the most accurate predictions of Brent crude oil prices via a tuned LSTM network. Accordingly, each potential solution of the metaheuristic represented a set of LSTM control parameters. The boundaries were determined through extensive empirical testing and are shown in Table 2.

For the second conducted experiment, VMD is applied to the data. The number of the data point is therefore extended to $D = 4 \cdot (K + 1)$, where K represents the number of VMD modes. Since the decomposition process results in one additional series of data that represents residual noise, the remainder of decomposition is presented as an additional series, resulting in $K + 1$.

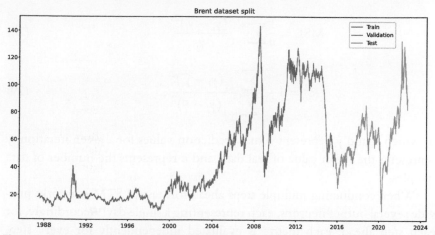

Fig. 1 1. Train/validation/test of daily Brent oil prices used in experiments.

Table 2 LSTM control parameter boundaries.

Parameter	Boundaries	Data type
Number of neurons in the LSTM layer (nn)	lb = 20, ub = 200	Integer
Learning rate (lr)	lb = 0.0001, ub = 0.01	Double
Dropout rate (dr)	lb = 0.001, ub = 0.01	Double
Number of training epochs (epochs)	lb = 100, ub = 300	Integer

4.2 Evaluation metrics, basic setup and opponent methods

To assess, the performance of the suggested method four evaluation metrics are utilized. These include root mean square error (RMSE) calculated using Eq. (26), mean absolute error (MAE) determined according to Eq. (27), mean square error (MSE) computed using Eq. (28) and coefficient of determination (R^2) determined using Eq. (29)

$$\text{RMSE} = \sqrt{1/n \sum_{i=1}^{n} (y_i - \widehat{y_i})^2} \tag{26}$$

$$\text{MAE} = \frac{1}{n} \sum_{i=1}^{n} |y_i - \widehat{y_i}| \tag{27}$$

$$\text{MSE} = \frac{1}{n} \sum_{i=1}^{n} \frac{(y_i - \widehat{y}_i)^2}{y_i} \tag{28}$$

$$R^2 = 1 - \frac{\sum_{i=1}^{n} (y_i - \widehat{y}_i)^2}{\sum_{i=1}^{n} (y_i - \bar{y})^2}, \tag{29}$$

in which y_i and $\widehat{y}_i$ represent actual prediction values for a given iteration, $\bar{y}$ represents the mean value of real data, and n represents the number of data samples.

When conducting multiple steps ahead forecasts, LSTM networks provide several output neurons, each representing a single day. Accordingly, for each step ahead, each metric is evaluated independently for every step. When multiple steps are provided as output an overall metric is introduced. However, when a single output neuron is provided, in the case of a single step ahead prediction, the metrics are identical.

When performing multiple steps ahead predictions, the initial steps—1 predictions provide less than the sufficient number of output data points to determine an appropriate error rate. Therefore in cases when for instance a 3-step ahead prediction is made, the first two predictions for the first and second days ahead are disregarded, and metrics are only calculated after the 3rd day ahead. For this reason, the metrics are not the same as simple arithmetic mean.

The prediction task can be formulated as a minimization problem guided by an objective function. This research makes use of Eq. (30) that relies on the overall MSE metric.

$$Obj = min(MSE) \tag{30}$$

Due to the heavy computation demands of the described methods, for both experiments, testing has been conducted using a population size of six individual agents and through five iterations. Additionally, to provide valid groups for comparative analysis, due to the random nature innate to metaheuristic algorithms the shown results are averaged over 10 independent runs.

Additionally, an early stopping condition is introduced for both simulations as $patience = \frac{epochs}{3}$ the recurrent dropout rate for the LSTM networks has been kept at the suggested default 0.01 rate, the utilized optimizer is *adam* with loss function *mse*, the *batch size* is set to 16, and the applied activation function being *relu*.

To ensure a valid assessment, several contemporary metaheuristics have been put into a comparative analysis to demonstrate the improvements made. The algorithms tested include the novel HHO-DO, the original HHO [45], BA [47], SCA [13], FA [10], TLB [48], implemented independently for this study, and tested under identical conditions as the HHO-DO algorithm, with control parameters set to their default values recommended in the papers that initially introduced the algorithms.

5. Experimental results and discussion

This section presents the results attained during experimentation and provides a discussion of the outcomes. The first results attained from the LSTM simulation are covered, which use the original dataset without decomposition methods. Following this, results from the VMD-LSTM method are presented along with a comparison with contemporary metaheuristics to present the improvements made and validate enactments made with the use of the HHO-DO algorithm.

Furthermore, the presented results are shown using normalized data points in the range of (0, 1). Subsequently, results attained results might differ from those shown in previous studies due to differences in scales. Best performance indicators for each simulation are marked with bold.

5.1 Simulations with LSTM

The following section covers findings and results as well as a comparative analysis of the LSTM approach applied to data prediction. In Table 3 the worst, mean, median, standard deviation, and variance of the objective function (MSE) are shown and cover 10 independent runs for five-step ahead predictions.

Table 4 covers performance indicators $R2$, MAE, MSE, RMSE attained during the best performing execution for each prediction step individually.

Objective function metric results shown in Table 3 strongly indicate that the proposed LSTM-HHO-DO approach on average and median obtains better results compared to other contemporary algorithms, only being

Table 3 Objective function metrics for LSTM application.

Method	Best	Worst	Mean	Median	Std	Var
LSTM-HHO-DO	**2.346E-04**	2.463E-04	**2.392E-04**	**2.380E-04**	4.700E-06	2.210E-11
LSTM-HHO	2.467E-04	2.722E-04	2.658E-04	2.722E-04	1.110E-05	1.230E-10
LSTM-SCA	2.478E-04	2.619E-04	2.528E-04	2.507E-04	5.460E-06	2.980E-11
LSTM-FA	2.357E-04	**2.456E-04**	2.406E-04	2.405E-04	**3.520E-06**	**1.240E-11**
LSTM-BA	2.421E-04	2.559E-04	2.503E-04	2.515E-04	5.730E-06	3.280E-11
LSTM-TLB	2.410E-04	3.010E-04	2.607E-04	2.505E-04	2.450E-05	6.000E-10

Table 4 The R2, MAE, MSE, and RMSE metrics of the best performing LSTM approach.

	Error indicator	LSTM-HHO-DO	LSTM-HHO	LSTM-SCA	LSTM-FA	LSTM-BA	LSTM-TLB
One-step ahead	R2	**0.989239**	0.988916	0.988825	0.988771	0.989125	0.988837
	MAE	**0.010868**	0.010978	0.011016	0.010907	0.010914	0.011149
	MSE	**0.000238**	0.000249	0.000246	0.000243	0.000242	0.000244
	RMSE	**0.015428**	0.015780	0.015682	0.015580	0.015552	0.015634
Two-step ahead	R2	**0.989706**	0.989070	0.988966	0.989129	0.988986	0.989205
	MAE	**0.010667**	0.010907	0.010953	0.010719	0.011032	0.010983
	MSE	**0.000228**	0.000243	0.000243	0.000237	0.000243	0.000236
	RMSE	**0.015113**	0.015602	0.015604	0.015391	0.015575	0.015378
Three-step ahead	R2	**0.989688**	0.988703	0.988883	0.989659	0.988945	0.989041
	MAE	0.010665	0.011282	0.011076	**0.010631**	0.011160	0.011100
	MSE	0.000230	0.000247	0.000245	**0.000227**	0.000242	0.000239
	RMSE	0.015180	0.015720	0.015656	**0.015064**	0.015571	0.015475
Four-step ahead	R2	**0.989455**	0.989119	0.988461	0.989176	0.989311	0.989165
	MAE	**0.010822**	0.010953	0.011376	0.010934	0.010899	0.011055
	MSE	0.000235	0.000241	0.000253	0.000235	0.000235	0.000239
	RMSE	**0.015317**	0.015522	0.015906	0.015344	0.015334	0.015447

Continued

Table 4 The R2, MAE, MSE, and RMSE metrics of the best performing LSTM approach.—cont'd

	Error indicator	LSTM-HHO-DO	LSTM-HHO	LSTM-SCA	LSTM-FA	LSTM-BA	LSTM-TLB
Five-step ahead	R2	0.989063	0.988567	0.988498	**0.989196**	0.988729	0.988721
	MAE	0.011102	0.011317	0.011349	**0.010945**	0.011178	0.011290
	MSE	0.000242	0.000253	0.000251	**0.000236**	0.000248	0.000246
	RMSE	0.015543	0.015900	0.015852	**0.015370**	0.015760	0.015680
Overall Results	R2	**0.989431**	0.988876	0.988728	0.989188	0.989019	0.988994
	MAE	**0.010825**	0.011087	0.011154	0.010827	0.011037	0.011116
	MSE	**0.000235**	0.000247	0.000248	0.000236	0.000242	0.000241
	RMSE	**0.015317**	0.015705	0.015740	0.015351	0.015559	0.015523

outperformed by the LSTM-HHO-FA algorithm in the worst case, suggesting that further improvement is possible and enforcing the no free lunch theorem.

Further, this observation is solidified by results shown in Table 4 where predictions for each step are shown. The proposed approach performed the best overall. However, when casting three and five-step ahead predictions it has been marginally outperformed by the LSTM-FA.

5.2 Simulations with VMD-LSTM

In the second experiment, the original dataset has been subjected to VMD, and the outputs have been piped as inputs to a LSTM network. Results are grouped similarly to the previous section, with Table 5 showing the worst, mean, median, standard deviation, and variance of the objective function (MSE), and Table 6 covering comparative analysis results of the proposed approach and other contemporary algorithms for individual prediction steps.

The improvements introduced by the application of VMD can be seen in Table 5 where the proposed VMD-LSTM-HHO-DO outperformed all other contemporary algorithms according to all evaluation metrics emphasizing the accuracy improvements made by subjecting data to decomposition methods.

The proposed VMD-LSTM-HHO-DO algorithm outperformed all contemporary algorithms on average and overall metrics as shown in Table 6. However, when casting one and four-step ahead predictions, the VMD-LSTM-SCA algorithm performed slightly better, further affirming the no-free-lunch theorem.

5.3 Comparison with other well-known methods

Five-step ahead predictions cast some of the tested state-of-the-art algorithms used to improve a LSTM network as well as real price data are shown in Fig. 2, while predictions made by the best performing LSTM-HHO-DO model, alongside real-world data can be seen in Fig. 3.

The best prediction made by the best performing LSTM-HHO-DO model, alongside real-world data can be seen in Fig. 3.

Table 5 Objective function metrics for VMD-LSTM application.

Method	Best	Worst	Mean	Median	Std	Var
VMD–LSTM–HHO–DO	**1.750E-04**	**1.750E-04**	**1.750E-04**	**1.750E-04**	**0**	**0**
VMD–LSTM–HHO	1.845E-04	2.037E-04	1.884E-04	1.845E-04	7.680E-06	5.890E-11
VMD–LSTM–SCA	1.970E-04	2.145E-04	2.057E-04	2.053E-04	7.820E-06	6.120E-11
VMD–LSTM–FA	2.056E-04	2.305E-04	2.226E-04	2.236E-04	9.120E-06	8.310E-11
VMD–LSTM–BA	1.914E-04	2.027E-04	1.958E-04	1.914E-04	5.360E-06	2.880E-11
VMD–LSTM–TLB	1.939E-04	2.254E-04	2.035E-04	2.022E-04	1.160E-05	1.340E-10

Table 6 The R2, MAE, MSE, and RMSE metrics of the best performing VMD-LSTM approach.

	Error indicator	VMD-LSTM-HHO-DO	VMD-LSTM-HHO	VMD-LSTM-SCA	VMD-LSTM-FA	VMD-LSTM-BA	VMD-LSTM-TLB
One-step ahead	R2	0.991313	0.991233	**0.992671**	0.990029	0.991466	0.992343
	MAE	0.009602	**0.008802**	0.008981	0.010450	0.009816	0.009138
	MSE	0.000192	0.000169	**0.000161**	0.000221	0.000191	0.000169
	RMSE	0.013866	0.012685	**0.012670**	0.014865	0.013824	0.012991
Two-step ahead	R2	**0.993067**	0.992557	0.990262	0.989065	0.991466	0.991418
	MAE	**0.008436**	0.009051	0.010879	0.010966	0.009838	0.009761
	MSE	**0.000157**	0.000169	0.000213	0.000241	0.000187	0.000192
	RMSE	**0.012513**	0.012702	0.014611	0.015520	0.013682	0.013842
Three-step ahead	R2	**0.992323**	0.992101	0.988879	0.991197	0.991039	0.991445
	MAE	**0.008892**	0.009802	0.011990	0.009958	0.009931	0.009645
	MSE	**0.000173**	0.000183	0.000239	0.000194	0.000200	0.000189
	RMSE	**0.013158**	0.013215	0.015473	0.013935	0.014151	0.013742
Four-step ahead	R2	0.991394	0.991174	**0.993357**	0.991559	0.991979	0.991835
	MAE	0.009561	0.009455	**0.008705**	0.009693	0.009437	0.009492
	MSE	0.000190	0.000172	**0.000148**	0.000186	0.000178	0.000182
	RMSE	0.013785	0.012796	**0.012166**	0.013648	0.013346	0.013497

Continued

Table 6 The R2, MAE, MSE, and RMSE metrics of the best performing VMD-LSTM approach.—cont'd

	Error indicator	VMD-LSTM-HHO-DO	VMD-LSTM-HHO	VMD-LSTM-SCA	VMD-LSTM-FA	VMD-LSTM-BA	VMD-LSTM-TLB
Five-step ahead	R2	**0.992753**	0.992142	0.989867	0.991673	0.991084	0.989215
	MAE	**0.008641**	0.009694	0.011544	0.009523	0.010058	0.011082
	MSE	**0.000163**	0.000186	0.000224	0.000186	0.000200	0.000238
	RMSE	**0.012763**	0.013319	0.014956	0.013630	0.014154	0.015429
Overall Results	R2	**0.992177**	0.991823	0.991023	0.990707	0.991406	0.991253
	MAE	**0.009027**	0.009361	0.010420	0.010118	0.009816	0.009824
	MSE	**0.000175**	0.000176	0.000197	0.000206	0.000191	0.000194
	RMSE	0.013228	**0.012946**	0.014037	0.014339	0.013835	0.013924

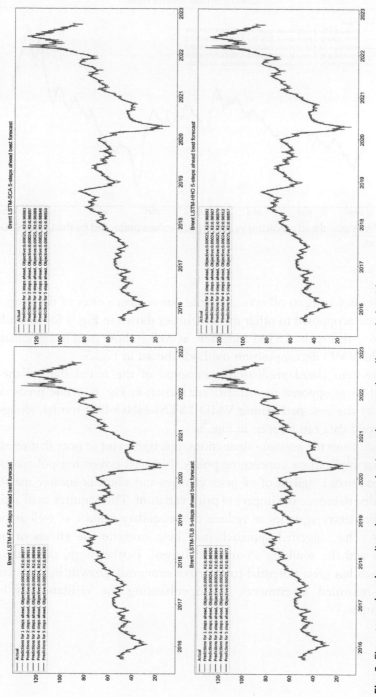

Fig. 2 Five steps ahead predicted vs actual price values cast by all tested algorithm.

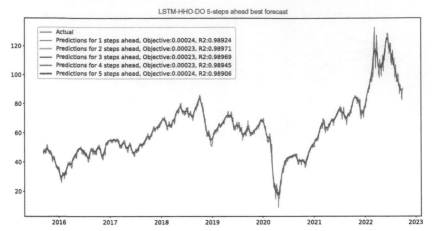

Fig. 3 Five steps ahead predicted vs actual price values presented by the LSTM-HHO-DA algorithm.

It is interesting to observe the rapid convergence rates of the proposed approach as opposed to other metaheuristics shown in Fig. 4 for algorithms used to optimize the LSTM approach, as well as those used in combination with the VMD decomposition methods shown in Fig. 5.

Five-stem ahead predictions cast some of the tested state-of-the-art algorithms as opposed to real data are shown in Fig. 6, while predictions made by the best performing VMD-LSTM-HHO-DO model, alongside real-world data can be seen in Fig. 7.

Aside from the previous deductions, it is important to note that research presents implications concerning policy. Namely, governing policymakers need to remain vigilant of oil price changes and adjust economic measures as needed to reduce the impact of price variation. This requires swift actions from monetary agencies to reduce the secondary impact as well as price shocks. The suggested approach may help mitigate the effects of price shocks and the resulting aftereffects thereof. Furthermore, the proposed approach has great potential to improve economic growth by stimulating better-informed investments, while enhancing the resistance of local economies.

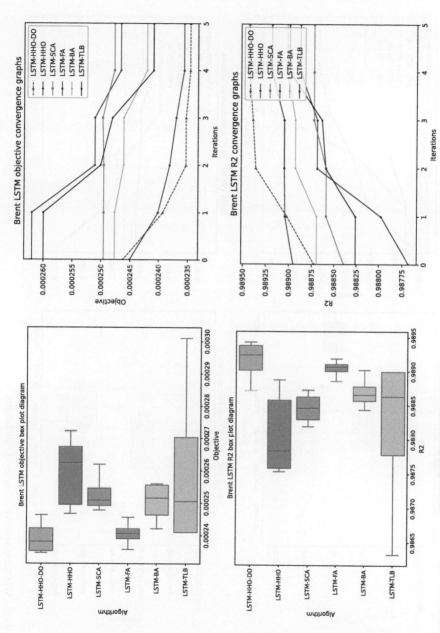

Fig. 4 LSTM Comparative analysis visualizations.

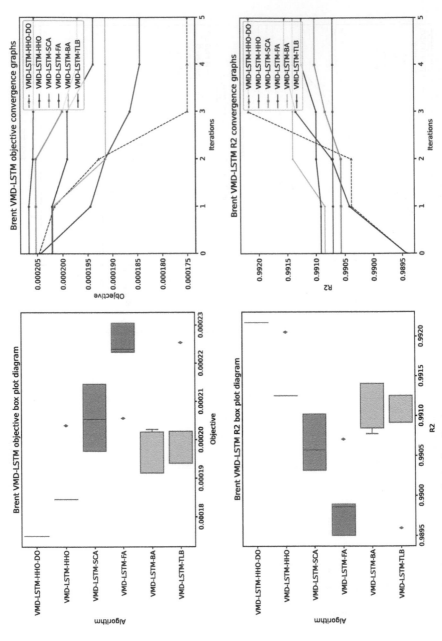

Fig. 5 VMD-LSTM Comparative analysis visualizations.

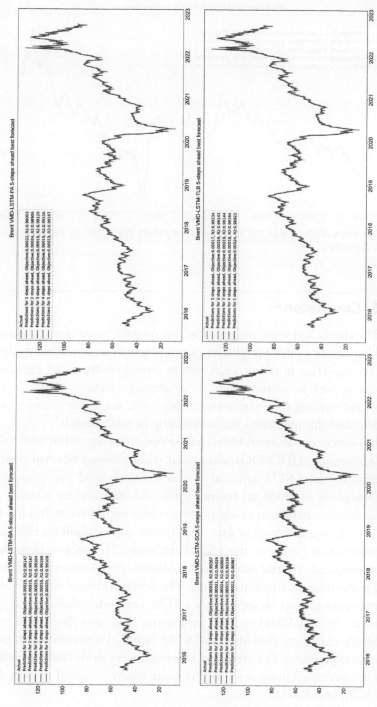

Fig. 6 Five steps ahead predicted vs actual price values cast by all tested algorithm.

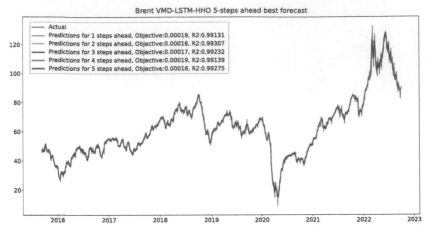

Fig. 7 Five steps ahead predicted vs actual price values presented by the VMD-LSTM-HHO-DA algorithm.

6. Conclusion

The conducted work presented in this research paper shows methods for tackling crude oil price prediction, a pressing economic subject made all the more important in recent years due to armed conflicts and pandemic outbreaks as well as several other socioeconomic factors. While several methods for tackling time-series forecasting exist, room for improvement also exists, and this motivated the conducting of this research.

In this work, an enhanced Harris hawks optimizer algorithm with a disputation operator (HHO-DO) was tasked with selecting optimal control parameters for an LSTM artificial neural network used for time-series price forecasting of crude oil trading price. Additionally, to account for the complexity inherent in crude oil price data, the VMD technique has been applied, to split complex data into simpler subcomponents that have been provided as inputs to the LSTM network. The proposed method has been validated on a real-world dataset consisting of Brent crude oil daily trading prices through two experiments. The first experiment used the original time-series data as an input to a LSTM network, while the second experiment utilized VMD to split time-series data into five signal subcomponents that were then used as LSTM inputs. Furthermore, in both conducted experiments five-step ahead forecasts were made and the predictions were evaluated using some of the most wildly accepted metrics $R2$, MAE, MSE, and $RMSE$.

The proposed LSTM-HHO-DO and VMD-LSTM-HHO-DO models have been subjected to a comparative analysis with other contemporary and well-performing metaheuristics. The proposed approach showed a general improvement in performance over other algorithms.

Future work will focus on expanding the conducted research and exploring the potential of additional signal decomposition methods, as well as different neural network architectures and their various combinations for tackling the pressing and important subject of crude oil price forecasting.

References

[1] K. Dragomiretskiy, D. Zosso, Variational mode decomposition, IEEE Trans. Signal Process. 62 (3) (2013) 531–544.

[2] M. Zivkovic, L. Jovanovic, M. Ivanovic, A. Krdzic, N. Bacanin, I. Strumberger, Feature selection using modified sine cosine algorithm with COVID-19 dataset, in: Evolutionary Computing and Mobile Sustainable Networks, Springer, 2022, pp. 15–31.

[3] R.S. Latha, B. Saravana Balaji, N. Bacanin, I. Strumberger, M. Zivkovic, M. Kabiljo, Feature selection using grey wolf optimization with random differential grouping, Comput. Syst. Sci. Eng. 43 (1) (2022) 317–332.

[4] Y. Singh, A. Biswas, Swaragram based residual neural architecture for raag identification in Indian classical music, in: 2021 12th International Conference on Computing Communication and Networking Technologies (ICCCNT), IEEE, 2021, pp. 1–6.

[5] D. Jovanovic, M. Marjanovic, M. Antonijevic, M. Zivkovic, N. Budimirovic, N. Bacanin, Feature selection by improved sand cat swarm optimizer for intrusion detection, in: 2022 International Conference on Artificial Intelligence in Everything (AIE), IEEE, 2022, pp. 685–690.

[6] S. Hochreiter, J. Schmidhuber, Long short-term memory, Neural Comput. 9 (8) (1997) 1735–1780.

[7] D. Sarkar, A. Biswas, Comparative performance analysis of recent evolutionary algorithms, in: Evolution in Computational Intelligence, Springer, 2022, pp. 151–159.

[8] D.H. Wolpert, W.G. Macready, No free lunch theorems for optimization, IEEE Trans. Evol. Comput. 1 (1) (1997) 67–82.

[9] S. Mirjalili, S.M. Mirjalili, A. Lewis, Grey wolf optimizer, Adv. Eng. Softw. 69 (2014) 46–61.

[10] X.-S. Yang, A. Slowik, Firefly algorithm, in: Swarm Intelligence Algorithms, CRC Press, 2020, pp. 163–174.

[11] S. Mirjalili, A. Lewis, The whale optimization algorithm, Adv. Eng. Softw. 95 (2016) 51–67.

[12] L. Abualigah, A. Diabat, S. Mirjalili, M. Abd Elaziz, A.H. Gandomi, The arithmetic optimization algorithm, Comput. Methods Appl. Mech. Eng. 376 (2021) 113609.

[13] S. Mirjalili, SCA: a sine cosine algorithm for solving optimization problems, Knowl. Based Syst. 96 (2016) 120–133.

[14] D. Sarkar, N. Mishra, A. Biswas, Genetic algorithm-based deep learning models: a design perspective, in: Proceedings of the Seventh International Conference on Mathematics and Computing, Springer, 2022, pp. 361–372.

[15] H. Niu, Y. Zhao, Crude oil prices and volatility prediction by a hybrid model based on kernel extreme learning machine, Math. Biosci. Eng. 18 (6) (2021) 8096–8122.

[16] A. Shabri, R. Samsudin, Crude oil price forecasting based on hybridizing wavelet multiple linear regression model, particle swarm optimization techniques, and principal component analysis, Sci. World J. 2014 (2014) 854520.

[17] W. Xie, L. Yu, S. Xu, S. Wang, A new method for crude oil price forecasting based on support vector machines, in: International Conference on Computational Science, Springer, 2006, pp. 444–451.

[18] N. Bacanin, M. Zivkovic, L. Jovanovic, M. Ivanovic, T.A. Rashid, Training a multilayer perception for modeling stock price index predictions using modified whale optimization algorithm, in: Computational Vision and Bio-Inspired Computing, Springer, 2022, pp. 415–430.

[19] A. Petrovic, I. Strumberger, T. Bezdan, H.S. Jassim, S.S. Nassor, Cryptocurrency price prediction by using hybrid machine learning and beetle antennae search approach, in: 2021 29th Telecommunications Forum (TELFOR), IEEE, 2021, pp. 1–4.

[20] G. Beni, J. Wang, Swarm intelligence in cellular robotic systems, in: Robots and Biological Systems: Towards a New Bionics? Springer Berlin Heidelberg, Berlin, Heidelberg, 1993, pp. 703–712.

[21] C.W. Reynolds, Flocks, herds and schools: a distributed behavioral model, SIGGRAPH Comput. Graph. 21 (4) (1987) 25–34, https://doi.org/10.1145/37402. 37406. 0097-8930.

[22] D. Karaboga, Artificial bee colony algorithm, Scholarpedia 5 (3) (2010) 6915.

[23] G. Dhiman, V. Kumar, Emperor penguin optimizer: a bio-inspired algorithm for engineering problems, Knowl. Based Syst. 159 (2018) 20–50.

[24] W. Deng, R. Chen, B. He, Y. Liu, L. Yin, J. Guo, A novel two-stage hybrid swarm intelligence optimization algorithm and application, Soft Comput. 16 (2012), https://doi.org/10.1007/s00500-012-0855-z.

[25] A. Mortazavi, V. Toğan, M. Moloodpoor, Solution of structural and mathematical optimization problems using a new hybrid swarm intelligence optimization algorithm, Adv. Eng. Softw. 127 (2019) 106–123, https://doi.org/10.1016/j.advengsoft.2018.11.004.

[26] K. Karthikeyan, S. Ramasamy, K. Shankar, S.K. Lakshmanaprabu, V. Varadarajan, M. Elhoseny, G. Manogaran, Energy consumption analysis of Virtual Machine migration in cloud using hybrid swarm optimization (ABC-BA), J. Supercomput. 76 (2020), https://doi.org/10.1007/s11227-018-2583-3.

[27] E. Houssein, A. Ewees, M. Elsayed Abd Elaziz, Improving twin support vector machine based on hybrid swarm optimizer for heartbeat classification, Pattern Recognit. Image Anal. 28 (2018) 243–253, https://doi.org/10.1134/S1054661818020037.

[28] S. Sumi, F. Zaman, H. Hirose, A rainfall forecasting method using machine learning models and its application to the Fukuoka city case, Int. J. Appl. Math. Comput. Sci. 22 (2012) 841–854, https://doi.org/10.2478/v10006-012-0062-1.

[29] M. Zivkovic, L. Jovanovic, M. Ivanovic, N. Bacanin, I. Strumberger, P.M. Joseph, XGBoost hyperparameters tuning by fitness-dependent optimizer for network intrusion detection, in: Communication and Intelligent Systems, Springer, 2022, pp. 947–962.

[30] M. Zivkovic, N. Bacanin, J. Arandjelovic, I. Strumberger, K. Venkatachalam, Firefly algorithm and deep neural network approach for intrusion detection, in: Applications of Artificial Intelligence and Machine Learning, Springer, 2022, pp. 1–12.

[31] D. Jovanovic, M. Antonijevic, M. Stankovic, M. Zivkovic, M. Tanaskovic, N. Bacanin, Tuning machine learning models using a group search firefly algorithm for credit card fraud detection, Mathematics 10 (13) (2022) 2272.

[32] A. Petrovic, N. Bacanin, M. Zivkovic, M. Marjanovic, M. Antonijevic, I. Strumberger, The AdaBoost approach tuned by firefly metaheuristics for fraud detection, in: 2022 IEEE World Conference on Applied Intelligence and Computing (AIC), IEEE, 2022, pp. 834–839.

[33] L. Jovanovic, M. Zivkovic, M. Antonijevic, D. Jovanovic, M. Ivanovic, H.S. Jassim, An emperor penguin optimizer application for medical diagnostics, in: 2022 IEEE Zooming Innovation in Consumer Technologies Conference (ZINC), IEEE, 2022, pp. 191–196.

[34] T. Bezdan, M. Zivkovic, N. Bacanin, A. Chhabra, M. Suresh, Feature selection by hybrid brain storm optimization algorithm for COVID-19 classification, J. Comput. Biol. 29 (2022) 515–529.

[35] M. Zivkovic, N. Bacanin, K. Venkatachalam, A. Nayyar, A. Djordjevic, I. Strumberger, F. Al-Turjman, COVID-19 cases prediction by using hybrid machine learning and beetle antennae search approach, Sustain. Cities Soc. 66 (2021) 102669.

[36] N. Bacanin, T. Bezdan, E. Tuba, I. Strumberger, M. Tuba, M. Zivkovic, Task scheduling in cloud computing environment by grey wolf optimizer, in: 2019 27th Telecommunications Forum (TELFOR), IEEE, 2019, pp. 1–4.

[37] T. Bezdan, M. Zivkovic, N. Bacanin, I. Strumberger, E. Tuba, M. Tuba, Multi-objective task scheduling in cloud computing environment by hybridized bat algorithm, J. Intell. Fuzzy Syst. 42 (1) (2022) 411–423.

[38] N. Bacanin, M. Antonijevic, T. Bezdan, M. Zivkovic, K. Venkatachalam, S. Malebary, Energy efficient offloading mechanism using particle swarm optimization in 5G enabled edge nodes, Clust. Comput. (2022) 1–12.

[39] M. Zivkovic, T. Zivkovic, K. Venkatachalam, N. Bacanin, Enhanced dragonfly algorithm adapted for wireless sensor network lifetime optimization, in: Data Intelligence and Cognitive Informatics, Springer, 2021, pp. 803–817.

[40] M. Zivkovic, N. Bacanin, E. Tuba, I. Strumberger, T. Bezdan, M. Tuba, Wireless sensor networks life time optimization based on the improved firefly algorithm, in: 2020 International Wireless Communications and Mobile Computing (IWCMC), IEEE, 2020, pp. 1176–1181.

[41] I. Strumberger, T. Bezdan, M. Ivanovic, L. Jovanovic, Improving energy usage in wireless sensor networks by whale optimization algorithm, in: 2021 29th Telecommunications Forum (TELFOR), IEEE, 2021, pp. 1–4.

[42] M. Zivkovic, T. Bezdan, I. Strumberger, N. Bacanin, K. Venkatachalam, Improved Harris hawks optimization algorithm for workflow scheduling challenge in cloud-edge environment, in: Computer Networks, Big Data and IoT, Springer, 2021, pp. 87–102.

[43] M. Salb, L. Jovanovic, M. Zivkovic, E. Tuba, A. Elsadai, N. Bacanin, Training logistic regression model by enhanced moth flame optimizer for spam email classification, in: Computer Networks and Inventive Communication Technologies, Springer, 2023, pp. 753–768.

[44] N. Bacanin, K. Alhazmi, M. Zivkovic, K. Venkatachalam, T. Bezdan, J. Nebhen, Training multi-layer perceptron with enhanced brain storm optimization metaheuristics, Comput. Mater. Contin. 70 (2022) 4199–4215.

[45] A.A. Heidari, S. Mirjalili, H. Faris, I. Aljarah, M. Mafarja, H. Chen, Harris hawks optimization: algorithm and applications, Future Gener. Comput. Syst. 97 (2019) 849–872.

[46] S. Talatahari, H. Bayzidi, M. Saraee, Social network search for global optimization, IEEE Access 9 (2021) 92815–92863.

[47] X.-S. Yang, A.H. Gandomi, Bat algorithm: a novel approach for global engineering optimization, Eng. Comput. 29 (2012) 464–483.

[48] R.V. Rao, V.J. Savsani, D.P. Vakharia, Teaching-learning-based optimization: a novel method for constrained mechanical design optimization problems, Comput. Aided Des. 43 (3) (2011) 303–315.

About the authors

Luka Jovanovic received his degree from the Faculty of Technical Sciences at Singidunum University. Throughout his studies, Luka actively contributed to various artificial intelligence research projects, resulting in over 50 peer-reviewed publications in esteemed journals and international conferences. Luka maintains a diverse range of research interests, covering topics such as reinforcement learning, natural language processing, distributed computing, and signal processing. In his current capacity at the Faculty of Technical Sciences, Singidunum University, Luka plays a crucial role in the realm of research. His responsibilities include spearheading research initiatives, publishing scholarly papers, conceptualizing innovative ideas, and implementing experiments through programming.

Milos Antonijevic has PhD in Computer Sciences (study program advanced security systems) from Singidunum University and master degree of Engineer of Organizational Sciences from Faculty of Organizational sciences, University of Belgrade (study program E-business). He started his career in education 14 years ago at High School of Graphics and Media in Belgrade. He currently works as an assistant professor at Singidunum University, Belgrade, Serbia and as certified ISO 27001 Auditor for various accreditation authorities. He is involved in scientific research in the field of computer science with focus on implementation of AI and optimization algorithms in various security systems.

Miodrag Zivkovic received his PhD degree from the School of Electrical Engineering, University of Belgrade in 2014. He currently works as a full professor at the Faculty of Informatics and Computing, Singidunum University, Belgrade, Serbia. He is involved in scientific research in the field of computer science and his specialty includes artificial intelligence, machine learning, deep learning, stochastic optimization algorithms, swarm intelligence, and human–computer inter-action. He has published more than 150 scien-

tific papers in high-quality journals, articles as book chapters, and international conferences indexed in Clarivate Analytics JCR, Scopus, WoS, IEEExplore, and other scientific databases.

Dr. Milos Dobrojevic obtained his BSc, MSc, and PhD degrees in 2000, 2003, and 2006, respectively, all from the Faculty of Mechanical Engineering, University of Belgrade, Serbia. As of 2000, Milos Dobrojevic was employed at Energoprojekt Visokogradnja, Belgrade, in the Construc-tion Equipment Division holding manage-rial positions first in general repair workshop, and later in tower cranes and concrete plants work unit. From 2007. He was employed in Innovation Center on Faculty of Mechanical

Engineering on positions of project manager and IT consultant, leading the realization of several international and domestic projects in the field of information technologies and industrial management. From 2008. He established Magma personal trademark (www.magma.rs) for CMS & ERP systems and several other software products that he developed. Since 2013, he is employed on University Sinergija, Bijeljina B&H, where he currently holds the position of associated professor in the field of IT on the Faculty of Computing and Informatics. From 2018, he works on University Singidunum in Belgrade, currently as associated professor in the field of IT on the Faculty of Technical Sciences. The research field of Milos

Dobrojevic currently covers Internet technologies, Internet of Things (IoT), data science, computer vision, technology development, and industrial resources management. He authored many web based software solutions, as well as numerous scientific and research articles in the field of IoT, AI, resources management, development of ERP systems and content management systems.

Mohamed Salb holds a PhD in Computer Science, specializing in Advanced Security Systems from Singidunum University, Belgrade, Serbia. Additionally, he obtained a master's degree in Information Technology from Singidunum University. With over a decade of experience in the field of education, Mohamed began his career at PRIMA International School in Belgrade, where he taught ICT and Mathematics. Currently, he serves as a Computer Science and Computing Teacher at PRIMA International School, where he develops curriculum, teaches coding and programming, and provides ICT support. In parallel to his teaching career, Mohamed is an accomplished freelance researcher, having published numerous research papers in the field of information technology. He has received recognition for his contributions, including the Libyan Innovation Award.

Ivana Strumberger started her university career in 2013 as teaching assistant at Faculty of Computer Science in Belgrade. She received her PhD degree from Singidunum University in 2020 from the domain of Computer Science (average grade: 9,93). She currently works as teaching assistant at Faculty of Informatics and Computing, Singidunum University, Belgrade, Serbia, however, she is in the process of being elected for assistant professor. She conducts research in the domain of computer science and her specialty includes swarm intelligence, machine learning, optimization and modeling, cloud computing, computer networks and distributed computing. She has published around 50 scientific papers in high quality journals and international conferences indexed in Clarivate Analytics JCR, Scopus, WoS, IEEExplore. She has also published 10 book chapters in Springer Lecture Notes in Computer Science series. She has also published one book from the domain of Cloud Computing. She is regular reviewer of many international state-of-the-art journals with high Clarivate Analytics and WoS impact factor such as Applied Soft Computing, Journal of Ambient Intelligence & Humanized Computing, Soft Computing, Swarm and Evolutionary Computation, etc. She has been included in prestigious list of Stanford University with best 2% world scientists for the year 2021.

Dr. Nebojsa Bacanin received his PhD degree from the Faculty of Mathematics, University of Belgrade (The University of Belgrade was in 2015 in 2016 ranked among 400 and 300, best universities in the world, respectively, according to the Shanghai list [http://www.shanghairanking.com/World-University-Rankings/University-of-Belgrade.html]) in 2015 (study program Computer Science, average grade 10,00). The external Committee Member for his PhD thesis defense was one of the greatest world scholars from the domain of artificial intelligence, specifically metaheuristics optimization (creator of a dozen state-of-the-art swarm intelligence algorithms), Professor Xin-She Yang, from Middlesex University, London, UK. He started University career in Serbia 18 years ago at Graduate School of Computer Science in Belgrade. He currently works as a Full Professor, Head of study program Applied Artificial Intelligence and as a Vice-Rector for Scientific Research at Singidunum University, Belgrade, Serbia. He is involved in scientific research in the field of computer science and his specialties includes stochastic optimization algorithms, swarm intelligence, machine learning, soft computing, optimization and modeling, image processing and cloud and distributed computing. He has published more than 320 scientific papers (more than 130 papers indexed in Clarivate Analytics SCIE) in high quality journals, articles as the book chapters and international conferences indexed in Clarivate Analytics JCR, Scopus, WoS, IEEExplore, and other scientific databases. He has also published four books from domains of Cloud Computing, Web Programming and Advanced Java Spring Programming. He is a member of numerous editorial boards, scientific and advisory committees of international conferences and journals and regular reviewer for international journals with high Clarivate Analytics and WoS impact factor. He also serves associate editor and as guest editor of many outstanding international journals indexed in Clarivate Analytics SCIE. He has also been included in the prestigious Stanford University list with 2% best world researchers from the domain of Artificial Intelligence, that encompasses whole career, for the years 2021 and 2022, while according to the AD Scientific Index" (Alper-Doger Scientific Index) he is currently ranked as the 2nd best researcher in Serbia from the domain of Computer Science. The research fields of Nebojsa Bacanin covers metaheuristics optimization, swarm intelligence, data science, artificial intelligence, and machine learning.

CHAPTER NINETEEN

Artificial neural network optimized with PSO to estimate the interfacial properties between FRP and concrete surface

Aman Kumar[a,b], **Harish Chandra Arora**[a,b], **Nishant Raj Kapoor**[a,b], **and Ashok Kumar**[a,b]

[a]Academy of Scientific and Innovative Research (AcSIR), Ghaziabad, Uttar Pradesh, India
[b]CSIR-Central Building Research Institute, Roorkee, Uttarakhand, India

Contents

Abstract

Deterioration of the concrete structure is a major problem in the construction industry. It reduces the strength, serviceability, and safety of the structures. Demolish these deteriorated structures is not the right solution, and it is uneconomical and produces a huge amount of waste. From a sustainability point of view, strengthening the deteriorated structures is an acceptable and possible solution. Strengthening with fiber-reinforced polymer (FRP), composites have been widely utilized for construction and rehabilitation purposes. The substrate bond between concrete and FRP composite is the crucial parameter deciding the influence of strengthening on the structural components. This bond is responsible to transfer the stresses from the concrete substrate to the FRP composite. There are various analytical models as well as standard guidelines available in the past studies to forecast the strength of the bond between the concrete and the FRP composite. The main disadvantage of these analytical models is that these models are valid only for limited datasets. In this chapter, a supervised machine learning (ML) model called artificial neural network (ANN) and optimized ANN with particle swarm optimization (PSO) was used to estimate the bond between the concrete and the FRP composite. The results of these ML models were compared with widely known standard guidelines such as fib, ACI, CNR-DT 200, and CS-TR-55-UK. It was found that the performance of PSO-ANN and ANN algorithms was good as compared to standard guidelines. The correlation coefficient of ANN and PSO-ANN models was 0.9777 and 0.9907, sequentially. The precision of the PSO-ANN algorithm was 1.33% higher as compared to the ANN model.

Notations
Acronyms

ABC	artificial bee colony
ACI	American Concrete Institute
ACO	ant colony optimization
ANFIS	adaptive neuro-fuzzy inference system
ANN	artificial neural network
CNR	National Research Center of Italy
EB	externally bonded
FRP	fiber-reinforced polymer
GPR	Gaussian process regression
INRC	Italian National Research Council
MAE	mean absolute error
MAPE	mean absolute percentage error
MARE	mean absolute relative error
ML	machine learning
NS	Nash–Sutcliffe efficiency index
NSM	near-surface mounted
PSO	particle swarm optimization
R	correlation coefficient
R^2	coefficient of determination
RC	reinforced concrete
RMSE	root mean squared error
TRC	textile-reinforced concrete

Symbols

b_c	width of concrete block
E_f	elastic modulus of FRP composite
f_{ct}	tensile strength of concrete
f_c'	compressive strength of concrete
f_f	tensile strength of FRP composite
L_f, L_b	bonded length
P_u	bond strength
t_f	thickness of FRP fiber
z	original value

Z^*	normalizing value
z_{max}	maximum value
z_{min}	minimum value
f_c	concrete tensile strength mean value
T_r	training set
V_a	validation set
T_e	testing set
b_f	width of FRP composite

1. Introduction

Concrete has been a popular structural component for more than 150 years due to its long life, affordability, simple upkeep, fire resistance, and rigidity [1]. The structures that were erected in the mid-to-late 1950s are showing indications of degradation, due to severe environmental conditions. The utmost reason for RC construction deterioration is the intrusion of CO_2 and/or chloride from the external environment [2]. The intrusion of ions may happen at the time of construction by using poor-quality water, poor-quality aggregates, and other types of materials. The postconstruction ingress of chloride ions or CO_2 is only possible to penetrate water inside the concrete through concrete pores and voids. Concrete buildings are degraded in many ways by reinforcement corrosion. Pitting corrosion lowers the area of the cross section of the reinforcement in a local location, lowering its ductility as well as yielding and ultimate capacities [3,4]. The corrosion increases the volume of the reinforcing bar, which further causes the spalling and losing the bond between the reinforcing bar and concrete. The various types of conventional and new classes of strengthening techniques like sprayed concrete, concrete, as well as steel jacketing, carbon-fiber-reinforced polymer (FRP) fabric strengthening, carbon-FRP laminates strengthening, and textile-reinforced concrete, strengthening are used to strengthen and retrofit the structural elements. These conventional techniques do have various limitations such as concrete jacketing increases the cross section as well as the dead load of the building, and steel jacketing has corrosion problems, etc. These limitations are overcome by the new class of strengthening materials. To escalate the shear strength, flexure strength, and axial strength of the structural elements, it is necessary to strengthen the deteriorated structural elements with a new class of repair composite materials.

Concrete structures strengthened with fabric composites are a widely used approach for enhancing their existing capacities (shear flexural and axial) [5]. In the last four decades, FRP composites have become key structural materials in the industry of repair and rehabilitation. Due to the various benefits of FRPs, such as high strength at lower weights, excellent corrosion resistance, and ease of handling, retrofitting or rehabilitating RC structures is the most prevalent application in civil infrastructure. These externally bonded (EB) or near-surface mounted (NSM) fabric strips on RC structures are become popular for practical applications [6]. Debonding is the general failure mode of FRP-retrofitted RC structures, which can be caused by the high interface stress concentration [7]. Extensive research has been done, and many methods for calculating the strength of bond between fiber polymers for reinforcement and concrete interface have been studied.

To estimate the strength of the bond between FRP and concrete surface is a very complex phenomena and is difficult to predict as it is based on various parameters like elastic modulus of FRP, f_{ck}, the width of concrete block, the width of FRP composite, the thickness of FRP system, bonded length, application type and surface conditions of the concrete block, etc.

The various worldwide used standard FRP-concrete bond strength prediction guidelines given by ACI 440.R-08, fib Bulletin, Italian National Research Council CNR-DT 200/2004, and CS-TR-55-UK. are referenced in the study to estimate the strength of the bond between concrete and FRP. However, the results obtained from these models are only valid for limited datasets. To overcome the limitations of these empirical models, several machine-learning (ML) models are available in the literature.

In this study, an optimized particle swarm optimization (PSO)-artificial neural network (ANN) approach has been utilized to determine the strength of the bond between the FRP and concrete surface. The performance of the proposed ML-based model with other analytical has been compared, and based on the suggested results, a more reliable and cost-effective model suggested is the primary objective of this study.

In this chapter, ANN and optimized ANN with PSO algorithm techniques are used to forecast the strength of the bond between FRP and the concrete surface. Section 2 explains the use of ML in civil engineering applications. Section 3 is about the collection of experimental data from the previous studies, the processing of data for the ML algorithms, and the performance metrics used to measure the performance of the analytical and ML models. The formulation of the analytical models is presented in Section 4 with an overview of the ML models (ANN and PSO-ANN). Results obtained from analytical and ML models are explained in Section 5 and the final highlights of this study are mentioned in Section 6.

2. Related work

The applicability of ML algorithms in the concrete engineering area is described as follows:

Jahed et al. [8] examined the strength of bond between the FRP and concrete using the artificial bee colony (ABC)-ANN and ANN. On comparison, it was found that the ANN model has poor precision and outcomes show that ABC-ANN can outperform it. The model's utility is limited because the author only used 150 experimental values in his dataset. Kumar et al. [9] optimized ANN with ABC to find out the strength of the bond. The worked-out results were compared with analytical models, ANN, and Gaussian process regression (GPR) models. The R-values of ANN-ABC and GPR models are 0.9514 and 0.9618, sequentially. The performance of GPR and optimized ANN model was higher as compared to existing analytical models. Pei and Wei [10] used an adaptive neuro-fuzzy inference system (ANFIS) model optimized with ant colony optimization (ACO) to analyze the FRP-to-concrete bond strength. The accuracy of the ACO-ANFIS model was higher as compared to analytical models. The R^2, NS, RMSE, MAE, and MARE of the ACO-ANFIS model were 0.97, 0.97, 1.29 kN, 0.81 kN, and 0.053, respectively.

The shear strength contribution of the beams strengthened with fabric composite reinforcement was explored by Naderpour and Alavi [11] with an artificial intelligence-based ANFIS model. When compared to the results directly obtained from existing standard equations, the study shows that the ANFIS model could forecast the shear influence of FRP in RC beams maintaining a higher degree of precision. Naderpour et al. [12] estimated the capacity of circular RC columns strengthened with fabric composite reinforcement with the ANFIS algorithm. The coefficient of determination of the ANFIS model was achieved up to 0.99, which is much higher as compared to standard models.

In 2021, Feng et al. [13] performed research to estimate the properties of self-consolidating concrete by utilizing 26 mix designs.ANFIS-PSO and ANFIS-DEO models showed more precise results whereas the ANFIS-ACO model was not able to predict with good accuracy. In another study

Basarir et al. [14] carried out the prediction to estimate the bending moments of concrete-filled steel tubes by the ANFIS model. A huge database was built and populated from the literature to train and test MLR, MNLR, and ANFIS models. The results show that the ANFIS model beats other standard techniques in forecasting the final pure bending of concrete-filled steel tubes with a high degree of precision. Pandit and Panda [15] forecasted the earthquake magnitude using ANN and ANFIS. When compared to ANFIS, the results indicated that ANN is more effective at predicting earthquake magnitude. Pandit and Biswal [16] also predicted the earthquake magnitude using ANFIS. Kar et al. [17] bring forward a model to calculate the shear strength contribution of EB FRP composites using ANFIS. The ANFIS estimations were compared with the actual findings of the testing and six frequently used design codal provisions estimations. The test results show that the ANFIS model is superior to other standard codal guidelines.

A neuro-fuzzy approach was used by Kar and Biswal [18] to predict the shear involvement of end-anchored FRP U-jackets. The current ANFIS model was developed and evaluated using 107 sets of data. It has been determined that the ANFIS algorithm was more highly effective than the considered design recommendations based on numerous statistical parameters. Naderpour and Mirrashid [19] estimated the shear capacity of RC beams using ANFIS. ANFIS was created using the subclustering method. In the training and test phases, the suggested ANFIS has a correlation coefficient of 0.98 and 0.94, respectively. The findings showed that the proposed neuro-fuzzy system could accurately forecast the RC beam's shear strength reinforced with steel stirrups. The strength and ultimate strain of concrete cylinders reinforced with FRP were examined by Mansouri et al. [20] with soft-computing techniques. In calculating the ultimate strain of FRP-strengthened concrete, the performance of the ANFIS-SC model was better than ANFIS-FCM and RBNN models, but M5Tree provided the poorest strength and strain predictions.

Mansouri et al. [21] also used artificial intelligence to predict the performance concrete with FRP. Four algorithms named ANFIS, ANN, multivariate adaptive regression splines, and M5 model tree were used to estimate the strength of FRP-wrapped concrete. In terms of forecasting the strength enhancement ratio and hoop strain reduction factor, the M5Tree model outperforms the other models, while the ANN model offered the best precise estimations of the strain enhancement ratio.

The punching shear of the slab column connections model was developed Naderpour and Mirrashid [22] with ANFIS. In this model, a neuro–fuzzy system uses five factors as inputs to estimate the considered output. For the training and testing datasets, the correlation coefficients between measured and estimated values were 0.9901 and 0.98022, respectively. The compressive of self-compacting concrete was predicted by Vakhshouri and Nejadi [23] using the ANFIS model with 55 mix design datasets. The influence of aggregate volume and the maximum aggregate size in the design mix is not much effective.

Naderpour and Mirrashid [24] estimated the moment capacity of spirally reinforced RC columns with ANFIS. The correlation coefficients for the training and testing phases were 0.99 and 0.98, respectively, according to the results of regression plots for standardized data. Nedushan [25] proposed a predicted model to worked out the elastic modulus of high and normal f_{ck} of concrete using nonlinear regression and ANFIS models. The ANFIS model performs better than nonlinear regression methods and other predictive models reported in the literature, according to the results. Cao et al. [26] identified the contribution of shear resistance of RC beams strengthened with FRP and the model generated using ANFIS. Six input parameters were employed to describe the bars' geometric and mechanical parameters, as well as shear characteristics. Sharafati et al. [27] optimize the ANFIS model with nature-inspired algorithms to forecast the f_{ck} of foamed concrete. The suggested predictive models' performance is measured using performance criteria. An ANFIS-PSO-based model shows good accuracy among other models. The shear strength of titled angle connectors was studied by Shariati et al. [28] and predicted using ANN, ANFIS, and extreme ML algorithms. The paper's findings reveal that slip is the greatest important component in tilted shear strength connectors, followed by the inclination angle. It has been shown that ELM takes less time and achieves significantly higher performance indices than ANFIS and ANN. Chung et al. [29] investigated the performance of carbon FRP RC members using a neuro–fuzzy logic system. A number of layers, retrofit ratio, the thickness of FRP system, existing concrete strength, stiffness, specimen size, and ultimate strength of fiber are utilized as input data to forecast stiffness, strain, and post-yielding modulus strength. When compared to previous researchers' proposed constitutive models, the proposed ANFIS models have a higher level of precision in predicting the constitutive characteristics of concrete strengthened with FRP. Wei et al. [30] used a hybrid ANFIS-firefly algorithm (FFA) to calculate the f_{ck} of high-strength concrete. For the training datasets, the correlation coefficient of the ANFIS–FFA algorithm is 1 and 0.9965 for the testing database.

3. Experimental setup

3.1 Data Bank

The data were collected from the previous experimental studies based on FRP-concrete bond strength. A random literature survey was done to collect the data from year 2010 to 2021. In the previous studies, it was found that the commonly used FRP-concrete bond strength prediction parameters are elastic modulus of FRP composite (E_f), concrete compressive strength (f'_c), tensile strength of FRP composite (f_f), width of concrete specimen (b_c), thickness of FRP fiber (t_f), width of FRP composite (b_f), and bonded length (L_f), and the same datasets have been gathered from the literature. The total number of 744 datasets were gathered from the literature, but after removing outliers, only 643 datasets were selected and also used in this study. Table 1 displays specific information about the input and output parameters. The statistical parameters of the collected database are presented in Table 2. Before processing the data in ANN, it was standardized between 0 and 1 using Eq. (1) [9].

$$Z^* = \frac{(z - z_{\min})}{z_{\max} - z_{\min}} \tag{1}$$

where $Z^* =$ normalizing value, $z =$ original value in the dataset, $z_{\max} =$ maximum value, and $z_{\min} =$ minimum value.

The frequency distribution of the input and output parameters is shown in Fig. 1. The majority of the compressive strength of concrete in Fig. 1A ranges in the value from 20 to 50 MPa. According to Fig. 1B, the predominant width measurement for concrete blocks is 150 mm. The maximum frequency of the elastic modulus of concrete ranges from 200 to 300 GPa, as seen in Fig. 1C. Similarly, the maximum frequency of the tensile strength of FRP composite is at 4000 GPa (Fig. 1D). Fig. 1E illustrates that the higher frequency values are found at a thickness of 0.2 mm of the FRP composite. According to Fig. 1F and G, the maximum frequencies for the FRP composite's width and bonded length are 50 and 200 mm, respectively. The majority of the bond strength value is between 5 and 20 kN (Fig. 1D). The correlation plot among the selected parameters is depicted in Fig. 2.

Table 1 Collected database.

References	E_f (GPa)	t_f (mm)	f_f (MPa)	L_f (mm)	b_f (mm)	f'_c (MPa)	b_c (mm)	P_u (kN)
Dai et al. [31]	74–230	0.12	1550–3550	330	100	35	200	28.6–43.2
Hosseini et al. [32]	238	0.131	4300	20–250	48	36.5–41.1	150	7.54–10.12
Li et al. [33]	270	0.167	3500	300	50	27–61	100	11.03–20.87
Chen et al. [34]	242	0.167	3513	300	50	42.2–72	100	12.2–20
Yuan et al. [35]	73	0.12	1333	200	40	39.68–50.98	150	10.12–12.96
Mostofinejad et al. [36]	230	0.166	3900	200	50	24.7–40.4	150	10.2–14.98
Heydari [37]	230–238	0.131–0.26	3900–4300	150	48	20–43	150	10.54–15.69
Dai et al. [38]	74–230	0.11–0.118	1550–3550	330	100	35	100	28.6–42.9
Yun et al. [39]	230	0.167	3400	300	50	36.9–48.6	150	14.4–20.1
Ueno et al. [40]	23.9–425.1	0.08–1.8	234–6700	50–700	10–100	16.2–74.5	80–500	2.4–46.35
Zhang et al. [41]	256	2	4100	180	20	29.3	100	7.53–9.54
Yuan et al. [42]	73–210	0.12–0.287	1400–2450	200	40	30.14	150	3.73–17.66
Moghaddas et al. [43]	76–230	0.11–0.34	2300–3900	200	30–60	22.7–48.2	150	3.9–24.5
Ceroni et al. [44]	80–230	0.166–0.48	2560–4830	200–300	100	15–20	250	12.52–22.81

Table 2 Statistical analysis of the collected dataset.

Parameters	Symbol	Type	Mean	Min.	Max.	Kurtosis	Skewness	Std.	Units
Concrete	f'_c	Input	35.9498	15	74.5000	3.7075	0.9639	12.6880	MPa
	b_c	Input	142.5163	80	500	16.4762	2.8153	51.1970	mm
FRP	E_f	Input	195.9706	23.9000	425.1000	2.8395	−0.5929	76.9364	GPa
	f_f	Input	3.4237e+03	234	6700	3.9376	−0.8142	890.2562	MPa
	t_f	Input	0.3470	0.0800	2	7.5594	2.2822	0.3904	mm
	b_f	Input	50.4541	10	100	3.8667	1.1189	20.9713	mm
	L_f	Input	195.6753	20	700	4.9498	0.4133	76.5265	mm
Bond strength	P_u	Output	14.0914	2.4000	46.3500	7.2368	1.8459	6.7304	kN

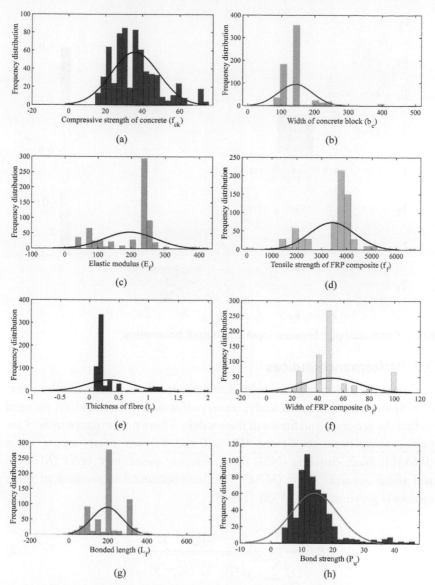

Fig. 1 Frequency distribution (A) $f_c{}'$, (B) b_c, (c) E_f, (D) f_f, (E) t_f, (F) b_f, (G) L_f and (H) P_u.

The correlation between the FRP composite tensile strength and the elastic modulus is good, with an R-value of 0.64. Similarly, with a value of 0.6, the correlation coefficient of bond strength and width of the FRP composite is good. The correlation coefficient between the elastic modulus of FRP composite and the thickness of FRP fiber is the lowest.

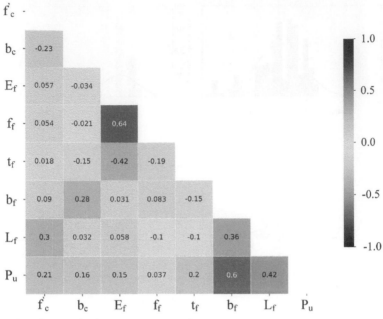

Fig. 2 Correlation plot between input and output parameters.

3.2 Performance indices

Performance indices are very helpful to compare the accuracy and the errors of different models. In this study, seven performance formulations are used to find the accuracy and fitness of the models. These performance indices are person correlation coefficient (R), a20-index, root mean square error (RMSE), Nash–Sutcliffe (NS), mean absolute percentage error (MAPE), and mean absolute error (MAE). The mathematical expression of these indices is given in Eqs. (2)–(8) [45–48]:

$$R = \frac{\sum_{i=1}^{N}(x_i - \overline{x})(y_i - \overline{y})}{\sqrt{\sum_{i=1}^{N}(x_i - \overline{x})^2(y_i - \overline{y})^2}} \tag{2}$$

$$\text{MAE} = \frac{1}{N}\sum_{i=1}^{N}|x_i - y_i| \tag{3}$$

$$\text{MAPE} = \frac{1}{N}\sum_{I=1}^{N}\left|\frac{x_i - y_i}{x_i}\right| \times 100 \tag{4}$$

$$\text{MSE} = \frac{\sum_{i=1}^{N}(x_i - y_i)^2}{N} \tag{5}$$

$$\text{RMSE} = \sqrt{\frac{\sum_{i=1}^{N}(x_i - y_i)^2}{N}} \tag{6}$$

$$\text{NS} = 1 - \frac{\sum_{i=1}^{N}(x_i - y_i)^2}{\sum_{i=1}^{N}(x_i - \bar{y_i})^2} \tag{7}$$

$$\text{a20-index} = \frac{m20}{N} \tag{8}$$

where the number of samples in the dataset is represented as N, x_i, $\bar{x}$, y_i, and $\bar{y}$ are represented as experimental value, mean of experimental values, forecasted value, and mean of forecasted values, respectively.

4. Preliminaries

4.1 Prediction of bond strength using standard guidelines

In this study, commonly used codal/standard guidelines such as fib Bulletin, CNR-DT 200, ACI 440.R-08, and CS-TR-55-UK have been used to predict the bond strength of FRP and concrete. The concerned formulation equations of these guidelines are presented in Eq. (9)–(22).

(a) fib Bulletin: The fib Bulletin [49] model was given by Neubauer and Rostasy in 1997 [50]. The bond strength calculation expression is given in Eq. (9):

$$P_u = \begin{cases} 0.64 k_p b_f \sqrt{0.53 E_f t_f (f_c')^{0.5}}, & L_b \geq L_e \\ 0.64 \dfrac{L_b}{L_e}\left(2 - \dfrac{L_b}{L_e}\right) k_p b_f \sqrt{0.53 E_f t_f (f_c')^{0.5}}, & L_b < L_e \end{cases} \tag{9}$$

where the geometric factor is represented as k_p and expressed in Eq. (10):

$$k_p = \sqrt{\frac{1.125\left(2 - \dfrac{b_f}{b_c}\right)}{1 + \dfrac{b_f}{400}}} \geq 1 \tag{10}$$

Effective length is represented by L_e and calculated using Eq. (11):

$$L_e = \sqrt{\frac{E_f t_f}{1.06 (f_c')^{0.5}}} \tag{11}$$

(b) ACI 440.R-08: In 2001, ACI [51] model was introduced to estimate the strength of the bond, which was given by Chen and Teng [52]. The bond strength calculation formula is expressed in Eq. (12):

$$P_u = 0.427\beta_p\beta_L\sqrt{f_c'}L_e b_f \tag{12}$$

Effective length (Le) is worked out utilizing Eq. (13):

$$L_e = \sqrt{\frac{E_f t_f}{\sqrt{f_c'}}} \tag{13}$$

Geometric parameters β_p and β_L are calculated using Eqs. (14) and (15), respectively:

$$\beta_p = \left[\frac{2 - \frac{b_f}{b_c}}{1 + \frac{b_f}{b_c}}\right]^{0.5} \tag{14}$$

$$\beta_L = \begin{cases} 1, & L \geq L_e \\ \sin\left(\frac{\pi L_b}{2L_e}\right), & L < L_e \end{cases} \tag{15}$$

(c) CNR-DT 200/2004: In 2004, INRC developed a new analytical model to estimate the bond strength [53]. The analytical model to calculate the bond strength is presented in Eq. (16):

$$P_u = \begin{cases} b_f\sqrt{2E_f t_f k_f}, & L_b \geq L_e \\ b_f\sqrt{2E_f t_f k_f \frac{L_b}{L_e}\left(2 - \frac{L_b}{L_e}\right)}, & L_b < L_e \end{cases} \tag{16}$$

where the specific fracture energy is represented by k_f and calculated using Eq. (17):

$$k_f = 0.03k_b\sqrt{f_c' f_c} \tag{17}$$

Geometric coefficient is represented by k_b and calculated using Eq. (18).

$$k_b = \sqrt{\frac{2 - \frac{b_f}{b_c}}{1 + \frac{b_f}{400}}} \geq 1 \tag{18}$$

Effective length (L_e) is calculated using Eq. (19):

$$L_e = \sqrt{\frac{E_f t_f}{2f_c}} \qquad (19)$$

where f_c is the concrete tensile strength mean value

(d) CS-TR-55-UK: In early 2000, the CS-TR-55-UK introduced a new analytical model to estimate the bond strength [54]. The analytical model to calculate bond strength is expressed in Eq. (20):

$$P_u = \begin{cases} 0.5k_b b_f \sqrt{E_f t_f f_{ct}}, & L_b \geq L_e \\ 0.5k_b b_f \sqrt{E_f t_f f_{ct}} \dfrac{L_b}{L_e} \left(2 - \dfrac{L_b}{L_e}\right), & L_b < L_e \end{cases} \qquad (20)$$

Geometric coefficient is represented by k_b and calculated using Eq. (21):

$$k_b = 1.06 \sqrt{\frac{2 - \dfrac{b_f}{b_c}}{1 + \dfrac{b_f}{400}}} \geq 1 \qquad (21)$$

Effective length (L_e) is calculated using Eq. (22):

$$L_e = 0.7 \sqrt{\frac{E_f t_f}{f_{ct}}} \qquad (22)$$

4.2 Computational methods

Artificial intelligence is one of the most trending tools in the computer industry to address the challenging issues in engineering, medical, and agricultural disciplines [55]. Artificial intelligence is mainly classified into three categories: (i) ML, (ii) deep learning (DL), and (iii) reinforcement learning (RL). ML is a data analytics approach that trains computers to learn from the experience in the same way that people and animals do. ML algorithms employ computer approaches directly from the data, rather than relying on model-based predefined equations in order to learn knowledge. DL is a type of learning that is very specialized [56]. DL is mostly deployed to perform the classification tasks in the form of images, sound, and texts. Sometimes, DL algorithms enhance the performance of models more than human-level performance [57]. RL is a sort of ML approach in which a computer agent learns to accomplish a job by interacting with a dynamic environment in a trial-and-error manner. Without human interaction or being explicitly programmed to complete the goal, this learning technique allows the system to make an amount of choices that increase a task's reward measure [58, 59]. In this chapter, a ML–based supervised learning approach has been used to create a correlation and develop a prediction model between the FRP and concrete interface bond strength.

4.2.1 Artificial neural network

In 1944, Walter Pitts and Warren McCullough developed new types of networks called neural networks. One of the most extensively used ML approaches is the ANN model, which is inspired by biological neurons. The ANN is one of the most extensively used statistical models for detecting the relationship between input and output via a set of interconnected data structures with multiple neurons capable of enormous calculations for information representation and data processing. The ANN model might be trained to forecast the required output from the supplied input. ANN is a type of artificial intelligence that operates in the same way as the human brain. ANN is made up of a sequence of linked neurons stacked in layers, just like the human brain. The weights linking the neurons determine the capacity of ANN structures to process provided information. The ANN structure can be either feedforward or recurrent; however, feedforward is the most commonly employed in engineering and also utilized in this chapter. The feedforward network is made up of three layers: input layer (IL), hidden layer (HL), and output layer (OL) as shown in Fig. 3. The neurons in the same layer cannot be connected with each other, but they are connected to the adjacent layers. Neurons are interlinked with each and have different weight values. The final signal is generated, when the total sum of inputs reaches a predetermined threshold, and each neuron has its own transfer function (also known as the activation function). In this work, "tansig" activation function (AF) was used in between the IL and HL as expressed in Eq. (23). The "purlin" AF was used in between HL and OL as expressed in Eq. (24) [9, 60]. The final output of the FRP–concrete bond strength can be obtained by using Eq. (25). The selection of the best neuron based on the performance criteria is shown in Table 3:

$$y = \text{tansig}(x) = \frac{2}{(1 + e^{-2x})} - 1 \tag{23}$$

$$y = pu\,\text{lin}(x) = x \tag{24}$$

$$y = \sum_{i=1}^{n} w_{ij}x_j + b_j \tag{25}$$

where x_j, w_{ij}, and b_j are the input training node, connected weights between input and hidden layers, and bias of the output layer.

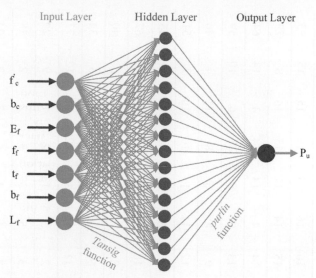

Fig. 3 Structure of artificial neural network.

4.2.2 Hybrid ANN (PSO)

PSO is an optimization approach proposed by Russell Eberhart and James Kennedy in 1995 after they discovered remarkable patterns among birds and fish in a social environment. The behavior of fish schools and bird flocks inspired this method, which led to the development of computer simulations. These algorithms then began to provide impressive results on optimization issues, and they have since served as a platform for additional study. The PSO method has a population (referred to as a swarm) of possible solutions, known as particles, that follow the optimal particles according to a simple formula and move through the space bounded by the function. The algorithmic rule stores the ultimate target value in global variables, as well as a global best (gBest) value, which refers to a particle's current best position among all the other particles in the swarm that is closest to the target/measured value. The hybrid process of the ANN model using PSO is shown in Fig. 4. The hyperparameters of the ANN were updated using PSO to fit according to the required results. First, PSO initializes the random particle position and vector and then updates the best position and velocity; if the criteria are satisfied, then it stops the process; otherwise, the loop process goes on until the criteria are stratified. The updated weights and biases updated in the ANN and output were calculated based on these collected weights and biases.

Table 3 ANN model performance.

Neurons	Values R T_r	V_a	T_e	Values MSE T_r	V_a	T_e	Rank R T_r	V_a	T_e	Rank MSE T_r	V_a	T_e	Total
3	0.9039	0.8692	0.8864	0.0047	0.0070	0.0065	16	16	15	16	15	15	93
4	0.9492	0.8899	0.9396	0.0024	0.0075	0.0028	15	14	11	15	16	8	79
5	0.9549	0.9596	0.9264	0.0022	0.0022	0.0046	14	4	14	14	4	14	64
6	0.9633	0.9416	0.9444	0.0019	0.0024	0.0031	13	9	8	13	6	10	59
7	0.9697	0.9524	0.9557	0.0015	0.0029	0.0021	9	8	6	8	11	4	46
8	0.9661	0.9281	0.9712	0.0016	0.0029	0.0028	12	11	1	10	10	7	51
9	0.9746	0.9556	0.9406	0.0013	0.0026	0.0034	5	6	10	5	8	11	45
10	0.9675	0.9227	0.9412	0.0016	0.0031	0.0041	10	12	9	11	12	13	67
11	0.9701	0.9534	0.9599	0.0016	0.0023	0.0022	8	7	5	9	5	6	40
12	0.9728	0.9656	0.9517	0.0014	0.0026	0.0021	6	2	7	7	7	5	34
13	0.9665	0.8843	0.9346	0.0017	0.0067	0.0036	11	15	13	12	14	12	77
14	0.9840	0.9638	0.936	0.0009	0.0016	0.0029	2	3	12	2	1	9	29
15	0.9827	0.9589	0.9667	0.0009	0.0020	0.0018	3	5	3	3	2	3	18
16	0.9714	0.9752	0.9629	0.0013	0.0020	0.0019	7	1	4	6	3	3	24
17	0.9859	0.9352	0.7902	0.0008	0.0027	0.0137	1	10	16	1	9	16	53
18	0.9789	0.9048	0.9707	0.0011	0.0047	0.0017	4	13	2	4	13	1	37

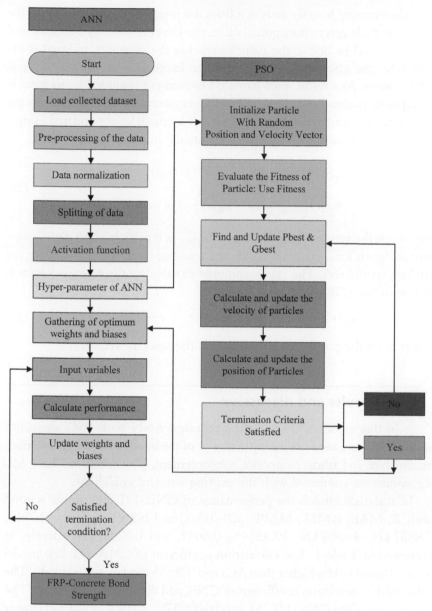

Fig. 4 Flow diagram of proposed ANN-PSO.

Each particle in a swarm keeps track of its own velocity and a particle best value (pBest), which represents the particle's best location in relation to the ultimate target pattern. Only if one of the particle's pBest values is discovered to be closer to the global target pattern than the current gBest value does the gBest value change. The value of a particle's velocity is estimated after determining how far away it is from the intended result. In general, the closer a particle gets to the optimal value, the lower its velocity becomes, and vice versa. When one of the particles reaches the objective and makes the cost zero, the gBest may finally complete its goal and return an optimal arrangement. As a result, each particle's movement is guided by its particle best (pBest) position and gBest position. Every iteration of this method is the conducted until a good solution is found. The particles are updated using a fitness function until the function is eventually minimized [61–68]:

$$
\begin{aligned}
v_{id}(t+1) = c_1 r_1 \Big(p_{\text{best, } id}(t) - x_{id}(t) \Big) + w v_{id}(t) \\
+ c_2 r_2 \Big(g_{\text{best, } id}(t) - x_{id}(t) \Big), \quad d = 1, 2, \ldots D
\end{aligned}
\tag{26}
$$

where w is the weight of the inertia, and $g_{(best, id)}$ is particle's best global position at the ith location. c_1 and c_2 are the learning rates, and r_1 and r_2 are the random coefficients. The expression used to calculate the position vector is shown in Eq. (27):

$$
x_{id}(t+1) = x_{id}(t) + v_{id}(t+1), \quad d = 1, 2, \ldots, D
\tag{27}
$$

where x_{id} is the position vector and v_{id} is the velocity vector.

5. Results and discussion

In this work, ANN and the optimized ANN with PSO algorithm approaches were used to identify behavior of the bond between the interface of concrete and fabric composite reinforcement. The results of these ML algorithms are compared with the existing standard guidelines.

In analytical models the performance of CNR–DT 200 model is good with R, MAE, RMSE, MAPE, a20-index, and NS values being 0.7919, 2.8807 kN, 4.6309 kN, 19.9314%, 0.6641, and 0.6224, respectively, as presented in Table 4. The correlation coefficient of CNR–DT 200 model is 12.79% and 6.02% higher than ACI and TR–55 models, respectively. The value of the correlation coefficient of CNR and fib models is the same. The MAE value of the CNR–DT 200 model is 8.37%, 20.01%, and 8.57% lower than fib, ACI, and TR–55 models, respectively. The RMSE value of the CNR–DT 200 model is 3.97%, 17.2%, and 7.52% lower than fib, ACI, and TR–55 models, respectively. The second best analytical model to forecast the bond strength is fib, based on R, RMSE, MAE, MAPE, a20-index, and NS performance indices. Fig. 5 depicts the scatter plot of the analytical models.

Table 4 Comparison results of analytical models and proposed models.

Model	R	RMSE (kN)	MAE (kN)	MAPE (%)	a20-index	NS	Std.
fib	0.7919	4.8225	3.1439	23.0109	0.5863	0.5905	7.1814
ACI	0.7021	5.5927	3.6014	24.4412	0.4899	0.4493	5.7409
CNR–DT 200	0.7919	4.6309	2.8807	19.9317	0.6641	0.6224	7.6592
CS-TR-55-UK	0.7469	5.0077	3.1507	23.3778	0.5817	0.5585	7.0548
ANN	0.9777	1.4955	1.1326	7.7993	0.9440	0.9918	6.5250
Hybrid ANN	0.9907	0.8289	0.6123	4.5253	0.9904	0.9951	6.7270

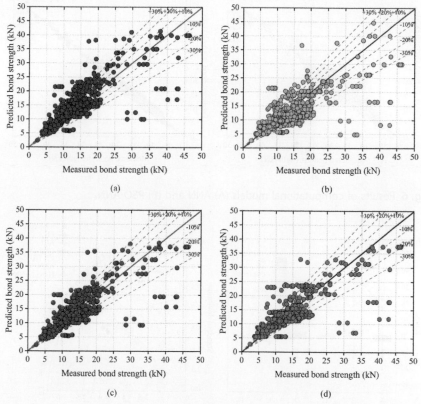

Fig. 5 Results of analytical models (A) fib, (B) ACI, (C) CNR-DT 200, and (D) CS-TR-55-UK.

The correlation coefficient of the PSO-ANN model is 0.9907, which is 25.1%, 41.11%, 25.1%, 32.64%, and 1.33% higher than fib, ACI, CNR, TR-55, and ANN models, respectively. The MAE and RMSE values of the PSO-ANN model are 0.6123 and 0.8289, respectively. The standard deviation (SD) of the original dataset is 6.7304 and SD of the PSO-ANN model is 6.7270, which is very close to the original value, as shown in Table 4. The SD of the ANN model is 6.5250, which is 3.15% higher than the original SD value. Fig. 6 shows the scatter plot of the ANN and PSO-ANN models. As shown in Fig. 6, the scattering of the dataset is almost observed in all the analytical plots, and the values crossed the error range of +30% to −30%. In Fig. 6, the developed ANN model shows less scattering of the predicted values, and even in the case of the PSO-ANN model, all the dataset lies within the error range of +30% to −30% and completely sticks to the linear line. Fig. 7 shows the graphical representation of the

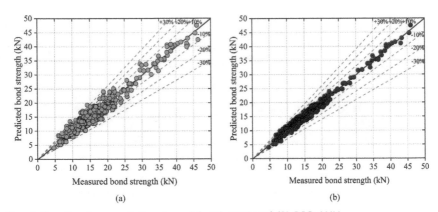

Fig. 6 Results of computational models (A) ANN and (B) PSO-ANN.

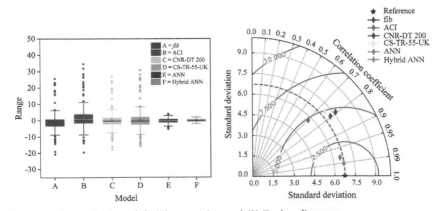

Fig. 7 Analysis of all models (A) error plot and (B) Taylor diagram.

fitness of the models. In Fig. 7A, all the analytical models have a higher range of errors, but these errors get reduced in the ANN and PSO-ANN models. The PSO-ANN model has the lowest error and completely sticks to the zero line range value. Fig. 7B is the Taylor plot, which is drawn in between the R, std., and the RMSE values. The experimental line on this plot is represented by the green dotted line. If the diamond-shaped fill precisely intersects or aligns with this line, it indicates the model's absolute accuracy. As depicted in Fig. 7B, the ACI model behind the *green dotted line* and the other three models fib, CNR-DT 200, and CS-TR-55-UK crossed the line, but the hybrid PSO-ANN model directly lies on the line, and also the best-predicted model based on the Taylor diagram analysis.

6. Conclusion

In this chapter, PSO was used to optimize an ANN for evaluating the strength of the bond between concrete and fabric composite reinforcement based on the collected input parameters. The seven input parameters were used to develop the prediction model, whereas the strength of the bond is defined as the output parameter. The predictions made by ANN and PSO-ANN show an excellent relationship with the tested outcomes. In comparison to the ANN model, the PSO-ANN model performs 1.33% better. The output of these ML algorithms was evaluated with the four commonly accepted standard recommendations. The predicted outcomes shows that the PSO-ANN model was more precise and reliable. The root mean square error of analytical models lies between 4.6309 and 5.5927 kN, and in ANN and PSO-ANN, its value is 1.4955 and 0.8289 kN, respectively. Taylor diagram and error plots were also used to show the performance of analytical as well as ML models.

The developed ML model can be used as an additional method to estimate the strength of the bond between concrete and fabric composite reinforcement. This model is trained and tested within the input range parameters; however, beyond the limit of input parameters, the models' accuracy may vary, or in an extreme case, the model will be unable to predict the expected results.

References

[1] R. Al-Mahaidi, R. Kalfat, Chapter 1—Introduction, in: R. Al-Mahaidi, R. Kalfat (Eds.), Rehabilitation of Concrete Structures With Fiber-Reinforced Polymer, Butterworth-Heinemann, 2018, pp. 1–5, ISBN: 978-0-12-811510-7, https://doi.org/10.1016/B978-0-12-811510-7.00001-X.

[2] M. Gotame, C.L. Franklin, M. Blomfors, J. Yang, K. Lundgren, Finite element analyses of FRP-strengthened concrete beams with corroded reinforcement, Eng. Struct. 257 (2022) 114007, https://doi.org/10.1016/j.engstruct.2022.114007.

[3] I. Fernandez, M.F. Herrador, A.R. Marí, J.M. Bairán, Structural effects of steel reinforcement corrosion on statically indeterminate reinforced concrete members, Mater. Struct. 49 (12) (2016) 4959–4973.

[4] E. Chen, C.G. Berrocal, I. Fernandez, I. Löfgren, K. Lundgren, Assessment of the mechanical behaviour of reinforcement bars with localised pitting corrosion by Digital Image Correlation, Eng. Struct. 219 (2020) 110936.

[5] R. Al-Mahaidi, R. Kalfat, Methods of structural rehabilitation and strengthening, in: Rehabilitation of Concrete Structures With Fiber-Reinforced Polymer, Elsevier, Amsterdam, The Netherlands, 2018, pp. 7–13, https://doi.org/10.1016/B978-0-12-811510-7.00002-1.

[6] J. Wang, Assessing the durability of the interface between fiber-reinforced polymer (FRP) composites and concrete in the rehabilitation of reinforced concrete structures, in: Developments in Fiber-Reinforced Polymer (FRP) Composites for Civil Engineering, Elsevier, 2013, pp. 148–173, https://doi.org/10.1533/9780857098955.1.148.

[7] Z. Chao, J. Wang, Viscoelastic analysis of FRP strengthened reinforced concrete beams, Compos. Struct. 93 (12) (2011) 3200–3208, https://doi.org/10.1016/j.compstruct.2011.06.006.

[8] J. Gao, M. Koopialipoor, D.J. Armaghani, A. Ghabussi, S. Baharom, A. Morasaei, A. Shariati, M. Khorami, J. Zhou, Evaluating the bond strength of FRP in concrete samples using machine learning methods, Smart Struct. Syst. 26 (4) (2020) 403–418.

[9] A. Kumar, H.C. Arora, M.A. Mohammed, K. Kumar, J. Nedoma, An optimized neuro-bee algorithm approach to predict the FRP-concrete bond strength of RC beams, IEEE Access 10 (2021) 3790–3806.

[10] Z. Pei, Y. Wei, Prediction of the bond strength of FRP-to-concrete under direct tension by ACO-based ANFIS approach, Compos. Struct. 282 (2022) 115070, https://doi.org/10.1016/j.compstruct.2021.115070.

[11] H. Naderpour, S.A. Alavi, A proposed model to estimate shear contribution of FRP in strengthened RC beams in terms of adaptive neuro-fuzzy inference system, Compos. Struct. 170 (2017) 215–227.

[12] H. Naderpour, K. Nagai, M. Haji, M. Mirrashid, Adaptive neuro-fuzzy inference modelling and sensitivity analysis for capacity estimation of fiber reinforced polymer-strengthened circular reinforced concrete columns, Expert Syst. 36 (4) (2019) e12410.

[13] Y. Feng, M. Mohammadi, L. Wang, M. Rashidi, P. Mehrabi, Application of artificial intelligence to evaluate the fresh properties of self-consolidating concrete, Materials 14 (17) (2021) 4885.

[14] H. Basarir, M. Elchalakani, A. Karrech, The prediction of ultimate pure bending moment of concrete-filled steel tubes by adaptive neuro-fuzzy inference system (ANFIS), Neural Comput. Applic. 31 (2) (2019) 1239–1252.

[15] A. Pandit, S. Panda, Prediction of earthquake magnitude using soft computing techniques: ANN and ANFIS, Arab. J. Geosci. 14 (13) (2021) 1–10.

[16] A. Pandit, K.C. Biswal, Prediction of earthquake magnitude using adaptive neuro fuzzy inference system, Earth Sci. Inf. 12 (4) (2019) 513–524.

[17] S. Kar, A.R. Pandit, K.C. Biswal, A neuro-fuzzy approach to estimate the shear contribution of externally bonded FRP composites, Asian J. Civ. Eng. 22 (2) (2021) 351–367.

[18] S. Kar, K.C. Biswal, A neuro-fuzzy approach to predict the shear contribution of end-anchored FRP U-jackets, Comput. Concr. 26 (5) (2020) 397–409.

[19] H. Naderpour, M. Mirrashid, Shear strength prediction of RC beams using adaptive neuro-fuzzy inference system, Sci. Iran. 27 (2) (2020) 657–670.

[20] I. Mansouri, O. Kisi, P. Sadeghian, C.-H. Lee, J.W. Hu, Prediction of ultimate strain and strength of FRP-confined concrete cylinders using soft computing methods, Appl. Sci. 7 (8) (2017) 751.

[21] I. Mansouri, T. Ozbakkaloglu, O. Kisi, T. Xie, Predicting behavior of FRP-confined concrete using neuro fuzzy, neural network, multivariate adaptive regression splines and M5 model tree techniques, Mater. Struct. 49 (10) (2016) 4319–4334.

[22] H. Naderpour, M. Mirrashid, A neuro-fuzzy model for punching shear prediction of slab-column connections reinforced with FRP, J. Soft Comput. Civil Eng. 3 (1) (2019) 16–26.

[23] B. Vakhshouri, S. Nejadi, Prediction of compressive strength of self-compacting concrete by ANFIS models, Neurocomputing 280 (2018) 13–22.

[24] H. Naderpour, M. Mirrashid, Moment capacity estimation of spirally reinforced concrete columns using ANFIS, Complex Intell. Syst. 6 (1) (2020) 97–107.

[25] B. Ahmadi-Nedushan, Prediction of elastic modulus of normal and high strength concrete using ANFIS and optimal nonlinear regression models, Construct. Build Mater. 36 (2012) 665–673.

[26] Y. Cao, Q. Fan, S.M. Azar, R. Alyousef, S.T. Yousif, K. Wakil, K. Jermsittiparsert, L.S. Ho, H. Alabduljabbar, A. Alaskar, Computational parameter identification of strongest influence on the shear resistance of reinforced concrete beams by fiber reinforcement polymer, Structures 27 (2020) 118–127. Elsevier.

[27] A. Sharafati, H. Naderpour, S.Q. Salih, E. Onyari, Z.M. Yaseen, Simulation of foamed concrete compressive strength prediction using adaptive neuro-fuzzy inference system optimized by nature-inspired algorithms, Front. Struct. Civil Eng. 15 (1) (2021) 61–79.

[28] M. Shariati, M.S. Mafipour, P. Mehrabi, A. Shariati, A. Toghroli, N.T. Trung, M.N.A. Salih, A novel approach to predict shear strength of tilted angle connectors using artificial intelligence techniques, Eng. Comput. 37 (3) (2021) 2089–2109.

[29] L. Chung, M.-W. Hur, T. Park, Performance evaluation of CFRP reinforced concrete members utilizing fuzzy technique, Int. J. Concr. Struct. Mater. 12 (1) (2018) 1–11.

[30] Y. Wei, A. Han, X. Xue, A data-driven study for evaluating the compressive strength of high-strength concrete, Int. J. Mach. Learn. Cybern. 12 (12) (2021) 3585–3595.

[31] J. Dai, T. Ueda, Y. Sato, Development of the nonlinear bond stress-slip model of fiber reinforced plastics sheet-concrete interfaces with a simple method, J. Compos. Constr. 9 (1) (2005) 52–62.

[32] A. Hosseini, D. Mostofinejad, Effective bond length of FRP-to-concrete adhesively-bonded joints: experimental evaluation of existing models, Int. J. Adhes. Adhes. 48 (2014) 150–158.

[33] W. Li, J. Li, X. Ren, C.K.Y. Leung, F. Xing, Coupling effect of concrete strength and bonding length on bond behaviors of fiber reinforced polymer-concrete interface, J. Reinf. Plast. Compos. 34 (5) (2015) 421–432.

[34] C. Chen, X. Li, D. Zhao, Z. Huang, L. Sui, F. Xing, Y. Zhou, Mechanism of surface preparation on FRP-concrete bond performance: a quantitative study, Compos. Part B Eng. 163 (2019) 193–206.

[35] C. Yuan, W. Chen, T.M. Pham, H. Hao, Effect of aggregate size on bond behaviour between basalt fibre reinforced polymer sheets and concrete, Compos. Part B Eng. 158 (2019) 459–474.

[36] D. Mostofinejad, K. Sanginabadi, M.R. Eftekhar, Effects of coarse aggregate volume on CFRP-concrete bond strength and behavior, Construct. Build Mater. 198 (2019) 42–57.

[37] M.H. Mofrad, D. Mostofinejad, A. Hosseini, A generic non-linear bond-slip model for CFRP composites bonded to concrete substrate using EBR and EBROG techniques, Compos. Struct. 220 (2019) 31–44.

[38] J.-G. Dai, Y. Sato, T. Ueda, Improving the load transfer and effective bond length for FRP composites bonded to concrete, Proc. Jpn. Concr. Inst. 24 (2) (2002) 1423–1428.

[39] Y. Yun, Y.-F. Wu, Durability of CFRP-concrete joints under freeze-thaw cycling, Cold Reg. Sci. Technol. 65 (3) (2011) 401–412.

[40] S. Ueno, H. Toutanji, R. Vuddandam, Introduction of a stress state criterion to predict bond strength between FRP and concrete substrate, J. Compos. Constr. 19 (1) (2015) 04014024.

[41] P. Zhang, D. Lei, Q. Ren, J. He, H. Shen, Z. Yang, Experimental and numerical investigation of debonding process of the FRP plate-concrete interface, Construct. Build Mater. 235 (2020) 117457, https://doi.org/10.1016/j.conbuildmat.2019.117457.

[42] C. Yuan, W. Chen, T.M. Pham, H. Hao, J. Cui, Y. Shi, Interfacial bond behaviour between hybrid carbon/basalt fibre composites and concrete under dynamic loading, Int. J. Adhes. Adhes. 99 (2020) 102569, https://doi.org/10.1016/j.ijadhadh.2020.102569.

[43] A. Moghaddas, D. Mostofinejad, A. Saljoughian, E. Ilia, An empirical FRP-concrete bond-slip model for externally-bonded reinforcement on grooves, Construct. Build Mater. 281 (2021) 122575, https://doi.org/10.1016/j.conbuildmat.2021.122575.

[44] F. Ceroni, A. Garofano, M. Pecce, Modelling of the bond behaviour of tuff elements externally bonded with FRP sheets, Compos. Part B Eng. 59 (2014) 248–259, https://doi.org/10.1016/j.compositesb.2013.12.007.

[45] N.R. Kapoor, A. Kumar, A. Kumar, A. Kumar, M.A. Mohammed, K. Kumar, S. Kadry, S. Lim, Machine learning-based CO_2 prediction for office room: a pilot study, Wirel. Commun. Mob. Comput. 2022 (2022) 1–16.

[46] A. Kumar, H.C. Arora, K. Kumar, M.A. Mohammed, A. Majumdar, A. Khamaksorn, O. Thinnukool, Prediction of FRCM-concrete bond strength with machine learning approach, Sustainability 14 (2) (2022) 845.

[47] A. Kumar, H.C. Arora, N.R. Kapoor, M.A. Mohammed, K. Kumar, A. Majumdar, O. Thinnukool, Compressive strength prediction of lightweight concrete: machine learning models, Sustainability 14 (4) (2022) 2404.

[48] K. Kumar, R.P. Saini, Development of correlation to predict the efficiency of a hydro machine under different operating conditions, Sustainable Energy Technol. Assess. 50 (2022) 101859, https://doi.org/10.1016/j.seta.2021.101859.

[49] T. Triantafillou, S. Matthys, K. Audenaert, G. Balázs, M. Blaschko, H. Blontrock, C. Czaderski, E. David, A. Di Tomasso, W. Duckett, et al., Externally Bonded FRP Reinforcement for RC Structures, International Federation for Structural Concrete (fib), 2001. Bulletin FIB 14.

[50] U. Neubauer, F.S. Rostasy, Design aspects of concrete structures strengthened with externally bonded CFRP-plates, in: Proceedings of the Seventh International Conference on Structural Faults and Repair, 8 July 1997. Volume 2: Concrete and Composites, 1997.

[51] C.E. Bakis, A. Ganjehlou, D.I. Kachlakev, M. Schupack, P. Balaguru, D.J. Gee, V.M. Karbhari, D.W. Scott, C.A. Ballinger, T.R. Gentry, et al., Guide for the Design and Construction of Externally Bonded FRP Systems for Strengthening Concrete Structures, Citeseer, 2002. Reported by ACI Committee 440.

[52] J.F. Chen, J.G. Teng, Anchorage strength models for FRP and steel plates bonded to concrete, J. Struct. Eng. 127 (7) (2001) 784–791.

[53] M.A. Aiello, L. Ascione, A. Baratta, F. Bastianini, U. Battista, A. Benedetti, V.P. Berardi, A. Bilotta, A. Borri, B. Briccoli, et al., Guide for the Design and Construction of Externally Bonded FRP Systems for Strengthening Existing Structures, 2014.

[54] C. Arya, J.L. Clarke, E.A. Kay, P.D. O'regan, TR 55: design guidance for strengthening concrete structures using fibre composite materials: a review, Eng. Struct. 24 (7) (2002) 889–900.

[55] A. Kumar, N. Mor, An approach-driven: use of artificial intelligence and its applications in civil engineering, in: Artificial Intelligence and IoT, Springer, 2021, pp. 201–221.

[56] N.R. Kapoor, A. Kumar, A. Kumar, Machine learning algorithms for predicting viral transmission probability in naturally ventilated office rooms, in: 2nd International Conference on i-Converge 2022: Changing Dimensions of the Built Environment, School of Architecture Planning and Design, DIT University, Dehradun, India, 2022, p. 79.

[57] N.R. Kapoor, A. Kumar, H.C. Arora, A. Kumar, Structural health monitoring of existing building structures for creating green smart cities using deep learning, in: Recurrent Neural Networks, CRC Press, pp. 203–232.

[58] K. El Bouchefry, R.S. de Souza, Chapter 12—Learning in big data: introduction to machine learning, in: P. Škoda, F. Adam (Eds.), Knowledge Discovery in Big Data From Astronomy and Earth Observation, Elsevier, 2020, pp. 225–249, ISBN: 978-0-12-819154-5, https://doi.org/10.1016/B978-0-12-819154-5.00023-0.

[59] K. Kumar, A. Kumar, N. Kumar, M.A. Mohammed, A.S. Al-Waisy, M.M. Jaber, R. Shah, M.N. Al-Andoli, Dimensions of Internet of Things: technological taxonomy architecture applications and open challenges–a systematic review, Wirel. Commun. Mob. Comput. 2022 (2022) 1–23.

[60] A. Kumar, N. Mor, Prediction of accuracy of high-strength concrete using data mining technique: a review, in: Proceedings of International Conference on IoT Inclusive Life (ICIIL 2019), NITTTR Chandigarh, India, Springer, 2020, pp. 259–267.

[61] S.O. Fadlallah, T.N. Anderson, R.J. Nates, Artificial neural network-particle swarm optimization (ANN-PSO) approach for behaviour prediction and structural optimization of lightweight sandwich composite heliostats, Arab. J. Sci. Eng. 46 (12) (2021) 12721–12742.

[62] A. Biswas, B. Biswas, A. Kumar, K.K. Mishra, Particle swarm optimisation with time varying cognitive avoidance component, Int. J. Comput. Sci. Eng. 16 (1) (2018) 27–41.

[63] A. Biswas, B. Biswas, Swarm intelligence techniques and their adaptive nature with applications, in: Complex System Modelling and Control Through Intelligent Soft Computations, Springer, 2015, pp. 253–273.

[64] A. Biswas, A. Kumar, K.K. Mishra, Particle swarm optimization with cognitive avoidance component, in: 2013 International Conference on Advances in Computing, Communications and Informatics (ICACCI), IEEE, 2013, pp. 149–154.

[65] A. Biswas, A.V. Lakra, S. Kumar, A. Singh, An improved random inertia weighted particle swarm optimization, in: 2013 International Symposium on Computational and Business Intelligence, IEEE, 2013, pp. 96–99.

[66] D. Sarkar, N. Mishra, A. Biswas, Genetic algorithm-based deep learning models: a design perspective, in: Proceedings of the Seventh International Conference on Mathematics and Computing, Springer, 2022, pp. 361–372.

[67] A. Biswas, K.K. Mishra, S. Tiwari, A.K. Misra, Physics-inspired optimization algorithms: a survey, J. Optim. 2013 (2013) 1–17.

[68] A. Biswas, B. Biswas, K.K. Mishra, An atomic model based optimization algorithm, in: 2016 2nd International Conference on Computational Intelligence and Networks (CINE), IEEE, 2016, pp. 63–68.

About the authors

Er. Aman Kumar hails from Bilaspur, Himachal Pradesh, India. He holds a Master of Engineering degree in construction technology and management from the prestigious National Institute of Technical Teacher's Training and Research Institute in Chandigarh, India. Currently, he is fervently pursuing a Ph.D. in engineering sciences, specializing in structural engineering, at the renowned CSIR-Central Building Research Institute in Roorkee, India. His academic journey is underscored by a deep passion for various facets of civil engineering, including sustainability development, nondestructive testing, concrete technology, and strengthening techniques such as fiber-reinforced polymer and fiber-reinforced cementitious matrix. He is also deeply engaged in exploring corrosion protection techniques for structural design, as well as the cutting-edge domains of artificial intelligence and the Internet of Things. Now his focus to solve the complex structural engineering problems with machine learning algorithms. Aman Kumar's dedication to the field is evident through his extensive research endeavours and comprehensive technical surveys. His scholarly pursuits have culminated in numerous research papers and book chapters, which have been featured in esteemed international scientific publications.

Dr. Harish Chandra Arora currently holds the esteemed position of Principal Scientist in Structural Engineering Group at CSIR-Central Building Research Institute in Roorkee, India. With a distinguished career spanning more than 29 years, Dr. Arora is a renowned figure in the field of structural engineering. Dr. Arora is also functioning as an Associate Professor in Academy of Scientific and Innovative Research (AcSIR), Ghaziabad, India. His contemporary research areas include structural composites, structural corrosion, distress diagnosis, seismic evaluation and repair & retrofitting of structures and machine learning applications in structural engineering etc. Dr. Arora's exceptional contributions to the field have garnered recognition in both national and international academic journals. Beyond his scholarly achievements, he has made a significant impact on the education and development of future engineers, having supervised and guided over 100[+] students in their pursuit of Bachelor of Technology and Masters of Technology degrees. Additionally, he continues to mentor and support research scholars pursuing doctoral programs at the Central Building Research Institute in Roorkee, India. Furthermore, Dr. Arora actively contributes to the scholarly community as a reviewer for journals published by Springer Nature and Elsevier. His commitment to maintain the quality and rigour of academic publications is highly regarded. Beyond his academic pursuits, Dr. Arora has undertaken numerous consultancy and Research & Development projects within the field of structural engineering, further showing his dedication to advancing the science and practice of sustainable construction.

Dr. Nishant Raj Kapoor hails from Kota, Rajasthan, India. Dr. Kapoor is an accomplished alumnus, having earned degrees from AcSIR (CSIR–CBRI, Roorkee), NITTTR in Bhopal, and RTU in Kota, India. His diverse research interests encompass real-time challenges in the built environment, with a focus on areas such as COVID-19, indoor human comfort, indoor environmental quality, comfort perceptions, building energy efficiency, artificial intelligence, and environmental engineering. With a prolific scholarly record, Dr. Kapoor has authored over 45 research papers, review articles, patents, conference articles, and book chapters. These contributions have been published in esteemed peer-reviewed international journals and books, reflecting his commitment to advancing scientific knowledge. Dr. Kapoor played a vital role in formulating ventilation-related guidelines to curb the spread of SARS-CoV-2 in Indian office and residential buildings, showcasing his practical impact on public health. Additionally, he actively engages with the global scientific community by serving as an editor for international scientific book projects and contributing as a reviewer for SCI/ Scopus indexed international journals. Recognizing his outstanding contributions, Dr. Kapoor was honored with the Diamond Jubilee Best Technology Award in 2021, a testament to his excellence in the field. Further attesting to his prowess, he received the Best Paper Award in September 2022, underscoring the quality and impact of his research.

Dr. Ashok Kumar is a distinguished professional with a rich and diverse experience of more than 33 years in research, academia and industry. He superannuated as Outstanding Scientist and Professor from the prestigious building research lab of India (CSIR-CBRI, Roorkee). His educational journey includes earning a B.Sc., B.Arch. (Gold Medal), M.U.R.P. (Hons.), and a Ph.D. from the Indian Institute of Technology Roorkee (IITR). He has made significant contributions to research and academia, showcasing expertise in green and energy-efficient buildings, green retrofits, and affordable housing. Dr. Kumar has a proven track record of handling national and international collaborative R&D projects in areas such as building energy efficiency, school and healthcare buildings, and green affordable housing. Having led and participated in over 95 international and national projects, Dr. Kumar has left an indelible mark on the field. Notable projects include the design of the Medical College Complex in Haldwani, Uttarakhand, the concept design for prefab hospitals during the Covid-19 pandemic, and the development of guidelines on ventilation for residential and office buildings in the context of the SARS-CoV-2 virus. Dr. Kumar and his team received the Director's Diamond Jubilee Best Technology Award in 2021 for their work on HVAC Ducting System for Integration of Covid-19 Disinfection Solutions. He has also been honored with the National Design Award for Architectural Engineering in 2019 and various Best Paper Awards for his contributions to the field. He serves on the Governing Council of the Bureau of Energy Efficiency, Ministry of Power, representing architects of India. He is also involved in expert committees related to energy conservation, awards, and sustainable habitat. With over 120 publications in international and national journals, conferences, and book chapters, as well as holding two patents and three copyrights, Dr. Kumar's impact on the field is both broad and profound. His commitment to education is evident through his role as a visiting professor at the National University of Singapore. Dr. Kumar's commitment to research, education, and innovation sets a high standard in the industry.

CHAPTER TWENTY

Discovering the characteristic set of metaheuristic algorithm to adapt with ANFIS model

Aref Yelghi[a,c], Shirmohammad Tavangari[b,c], and Arman Bath[a,b,c]
[a]Faculty of Engineering, Computer Engineering, Zeytinburnu, Zeytinburnu, Istanbul, Turkey
[b]Electrical and Computer Engineering Faculty, Vancouver, British Columbia, Canada
[c]Department of Political Science, Vancouver, British Columbia, Canada

Contents

Abstract

In recent years, a number of applications based on Neural Networks, Neuro- Fuzzy, and optimization algorithms have been more common for solving regression and classification problems. In the adaptive neuro-fuzzy inference system (ANFIS), many researchers used the adaptation of metaheuristic algorithms with ANFIS to propose the best estimation model. In contrast, most researchers only focused on the experiment without demonstrating any mathematical characteristics or noting which parameters of the optimization algorithm are coordinated with ANFIS, during the run. This paper pro- vides an adaption of metaheuristic algorithms with ANFIS which has been performed by considering accuracy parameters in layer 1 and layer 4 for the estimation problem. It has integrated six well-known metaheuristic algorithms and extracted their properties. An experiment demonstrated that metaheuristic algorithms based on evolutionary computation are more stable than swarm intelligence methods in tuning parameters of ANFIS.

Advances in Computers, Volume 135
ISSN 0065-2458
https://doi.org/10.1016/bs.adcom.2023.11.009

1. Introduction

In the process of integrating the economies of countries into the world, the estimation of exchange rates becomes extremely significant. Many classical methods and models have been used to predict exchange rates. However, due to its significant inefficiencies and disadvantages, very accurate predictions have not been possible. ANFIS will be used to estimate the Tl–Dollar exchange rate in this study. The adaptive network-based fuzzy inference system (ANFIS) has been applied to real and complex problems in recent years. This intelligence technique is popular in artificial intelligence, and researchers are highly interested in the field. ANFIS is computationally less expensive, clear in processing, easy to implement, and outperforms statistics-based comparisons [1,2]. During training, the ANFIS parameters need to be adjusted. One of the two types of algorithms is a derivative-based optimization algorithm and the other is a metaheuristic algorithm [3]. In some recent work, Alagarsamy et al. [4] presented a Runge–Kutta (AARK) with ANFIS which is integrated with the self-adaptive discrete wavelet transform. In their work, the extracted features were applied to segmented regions, and then segmentation was used for ear regions. The other work by Harandizadeh et al. [5] suggested improved ANFIS approaches to estimate the bearing capacity of piles. The model was an integration of ANFIS and the group method of data handling. This model tuned the parameters of the ANFIS model by using the gravitational Search Algorithm. Talebizadeh et al. [6] developed the ANFIS and ANN(artificial neural network) models for forecasting the lake level fluctuations in Lake Urmia in Iran. By defining four inputs, lake levels, rainfall, evaporation, and inflow, to design models, what has been observed is that the results of the ANFIS model are better than those of ANN in both accuracy and uncertainty. In a study by Kiyak et al. [7], the ANFIS model is used to simulate the distribution of white color HP-LEDs in training subtraction clustering with a hybrid learning approach and the results indicate that the model can be used to predict the distribution of illumination. Zhou et al. [8] created hybrid FFA-ANFIS and GA-ANFIS models to predict the particle size distribution of a muck pile (fragmentation). In their work, they optimized the premise and consequent parameters of ANFIS in order to obtain the maximum performance result. Gholami et al. [9] Hydride AN- FIS model with differential evolution (DE) and singular value decomposition value (SVD) algorithms. They can predict shape profiles with respect to bank-observed profiles for threshold channels.

Karaboga et al. [10] proposed ANFIS and Artificial Bee Colony algorithms which include the advantage of speed convergence feature and are suitable to train ANFIS. Another work [11] uses the ABC algorithm for predicting and investigating how many tourists come to Turkey from nine different countries. Singh et al. The ANFIS model was applied to estimate the rock. This allows us to understand the magnitude and properties of deformation caused by the change in the stress field. This is discussed in [12]. In order to optimize the input-output data, the full factorial design method is used in the training ANFIS-based model which is presented by Buragohain et al. [13]. Based on renowned examples of data from a gas furnace and thermal power plant, their work indicated forecasting. Lei et al. In [14], it was shown that multiple AFIS mixtures (with GAs) can reliably distinguish between fault categories and severity, showing better classification performance than singular AFIS-dependent classifiers. Yasin et al. [15] proposed a unique combination of ANFIS with Firefly. The proposed model is applied to forecast the historical monthly streamflow in the Pahang River. Ahmedlou et al. [16] modelized ANFIS with two metaheuristic algorithms: biogeography-based optimization (BBO) and BAT algorithms (BA). The models applied GIS to map flood susceptibility in a region of Iran. By considering the accuracy and time complexity in obtaining the result, BBO was found to be superior to another method.

2. ANFIS model

The adaptive network-based fuzzy inference system (ANFIS) is one of the artificial intelligence methods based on artificial neural networks (ANN) and fuzzy logic (FL). The ANFIS framework provides the relation between inputs and outputs. This dataset contains 5 layers, with the exception of the first and last layers, which contain the independent variables and dependent variables of the dataset, respectively. The ANFIS Framework is shown in Fig. 1.

Rule 1:

Rule 2:

I˙F x is A1 and y B1 then z isf1(x,y) I˙F x is A2 and y B2 then z is f1(x,y)

The nodes in layers 1 and 4 indicate parameter sets that are tuned with an optimization algorithm. Layers 2 and 3 are fixed in the Framework. The first layer includes adaptive nodes with functions that are evaluated as shown in Eq. (1) and Eq. (2).

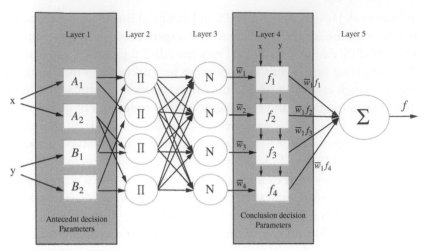

Fig. 1 ANFIS model for two inputs.

$$o_{1,i} = \mu_{A_x}(x) \text{ For i} = 1, 2 \tag{1}$$

$$o_{1,i} = \mu_{B_{i-2}}(x) \text{ For i} = 3, 4 \tag{2}$$

where x and y are inputs of ANFIS and A and B are fuzzy sets. Outputs of Sugeno Fuzzy and the fuzzification step mu(x), mu(y) the membership functions, which can be defined a bell shape equations are given in Eq. (3) and Eq.(4).

$$\mu(x) = \frac{1}{1 + \left(\frac{x-c_i}{a_i}\right)^{2b_i}} \tag{3}$$

Or

$$\mu(x) = \exp\left\{-\left(\frac{x-c_i}{a_i}\right)^2\right\} \tag{4}$$

Parameters are called premise parameters. They will be changed in the training step, as well, the bell-shaped functions will be changed. Each fixed node in the second layer is multiplied by input signals from the first layer. They showed their output of them in Eq. (5). A rule has power. It can be used with either the and-or or and–probability operators.

$$o_{2,i} = w_i = \mu_{A_i}(x) \cdot \mu_{B_i}(y) \text{For i} = 1, 2 \tag{5}$$

In the third layer, every node, normalize the input signal by considering all rules. The equation is given as Eq. (6).

$$o_{3,i} = \overline{w}_i = \frac{w_i}{\sum w_i} = \frac{w_i}{w_1 + w_2} \text{for i} = 1, 2 \qquad (6)$$

In the Fourth layer, each node is an adaptive layer which is calculated using Eq. (7).

$$o_{4,i} = \overline{w}_i \cdot f_i \text{ For i} = 1, 2 \qquad (7)$$

Rule 1: if x is A_1 and y is B_1 then $f_1 = p_1 x + q_1 y + r_1$.

Rule 2: if x is A_2 and y is B_2 then $f_2 = p_2 x + q_2 y + r_2$.

p_i, q_i, r_i Parameters called as consequent parameters.

The last layer in Eq. (8), compute the overall output and which will estimate the output during running [17,18].

$$o_{5,i} = f_{out} = \sum_i \overline{w}_i \cdot f_i = overalloutput \qquad (8)$$

3. Metaheuristics

It has been popular to use swarm and evolutionary computing methods to solve difficult problems and integrate them into complex systems.

This is in order to estimate and classify problems. Particle swarm optimization (PSO) is a population-based swarm optimization algorithm proposed by Eberhart and Kennedy [17] in 1995. In this study, we integrated those metaheuristics with ANFIS, as indicated in Fig. 2. The definition problem is presented in Table 1 which is taken from [22]. This table presents 7 real systems and based on their target, the model attempts to estimate their output. In the integration, we consider the decision variables for each real system. Decision variables are interfaces between ANFIS and each metaheuristic algorithm. The decision variables size for each system is presented in Fig.3.

4. Result analysis

4.1 Experimental setup

All approaches are implemented in Matlab R2018b and the system configuration includes an Intel(R) Core(TM) i5-8265U CPU @ 1.60GHz,

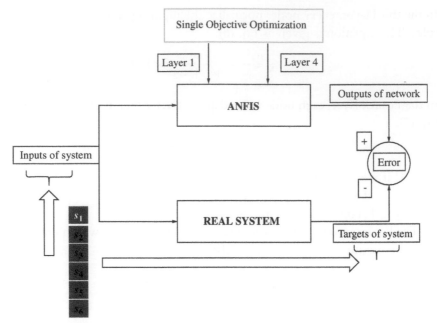

Fig. 2 General system.

Table 1 Problems used in application

Name	Definition	System
f1	Estimation of exchange rate (USD–YTL)	S1..S7
X	Buy	x(t), time based on day
Y	Sell	y(t), time based on day

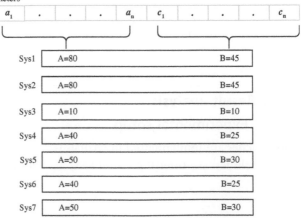

Fig. 3 Decision variables size for each system.

Table 2 Data sets used in application

System	Inputs of system	Target of system
s1	x(t-1),x(t-2),x(t-3),x(t-4),x(t-5),x(t-6),x(t-7)	x(t + 1)
s2	y(t-1),y(t-2),y(t-3),y(t-4),y(t-5),y(t-6),y(t-7)	y(t + 1)
s3	x(t)	y(t)
s4	x(t-1),x(t-2),x(t-4), x(t-6)	x(t + 1)
s5	x(t-1), x(t-3),x(t-5), x(t-7)	x(t + 1)
s6	y(t-1),y(t-2),y(t-4), y(t-6)	y(t + 1)
s7	y(t-1),y(t-3),y(t-5), y(t-7)	y(t + 1)

1.80 GHz in Windows 10. As mentioned before, the data set is taken from [22] and includes an estimation of the TL and Dollar exchange rate from 2007 to 2020. As mentioned before, it is then converted according to the 7 system, as illustrated in Table 2. As part of ANFIS, consider five cluster inputs based on the Gaussian with SUGENO Method. One example of the sys1 which is estimated by ANFIS and each metaheuristic algorithm is shown in Figs. 4–11.

4.2 Experimental result

As a result, in Table 3, a comparison of performance with 30 runs for the integrated ANFIS with optimization is presented. This comparison demonstrated that the PSO, HS, and GA algorithms with specific features are more powerful than the others. Table 4 demonstrates the comparison of performance for metaheuristic algorithms integrated based on the root-mean- square deviation (RMSE) for training and testing steps. For showing the performance of models, we only show an estimate of Sys 3 with ANFIS and Metaheuristic Algorithms. It is given in Table 5. On the basis of Table 6, we can observe which algorithms are most likely to adapt to ANFIS. It is presented in Table 7. Algorithms are divided into and can be divided into Probability and Probability Searching Methods and Categories. In this category. In m-based metaheuristics demonstrated a local search (Exploitation) capacity, whereas Evolutionary computing meta-heuristics demonstrated a considerable ability to explore the space problem (Figs. 12-15).

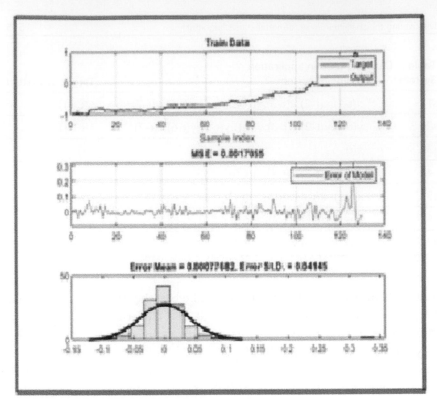

Fig. 4 PSO algorithm Train step for sys1.

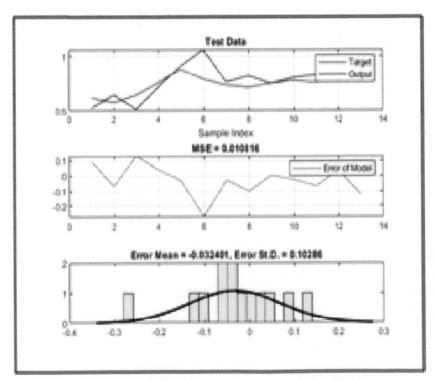

Fig. 5 PSO algorithm Test step for sys1.

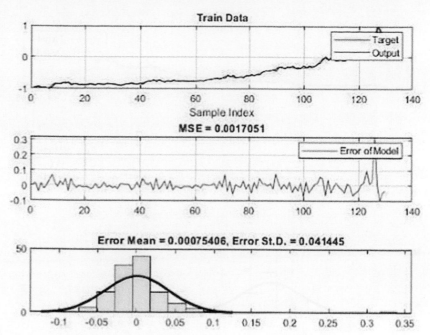

Fig. 6 GA algorithm Train step for sys1.

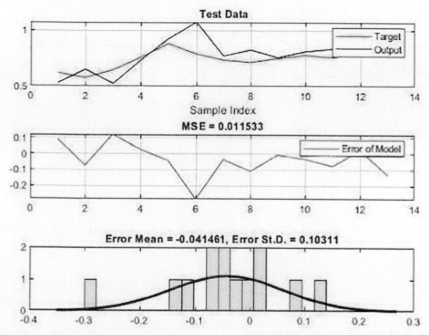

Fig. 7 GA algorithm Test step for sys1.

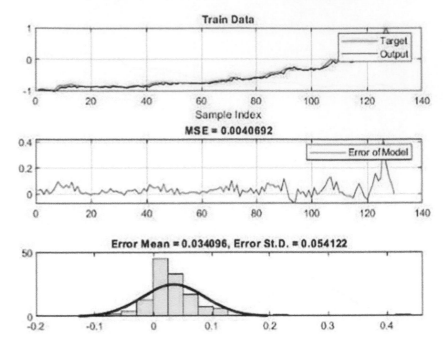

Fig. 8 FA algorithm Train step for sys1.

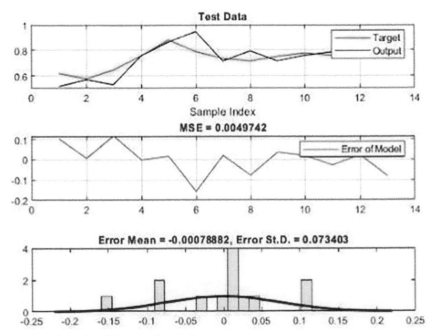

Fig. 9 FA algorithm Test step for sys1.

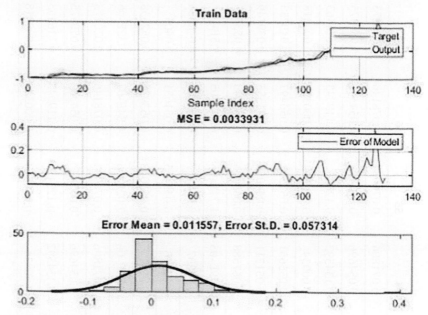

Fig. 10 HS algorithm Train step for sys1.

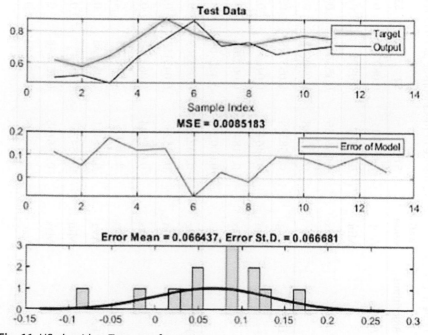

Fig. 11 HS algorithm Test step for sys1.

Table 3 Comparison of performance of metaheuristic algorithms in AN-FIS.

Algorithms	Statistic value	S1	S2	S3	S4	S5	S6	S7
ABC	Avg	0.093411	0.084738	0.056077	0.071806	0.076084	0.078146	0.070273
	Min	0.05471	0.0574	0.020318	0.054696	0.05247	0.05436	0.050575
	Max	0.156155	0.153648	0.094685	0.134592	0.12545	0.119043	0.116463
	Var	0.000688	0.000602	0.000627	0.00034	0.000564	0.000399	0.000318
BAT	Avg	0.058388	0.053988	0.015791	0.055651	0.051305	0.056497	0.049253
	Min	0.046428	0.046012	0.001864	0.051711	0.044059	0.051355	0.039149
	Max	0.118432	0.062266	0.067223	0.067588	0.065006	0.085545	0.053545
	Var	0.000278	3.42E-05	0.000252	1.79E-05	1.71E-05	6.53E-05	1.5E-05
HS	Avg	0.052288	0.053569	0.002084	0.05207	0.047757	0.052729	0.048563
	Min	0.045258	0.045155	0.001784	0.04966	0.041465	0.051631	0.0403
	Max	0.062544	0.062588	0.003628	0.054136	0.051048	0.053925	0.052638
	Var	2.4E-05	2.61E-05	2.13E-07	1.26E-06	7.02E-06	4.84E-07	7.32E-06
FA	Avg	0.093411	0.084738	0.056077	0.071806	0.076084	0.078146	0.070273
	Min	0.05471	0.0574	0.020318	0.054696	0.05247	0.05436	0.050575
	Max	0.156155	0.153648	0.094685	0.134592	0.12545	0.119043	0.116463
	Var	0.000688	0.000602	0.000627	0.00034	0.000564	0.000399	0.000318

GA	Avg	0.040524	0.04035	0.045128	0.051394	0.00174	0.05174	0.04511
	Min	0.035652	0.034967	0.039223	0.046765	0.001706	0.049644	0.036263
	Max	0.041348	0.041719	0.046579	0.052139	0.001755	0.052282	0.04673
	Var	2.04E-06	4.39E-06	4.21E-06	1.4E-06	1.66E-10	3.65E-07	7.54E-06
PSO	Avg	0.040497	0.041544	0.045737	0.051684	0.001724	0.05158	0.044453
	Min	0.036516	0.040554	0.039379	0.050555	0.001713	0.050449	0.036874
	Max	0.041298	0.041676	0.046691	0.052135	0.001768	0.052258	0.0468
	Var	1.94E-06	1.09E-07	3.04E-06	2.24E-07	3.31E-10	2.78E-07	1.11E-05

Table 4 The comparison of performance for metaheuristic algorithms with integrated ANFIS (RMSE).

Problem	PSO		GA		FA		HS		BAT		ABC	
System	Train	Test	Train	Test	Train	Test	Train	Test	Train	Test	Train	Test
S_1	0.04	0.104	0.041	0.103	0.093	0.206	0.052	0.118	0.058	0.13	0.093	0.206
S_2	0.042	0.102	0.04	0.107	0.086	0.219	0.054	0.134	0.054	0.159	0.086	0.219
S_3	0.002	0.003	0.002	0.003	0.056	0.24	0.002	0.002	0.016	0.3	0.056	0.24
S_4	0.052	0.075	0.051	0.083	0.072	0.107	0.052	0.07	0.056	0.102	0.072	0.107
S_5	0.046	0.101	0.045	0.105	0.078	0.118	0.048	0.088	0.051	0.095	0.076	0.162
S_6	0.052	0.086	0.052	0.08	0.078	0.118	0.053	0.078	0.056	0.087	0.078	0.118
S_7	0.044	0.089	0.045	0.113	0.07	0.121	0.049	0.082	0.049	0.094	0.07	0.121

Table 5 Estimate the Sys 3 with ANFIS and Metaheuristic Algorithms.

TARGET	ANFIS_ABC	ANFIS_BAT	ANFIS_FA	ANFIS_GA	ANFIS_HS	ANFIS_PSO
0.247601	0.201645	0.159004	0.210724	0.245103	0.245997	0.24519
0.328215	0.202485	0.258806	0.279599	0.325814	0.326708	0.325901
0.37428	0.216337	0.33773	0.318956	0.371934	0.372828	0.372022
0.754319	0.426681	0.764802	0.643654	0.752429	0.753323	0.752516
1	0.563821	1.010663	0.85028	0.994561	0.995455	0.994648
0.804223	0.45498	0.815535	0.686291	0.802393	0.803287	0.802479
0.616123	0.348316	0.6243	0.525582	0.614067	0.614961	0.614154
0.589251	0.333079	0.59696	0.502624	0.587164	0.588057	0.587251
0.616123	0.348316	0.6243	0.525582	0.614067	0.614961	0.614154
0.573896	0.324373	0.581318	0.489505	0.57179	0.572684	0.571877
0.642994	0.363553	0.651625	0.548541	0.640971	0.641865	0.641058
0.754319	0.426681	0.764802	0.643654	0.752429	0.753323	0.752516
0.877159	0.496339	0.889684	0.748607	0.875417	0.876311	0.875503
0.785029	0.444096	0.796022	0.669892	0.783176	0.78407	0.783263
0.731286	0.41362	0.741386	0.623975	0.729368	0.730262	0.729455
0.712092	0.402736	0.721874	0.607576	0.710152	0.711045	0.710239

Table 5 Estimate the Sys 3 with ANFIS and Metaheuristic Algorithms.—cont'd

TARGET	ANFIS_ABC	ANFIS_BAT	ANFIS_FA	ANFIS_GA	ANFIS_HS	ANFIS_PSO
0.746641	0.422327	0.756997	0.637094	0.744742	0.745636	0.744829
0.773512	0.437565	0.784315	0.660053	0.771646	0.772539	0.771732
0.754319	0.426681	0.764802	0.643654	0.752429	0.753323	0.752516
0.796545	0.448449	0.803827	0.676452	0.790863	0.791756	0.790949

Table 6 Characteristic of metaheuristic algorithms.

Algorithm	Probability distribution	Search characteristics	Category
PSO	Uniform	Guided search in search space	Swarm-based
GA	Roulette Wheel	Cross Over, Mutation, selection based on exponential	Evolutionary-based
FA	Gaussian, Uniform	Nonlinear attraction (EXP) in search space	Swarm-based
HS	Uniform	HMCR, PAR	Evolutionary-based
BAT	Uniform	Frequency-tuning	Swarm-based
ABC	Roulette Wheel	There Guided agent in search space	Swarm-based

Table 7 The obtained of Characteristic from metaheuristic algorithm.

Algorithm	Probability distribution	Search characteristics	Category
Hybridge (HS,GA)	Uniform	Crossover(local search)- PAR(Global search)	Evolutionary-based
Hybridge (GA,HS)	Uniform	HMCR(local search)- Mutation(Global search)	Evolutionary-based
Hybridge (PSO,HS)	Uniform	Decreasing the values of C1 and C2 parameters (local search)- Mutation(Global search)	Swarm-based
Hybridge (PSO,GA)	Uniform	crossover(Local search) Increasing the values of C1 and C2(Global search)	Swarm-based

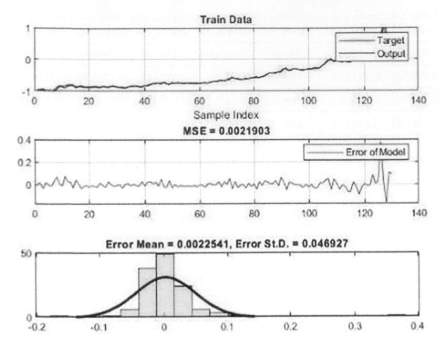

Fig. 12 BAT algorithm Train step for sys1.

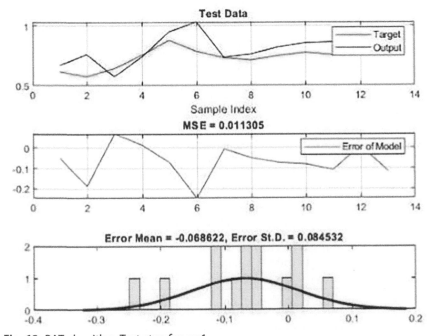

Fig. 13 BAT algorithm Test step for sys1.

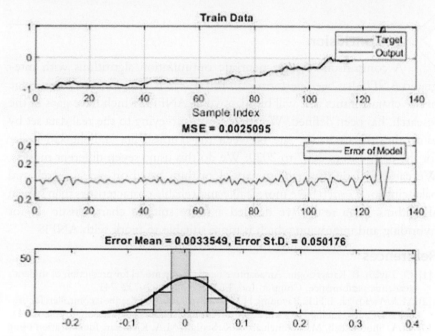

Fig. 14 ABC algorithm Train step for sys1.

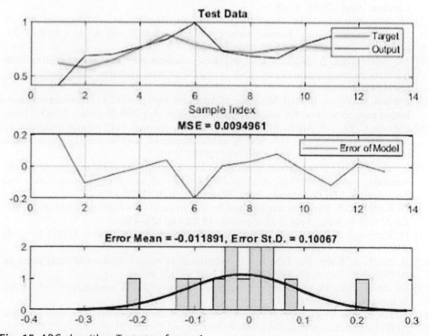

Fig. 15 ABC algorithm Test step for sys1.

5. Conclusion

A comparison of meta-heuristic optimization algorithms with integrated ANFIS has been performed in this research, and the basic knowledge of the characteristics that will be adaptive to ANFIS, which have gaps in the research, has been defined. We apply more surveying to the real data set by estimating the exchange rate between the Turkish lira and the US dollar for the period from 2007 to 2020. We do this using seven different models. We can conclude that evolutionary algorithms based on global search and balancing local search have more stable and reliable characteristics than swarm algorithms. As a result, we defined a more suitable characteristic set for hybriding and improving which is more suitable to work with ANFIS.

References

[1] O. Taylan, B. Karagözöğlu, An adaptive neuro-fuzzy model for prediction of student's academic performance, Comput. Ind. Eng. 57 (3) (2009) 732–741.
[2] M.A. Awadallah, E.H.E. Bayoumi, H.M. Soliman, Adaptive deadbeat controllers for brushless DC drives using PSO and ANFIS techniques, J. Electr. Eng. 60 (1) (2009) 3–11.
[3] M.A. Shoorehdeli, M. Teshnehlab, A.K. Sedigh, M.A. Khanesar, Identification using ANFIS with intelligent hybrid stable learning algorithm approaches and stability analysis of training methods, Appl. Soft Comput. 9 (2) (2009) 833–850.
[4] S.B. Alagarsamy, S. Kondappan, Earre cognition system using adaptive approach Runge–Kutta (AARK) threshold segmentation with ANFIS classification, Neural. Comput. Appl. (2018) 1–12.
[5] H. Harandizadeh, M.M. Toufigh, V. Toufigh, Application of improved ANFIS approaches to estimate bearing capacity of piles, Soft Comput. 23 (2018) 1–13. https://doi.org/10.1007/s00500-018-3517-y.
[6] M. Talebizadeh, A. Moridnejad, Uncertainty analysis for the- forecast of lake level fluctuations using ensembles of ANN and ANFIS models, Expert Syst. Appl. 38 (4) (2011) 4126–4135.
[7] I. Kiyak, V. Topuz, B. Oral, Modeling of dimmable high power LED illumination distribution using ANFIS on the isolated area, Expert Syst. Appl. 38 (9) (2011) 11843–11848.
[8] J. Zhou, C. Li, C.A. Arslan, M. Hasanipanah, H.B. Amnieh, Performance evaluation of hybrid FFA-ANFIS and GA-ANFIS model stopredict particle size distribution of a muck-pile after blasting, Eng. Comput. (2019) 1–10.
[9] A. Gholami, H. Bonakdari, I. Ebtehaj, S.H.A. Talesh, S.R. Khodashenas, A. Jamali, Analyzing bank profile shape of alluvial stable channels using robust optimization and evolutionary ANFIS methods, Appl. Water Sci. 9 (3) (2019) 40.
[10] D. Karaboga, E. Kaya, An adaptive and hybrid artificial bee- colony algorithm (aABC) for ANFIS training, Appl. Soft Comput. 49 (2016) 423–436.
[11] D. Karaboga, E. Kaya, Estimation of number of foreign visitors with ANFIS by using ABC algorithm, Soft Comput. (2019) 1–13.
[12] R. Singh, A. Kainthola, T.N. Singh, Estimation of elastic- constant of rocks using an ANFIS approach, Appl. Soft Comput. 12 (1) (2012) 40–45.
[13] M. Buragohain, C. Mahanta, A novel approach for ANFIS modelling based on full factorial design, Appl. Soft Comput. 8 (1) (2008) 609–625.
[14] Y. Lei, Z. He, Y. Zi, Q. Hu, Fault diagnosis of rotating machinery based on multiple ANFIS combination with GAs, Mech. Syst. Signal Process 21 (5) (2007) 2280–2294.

Printed and bound by CPI Group (UK) Ltd, Croydon, CR0 4YY

03/10/2024

01040431-0011